GCSE Chemistry

Graham Hill

Nigel Heslop, Graham Hill

The Publishers would like to thank the following for permission to reproduce copyright material:

Photo credits

The publishers would like to thank the following for permission to reproduce copyright material:
p.1 Corbis/Stephanie Maze; **p.2** *t* Rex Features/Nils Jorgensen, *b* Corbis/Chris Bland/Eye Ubiquitous; **p.5** Geoscience Features Picture Library; **p.10** Rex Features/Dave Penman; **p.12** *t* Corbis/Tim McGuire, *b* Rex Features/Sunset; **p.13** *tl* Alamy/Russ Merne, *cl* Alamy/Danita Delimont, *cr* Alamy/Greenshoots Communications; **p.15** *cl* Corbis/Philadelphia Museum of Art, *c* Photolibray.com, *cr* Rob Melnychuk/Taxi/Getty Images; **p.17** Geoscience Features Picture Library; **p.18** Alamy/Steve Atkins; **p.21** Alamy/Peter Bowater; **p.22** *t* Science Photo Library/Pascal Goetgheluck, *b* Alamy/Scott Camazine; **p.28** Getty Images/Hulton Archive; **p.31** all Nigel Heslop; **p.33** both Nigel Heslop; **p.34** Science Photo Library/Paul Rapson; **p.37** Still Pictures/Ray Pfortner; **p.39** Corbis/Peter Turnley; **p.40** *cl* Nigel Heslop, *cr* Nigel Heslop, *b* Bob Battersby; **p.41** all Nigel Heslop; **p.42** *t* Nigel Heslop, *b* Geoscience Features; **p.43** Nigel Heslop; **p.44** Bob Battersby; **p.45** *c* Photodisc, *bl* Nigel Heslop, *bc* Nigel Heslop, *br* Nigel Heslop; **p.46** Nigel Heslop; **p.49** *t* Still Pictures/Wolfgang Maria Weber, *b* Getty Images/UHB Trust; **p.50** *t* Science Photo Library/Jerry Mason, *c* Bob Battersby, *b* Nigel Heslop; **p.53** Ingram; **p.54** Ingram; **p.55** *tc* Photodisc, *tr* Rex Features/Burger/Phanie, *cl* Bob Battersby; **p.56** *t* Anthony Blake Photo Library/Tim Hill, *b* Bob Battersby; **p.57** Alamy/Marie-Louise Avery; **p.58** *t* Andrew Lambert, *c* Science Photo Library/Maximilian Stock Ltd., *b* Rex Features/The Travel Library; **p.61** Alamy/AGStockUSA, Inc; **p.65** Science Photo Library/Planetary Visions Ltd.; **p.66** *cl* Corbis/Jim Craigmyle, *cr* Getty Images/Wayne Levin; **p.74** Rex Features/RYB; **p.75** GeoScience Features; **p.76** *c* Getty Images/AFP, *b* Rex Features/Sipa Press; **p.79** Rex Features/Sipa Press; **p.80** Rex Features/The Travel Library; **p.84** *l* Science Photo Library/David Parker, *tr* Science & Society Picture Library/Science Museum, *br* Rex Features; **p.86** Corbis/Russell Boyce/Reuters; **p.90** University of Cambridge Cavendish Library; **p.92** Lorna Ainger; **p.93** *t* Alamy/Steve Atkins, *b* Alamy/Gunter Marx; **p.94** Rex Features/The Travel Library; **p.97** PurestockX; **p.99** PurestockX; **p.106** Corbis/Matthias Kulka; **p.107** *t* Alamy/Kevin Schafer/Peter Arnold Inc., *b* Science Photo Library/Sinclair Stammers; **p.108** *l* Science Photo Library/Philippe Plailly/Eurelios, *r* Science Photo Library/Kenneth Libbrecht; **p.109** *l* Corbis/Alan Goldsmith, *r* Corbis/David Samuel Robbins; **p.110** Science Photo Library/Charles D. Winters; **p.112** Corbis/Owaki – Kulla; **p.113** Corbis/Nik Wheeler; **p.116** *t* Hodder, *b* Corbis/Bruce Peebles; **p.117** Alamy/Frances Roberts; **p.119** *t* Still Pictures/Leonard Lessin, *b* Corbis/Paul McErlane/Reuters; **p.120** Welcome Trust **p.127** Rex Features/Woman's Weekly; **p.128** Science Photo Library/Andrew Lambert; **p.129** Science Photo Library/Colin Cuthbert; **p.132** *l* Alamy/David Hoffman, *r* Corbis/Jon Hicks; **p.133** Corbis/photocuisine; **p.136** Martyn f. Chillmaid; **p.137** *tl* PurestockX, *cl* PurestockX, *bl* Nigel Heslop, *bc* Nigel Heslop, *br* Nigel Heslop; **p. 138** Lorna Ainger; **p.142** Science Photo Library/Richard Folwell; **p.144** *l* Robert Opie, *c* Alamy/Dennis MacDonald, *r* Rex Features/Dan Talson; **p.145** *tl* Nigel Heslop, *cl* Nigel Heslop, *r* Martyn f. Chillmaid; **p.147** Science Photo Library/Emilio Segre Visual Archives/American Institute of Physics; **p.153** Science Photo Library/Alan Sirulnikoff; **p.155** Science Photo Library/Andrew Lambert; **p.158** Science Photo Library/Gusto; **p.159** Science Photo Library/Martyn f. Chillmaid; **p.160** both Nigel Heslop; **p.161** Science Photo Library/James Holmes, Hays Chemicals; **p.167** Alamy/LHB Photo; **p.169** Reproduced courtesy of the Library and Information Centre of The Royal Society of Chemistry; **p.170** ©IstockPhoto.com/Long Ha; **p.172** Science Photo Library; p.177 Science Photo Library/Veronique LePlat; **p.181** *l* Science Photo Library/Adrian Thomas, *c* Chizuko Kimura/Alamy; **p.182** Martyn f. Chillmaid; **p.183** © Image by Christie's Images/CORBIS, Boats in the Harbour, Collioure, 1905 by Andre Derain © ADAGP, Paris and DACS, London; **p.187** *l* Alamy/Andrew Duke, *r* Science Photo Library/Simon Fraser; **p.188** Science Photo Library/David Parker; **p.189** Martin Sookias photography; **p.195** *t* Alamy/Motoring Picture Library, *b* Science Photo Library/Martyn f. Chillmaid; **p.197** Science Photo Library; **p.198** *l* Science Photo Library, *r* Royal Society of Chemistry; **p.201** Alamy/John Goulter; **p.203** PurestockX; **p.204** Science Photo Library/Charles D. Winters; p.206 Martyn f. Chillmaid; p.207 *t* Science Photo Library/Martyn f. Chillmaid, *b* Chassenet/Photolibrary; **p.208** Last Resort Picture Library; **p.209** *t* Alamy/Westend 61, *b* PurestockX; **p.210** *c* Science Photo Library/Andrew Lambert Photography, *bl* Science Photo Library/Sheila Terry, *br* Anne Trevillion; **p.211** Botanica/Photolibrary; **p.212** *t* Science Photo Library/Sheila Terry, *b* Last Resort Picture Library; **p.215** Last Resort Picture Library; **p.220** Alamy/Bubbles Photolibrary; **p.233** *t* PA/EMPICS, *b* PurestockX; **p.238** l Corbis/© Jeffrey L. Rotman, *r* Science Photo Library/Philippe Psaila; **p.239** Martin Sookias photography; p.240 Corbis/© Bryan F. Peterson; **p.241** www.shoutpictures.com; **p.245** Andrew Lambert Photography; **p.246** all Andrew Lambert Photography; **p.247** Last Resort Picture Library; **p.249** Science Photo Library/Jerry Mason; **p.251** *t* Getty Images/Mario Tama, *b* Science Photo Library/Tek Image; **p.252** Science Photo Library/Geoff Tompkinson.

b = bottom, *c* = centre, *l* = left, *r* = right, *t* = top

Acknowledgements

British Water for permission to use the map on page 208 showing water hardness in England and Wales (© Hard water map area England and Wales, British Water). Data in Table 11.4 page 211 reproduced from www.mineralwaters.org © Pongü Text & Design GmbH. Data in Table 11.6 page 214 reproduced from the Drinking Water Inspectorate website: www.dwi.gov.uk. Crown copyright material is reproduced with the permission of the controller of HMSO.

Every effort has been made to trace all copyright holders, but if any have been inadvertently overlooked the Publishers will be pleased to make the necessary arrangements at the first opportunity.

Risk assessment

As a service to users, a risk assessment for this text has been carried out by CLEAPSS and is available on request to the Publishers. However, the Publishers accept no legal responsibility on any issue arising from this risk assessment: whilst every effort has been made to check the instructions for practical work in this book, it is still the duty and legal obligation of schools to carry out their own risk assessment.

Hachette's policy is to use papers that are natural, renewable and recyclable products and made from wood grown in sustainable forests. The logging and manufacturing processes are expected to conform to the environmental regulations of the country of origin.

Orders: please contact Bookpoint Ltd, 130 Milton Park, Abingdon, Oxon OX14 4SB.
Telephone: (44) 01235 827720. Fax: (44) 01235 400454. Lines are open 9am–5pm, Monday to Saturday, with a 24-hour message answering service. Visit our website at www.hoddereducation.co.uk

First published in 2007 by
Hodder Education,
part of Hachette Livre UK
338 Euston Road
London NW1 3BH

Impression number 5 4 3 2
Year 2011 2010 2009 2008

Cover photos Science Photo Library: pharmaceutical technician, TEKIMAGE; precipitation reaction, David Taylor; salt crystals, Andrew Syred.
Illustrations by Barking Dog Art
Typeset in Times 11.5pt by Fakenham Photosetting Limited, Fakenham, Norfolk

Printed in Italy

A catalogue record for this title is available from the British Library.

ISBN-13: 978 0 340 92800 4

Contents

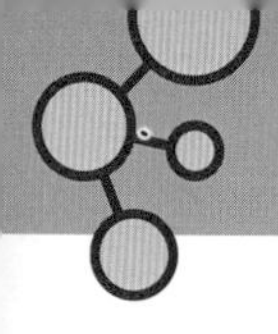

Introduction

Welcome to the AQA GCSE Chemistry Student's Book. This book covers all the Chemistry content as well as the key 'How science works' elements of the new specification.

Each chapter starts with a set of **learning objectives**. Don't forget to refer back to these when checking whether you have understood the material covered in a particular chapter. **Questions** appear throughout each chapter, which will help to test your knowledge and understanding of the subject as you go along. They will also help you develop key skills and understand how science works.

Activities are found throughout the book. These will take you longer to complete than the questions, but will show you many of the real-life applications and implications of science. At the end of each chapter a **summary** provides the important points and key words. You will find the summaries useful in reviewing the work you have completed and in revising for your examinations. Don't forget to use the **index** to help you find the topic you are working on.

You will find **exam questions** at the end of each chapter, to help you prepare for your exams. These include similar questions to those in the unit tests.

You will sit three written papers for GCSE Chemistry: Chemistry 1, Chemistry 2 and Chemistry 3. These match the three sections of this book.

The book includes both **Higher-tier** and **Foundation-tier** material. Learning objectives and summary points that are needed for the Higher-tier exam are shown by a tick in a coloured circle (e.g. ✓). In the text, the sections of the book that you must include if you are taking the Higher-tier exam are shown with a coloured stripe. Questions numbered inside a coloured circle (e.g. 2) would only be asked on the Higher-tier exam paper.

Finally, we would like to thank Gillian Lindsey, Becca Law and Anne Trevillion, all members of the Science Team at Hodder Murray, for their conscientious, perceptive and intelligent contributions to the production of this book.

Good luck with your studies!

Graham Hill and Nigel Heslop

Chapter 1
How do rocks provide useful materials?

At the end of this chapter you should:

- ✓ appreciate how atoms, as the smallest particles in elements, can join together to form molecules in compounds;
- ✓ be able to use symbols and formulae to write balanced equations for chemical reactions;
- ✓ appreciate how rocks provide stone for building, metals and other useful materials;
- ✓ know how limestone is used to manufacture quicklime, slaked lime, cement, concrete and glass;
- ✓ understand how metals can be extracted from their ores;
- ✓ have considered the social, economic and environmental impact of quarrying, mining and extracting metals;
- ✓ have considered the benefits and drawbacks of using metals and recycling metals.

Figure 1.1 Iron ore, like that being mined in this photograph, is useless. You can't grow anything in it, you can't eat it and you can't build with it. But, if it is heated with limestone, coke and air, it produces iron, and from iron we can make steel, which is one of our most useful materials.

1.1 What sorts of materials are there?

Materials that occur naturally, like iron ore, limestone, water and air are called **raw materials** or naturally occurring materials. Materials like iron and steel don't occur naturally, but we can make them using raw materials such as iron ore. Because of this, iron and steel can be called **manufactured materials**. These manufactured materials are useful products which we need for everyday modern life.

What are our most important raw materials?

Table 1.1 shows the five most important raw materials and some of the useful manufactured materials we can obtain from them.

The useful, manufactured materials are:
- either separated from the natural raw materials by processes such as distillation;
- or made from the raw materials by chemical reactions like iron from iron ore.

Raw material	Useful manufactured materials obtained from the raw material
Rocks	• Metals (iron, aluminium, copper) • Alloys (steel, brass) • Building materials (cement, glass)
Crude oil	• Fuels (petrol, diesel) • Plastics (polythene, PVC and polyester)
Air	• Nitrogen for making ammonia, nitric acid and fertilisers • Oxygen for breathing equipment
Seawater	• Table salt, sodium hydroxide, chlorine and hydrogen from brine (concentrated sodium chloride)
Plants	• Fruit and vegetables • Plant oils for cooking and medicines • Fuels from biomass materials

Table 1.1 The five most important raw materials

Figure 1.2 At one time, football boots were made of leather.

❶ What raw material does leather come from?

❷ Today, football boots are made from polymers such as PVC and polyester. What raw material are the polymers made from?

Figure 1.3 Most of the outside of this building is glass.

❸ a) Suggest two benefits of using glass for the building in Figure 1.3.
b) Suggest two drawbacks and risks.

❹ Copy and complete the following table. The first line has been done for you.

Manufactured material	Which raw material in Table 1.1 did the manufactured material come from?
Metals in your TV	Rocks
Plastics in your iPod	
Ammonia in cleaning fluid	
Olive oil for cooking	
Polyester in your shirt/blouse	
Glass for bottles and jars	

Raw materials are the naturally occurring substances, such as air, iron ore and limestone, that we use to make **manufactured materials**.

Activity – Using the natural resources on Geecee

Geecee (Figure 1.4) is an imaginary island off the west coast of Britain. The island has strong winds from the west, high mountains with fast flowing rivers, thick forests along the west coast and peat bogs in the south east. There are no supplies of coal, oil or natural gas. At present no one lives on Geecee.

Suppose you have been sent to study whether people could live on the island and develop a fishing industry.

1. What naturally occurring material would you use to build the first homes on Geecee? Explain your answer.
2. Describe two ways of producing heat to cook food.
3. Where would you build the port for the fishing fleet? Give two reasons for your choice of site.
4. Where would you build homes on the island? Explain your answer.
5. Which naturally occurring material(s) would you try to conserve? Explain your answer.

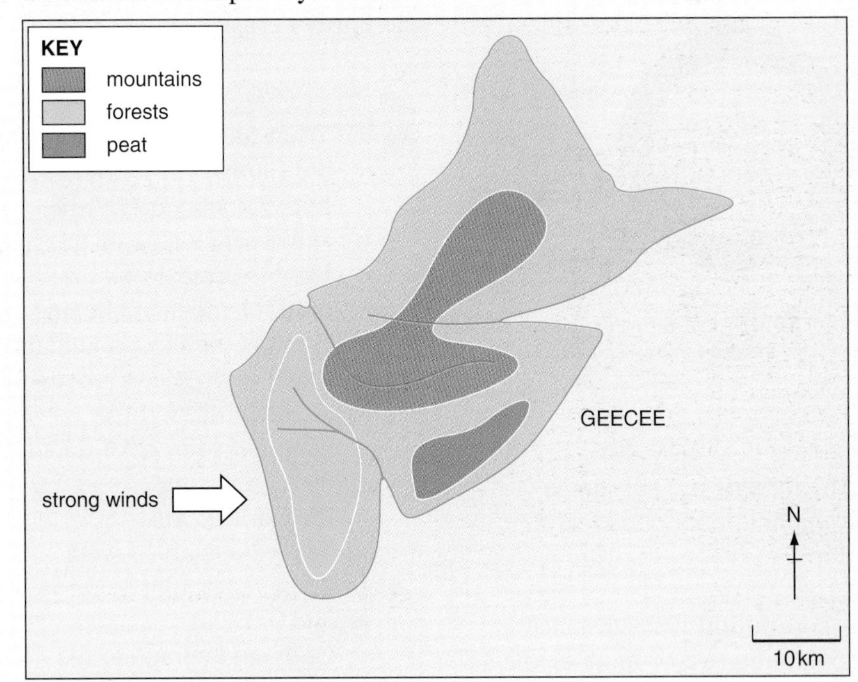

Figure 1.4 The imaginary island of Geecee

1.2 Elements, compounds and mixtures

When electricity passes through molten sodium chloride, it breaks down to form sodium and chlorine (Figure 1.5). We can summarise the chemical reaction by writing a word equation.

$$\text{sodium chloride} \rightarrow \text{sodium} + \text{chlorine}$$

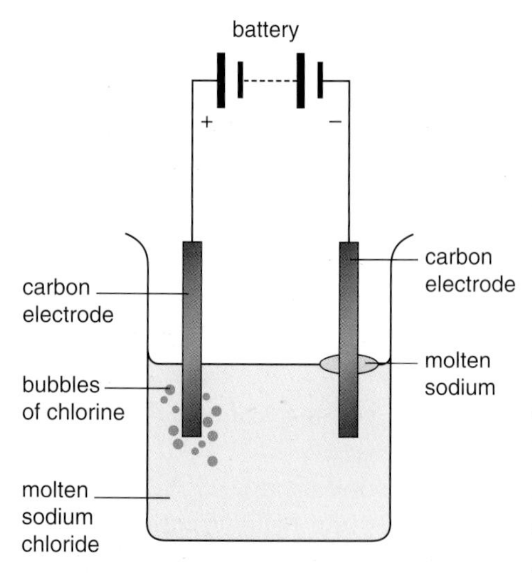

Figure 1.5 When electricity passes through molten sodium chloride, it breaks down into sodium and chlorine.

No matter how the sodium and chlorine are treated, they cannnot be broken down into simpler substances.

Substances, like sodium and chlorine, that cannot be broken down into simpler substances are called **elements**. Elements are the simplest possible materials.

Other elements include aluminium, iron, oxygen and carbon. But substances like sodium chloride and water are not elements because they can be broken down into simpler substances. Substances, like sodium chloride and water, which contain two or more elements chemically joined together are called **compounds**.

There are about 100 different elements and, although there are millions and millions of different substances in the Universe, they all contain one or more of these elements.

For example, water is made of hydrogen and oxygen. Sand is made of silicon and oxygen and limestone contains calcium, carbon and oxygen. So, elements are the simplest building blocks for all substances.

Figure 1.6 shows the percentages of the five most common elements in the Earth's crust.

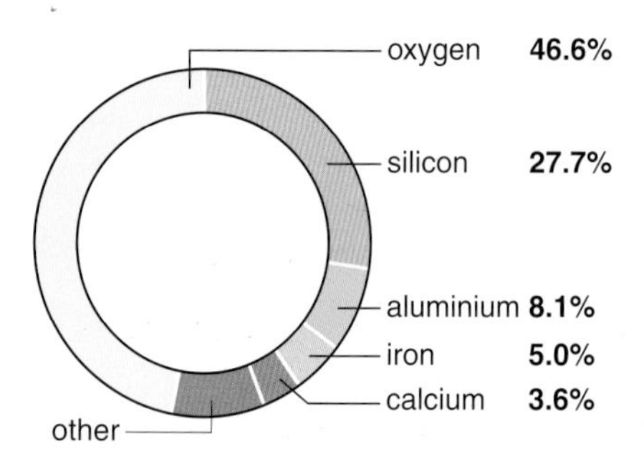

Figure 1.6 The percentages of the five most common elements in the Earth's crust

5 a) What total percentage of the Earth's crust do these five elements make up?
b) What percentage of the Earth's crust do all the other elements make up?
c) What is the most abundant metal in the Earth's crust?
d) A student found five different sources of data for the percentage of sodium in the Earth's crust.
The values given were 2.7%, 3.5%, 2.6%, 2.9% and 3.0%.
Copy and complete the following sentences.

The **range in the data** is from the minimum value of _____% to the maximum value of _____%. One of the values given is very different to the other four. This **anomalous value** is _____%. Anomalous (unusual or irregular) values are usually ignored in calculating an average or mean value. When this is done:
The mean value for the % of sodium in the Earth's crust = _____%.

An **element** is a substance containing only one kind of atom.

A **compound** is a substance containing two or more elements joined together chemically.

A **mixture** is two or more substances which are mixed together but not combined together chemically.

In an **alloy** other elements are mixed with metals to give the mixture particular properties.

Most of the materials that occur naturally and that we use every day are not pure elements or pure compounds. They are **mixtures** of substances. They may be:

- mixtures of elements, such as **alloys**, which are metals mixed with other elements (for example, mild steel is mainly iron with about 0.2% carbon);
- mixtures of compounds, such as seawater, which contains salt (sodium chloride) and water.

Figure 1.7 This photo of a gold crystal was taken through an electron microscope. Each yellow blob is a separate gold atom. The gold atoms are touching each other

What are the particles in elements?

The smallest particles of an element are **atoms**.

Electron microscopes can magnify objects more than a million times. Using electron microscopes, it is possible to identify atoms. Figure 1.7 shows an electron microscope photo of gold.

As all substances are made of elements and all elements are made of atoms, it follows that all substances are made of atoms. Each element contains only one sort of atom. So, as there are about 100 different elements, there are also about 100 different kinds of atom. Iron contains only iron atoms, copper contains only copper atoms and so on.

An **atom** is the smallest particle of an element. All substances are made of atoms.

Representing atoms with symbols

Atoms of each element can be represented by a chemical **symbol**. For example, O represents an atom of oxygen, Fe represents an atom of iron and C represents an atom of carbon. The symbol for an element can also be used as shorthand for the name of the element.

The names and symbols of all the common elements are shown in the Periodic Table on page 8.

6 a) Estimate the diameter of one gold atom in the photo.
b) In order to estimate the diameter of a gold atom more accurately, it is better to measure a line of four or five of them and then divide by four or five. Why is this more **reliable** than measuring the diameter of just one gold atom? (Hint: **Data** is more reliable if you can be sure it is **accurate**.)
c) Calculate the actual diameter of a gold atom. (Assume the magnification is 40 000 000.)

7 Why is it useful to represent elements using symbols? (Hint: Suppose you had to write the word 'magnesium' many times.)

8 Use the Periodic Table containing symbols on page 8 to answer the following questions.
a) What are the symbols for carbon, calcium, cobalt, copper, chlorine and chromium?
b) What elements are represented by N, Ni, P, K, Si, Ag, S and Na?

Sometimes, it is useful to draw pictures of atoms as coloured circles with the symbol in the centre (Figure 1.8). This helps scientists to understand the structure of substances and how they react with each other.

Data refers to a collection of measurements.

A **reliable** measurement or result is one that can be repeated.

An **accurate** measurement is one that is close to the **true value**, which is the value that would be obtained if there were no errors in the measurement.

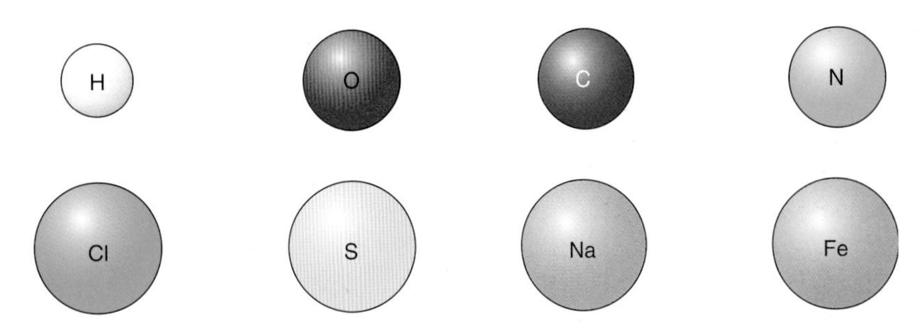

Figure 1.8 Pictures of atoms. Hydrogen atoms are usually shown as white circles, oxygen red, carbon black, nitrogen blue, chlorine green, sulfur yellow and metals grey.

Atoms of each element are represented by a chemical **symbol**, for example O represents an atom of oxygen.

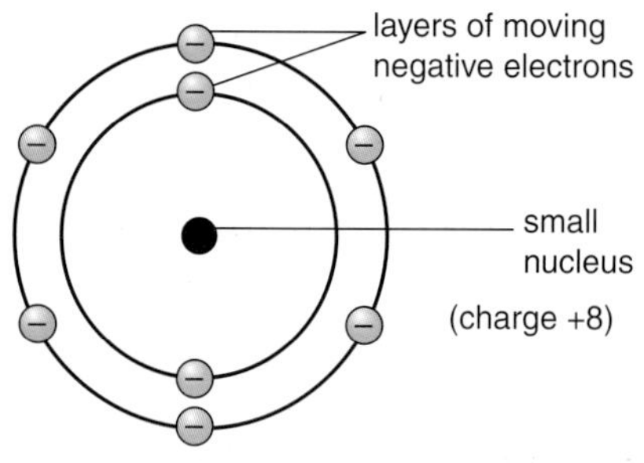

Figure 1.9 The nucleus and electrons in an atom of oxygen

How do atoms combine?

Atoms have a small positive **nucleus** surrounded by layers of moving negative **electrons**. The positive charge on the nucleus exactly cancels the negative charge on the electrons, so the overall charge on an atom is neutral. In Figure 1.9 the nucleus of the oxygen atom has a charge of +8, which exactly cancels the negative charge on the eight electrons moving around the nucleus.

In Chemistry 2 you will learn more about the particles inside the nucleus, called protons and neutrons (Section 5.1). Protons have a positive charge and neutrons have no charge. You will also learn how the number of protons in the nucleus, called the atomic number or proton number, determines which element an atom is.

When elements react, their atoms join with other atoms to form compounds. It is the electrons on the outside of the atoms that move in chemical reactions. So, making a compound involves:

- either sharing electrons to form **molecules**;
- or transferring (giving and taking) electrons to form charged particles called **ions**.

When atoms of non-metals join together, they share electrons and form molecules.

In a molecule of water, two atoms of hydrogen combine with one atom of oxygen. The two H atoms and the one O atom are held together in chemical bonds by the attraction of their positive nuclei for shared electrons.

Figure 1.10 shows how the electrons are shared in a molecule of water. You will find out more about this way of drawing electrons in atoms and molecules in Chemistry 2 (Section 5.6).

Atoms have a small central **nucleus** which has a positive charge.

Electrons are very small negatively charged particles that move around the nucleus.

The number of protons in an atom is called its **atomic number**.

A **molecule** is a particle containing two or more atoms joined by chemical bonds.

An **ion** is a charged particle formed from an atom by the loss or gain of one or more electrons.

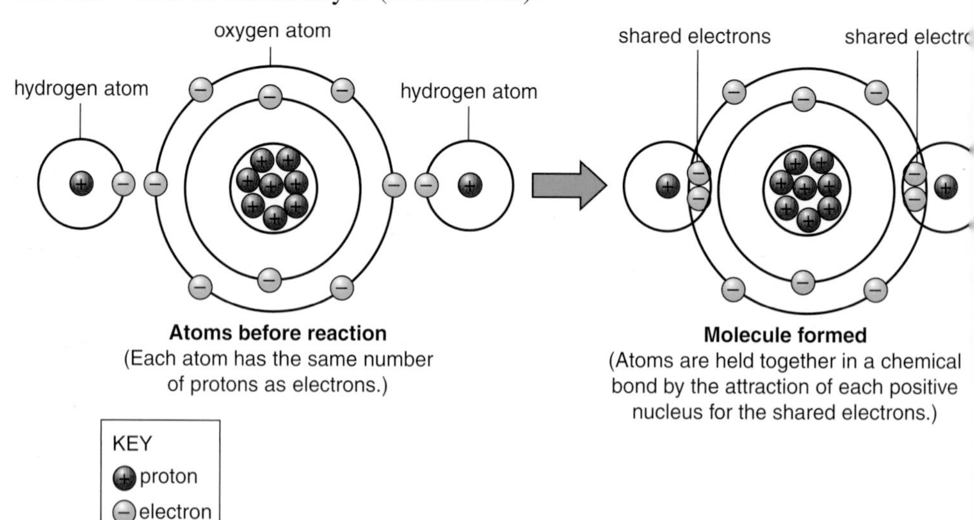

Figure 1.10 The formation of a molecule of water

The symbols for elements can also be used to represent molecules in compounds. So, water is written as H_2O – two hydrogen atoms (H) and one oxygen atom (O). Carbon dioxide is written as CO_2 – one carbon atom (C) and two oxygen atoms (O).

The **formula** of a compound shows the number of atoms of the different elements joined together in one molecule.

'H_2O' and 'CO_2' are called molecular formulae, or just **formulae** for short.

These formulae show the number of atoms of the different elements joined together in one molecule of a compound.

❾ How many atoms of the different elements are there in one molecule of
a) ammonia;
b) hydrogen chloride?

❿ Look at Figure 1.11. What is the formula of: a) ammonia; b) hydrogen chloride?

⓫ a) Draw a picture for a molecule of methane (natural gas), CH_4. All four hydrogen atoms are bonded to the carbon atom.
b) What are the advantages of writing 'CH_4' for methane?

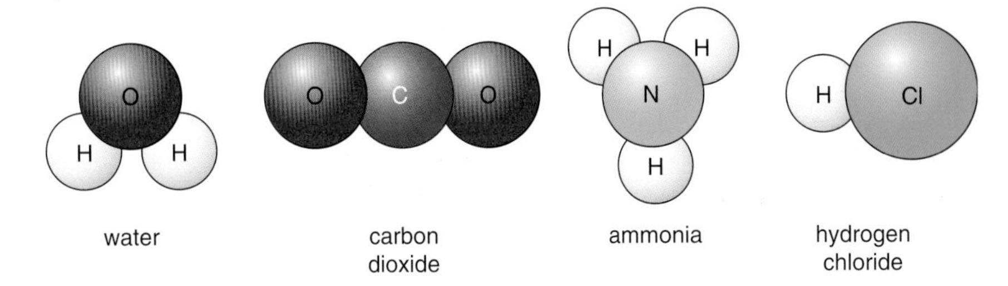

Figure 1.11 Molecules of water, carbon dioxide, ammonia and hydrogen chloride

When atoms of a metal join with atoms of a non-metal, they form ions.

So, compounds such as sodium chloride (NaCl), calcium oxide (CaO) and red iron oxide (Fe_2O_3) consist of ions, not molecules.

When these metal/non-metal compounds form, the metal atoms give up electrons to form positive ions and the non-metal atoms take electrons to form negative ions. Figure 1.12 shows the formation of a positive ion and a negative ion by the movement of one electron from a metal to a non-metal (you will need to use this type of diagram in Chemistry 2). In the crystals of these metal/non-metal compounds, the positive and negative ions are arranged in lattices. The opposite charges hold the ions together in the lattices by strong chemical bonds.

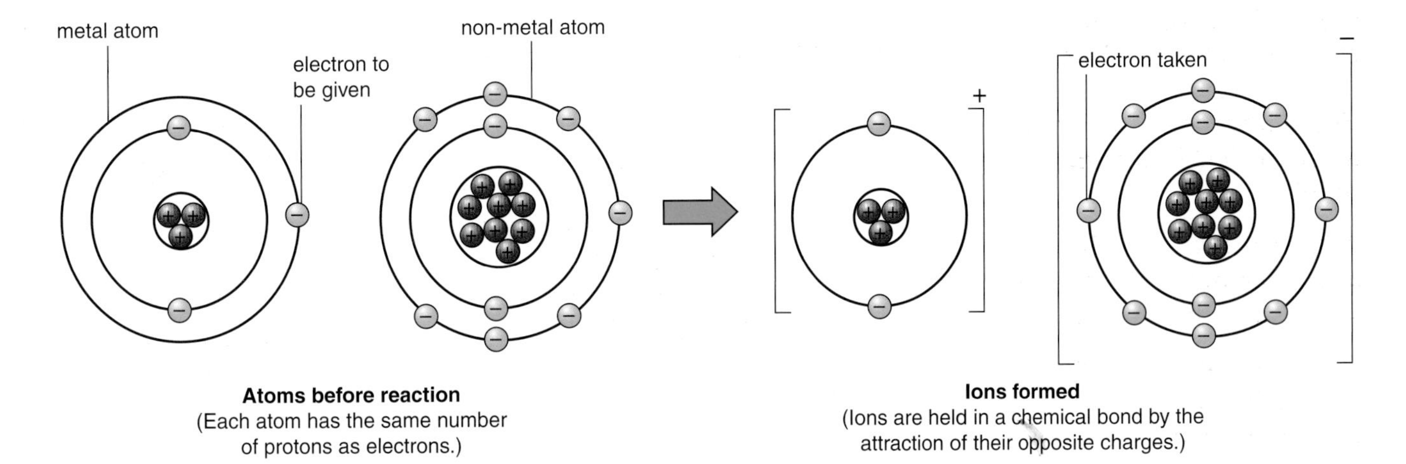

Figure 1.12 The formation of ions when atoms give and take electrons

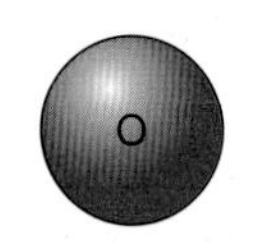

an oxygen atom

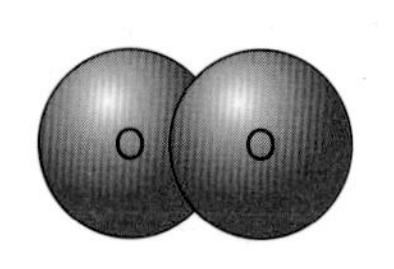

an oxygen molecule

Figure 1.13 An atom and a molecule of oxygen

Atoms and molecules of elements

Almost all elements can be represented by their symbols. For example, Fe for iron, C for carbon. But, this is not the case with hydrogen, oxygen, nitrogen and chlorine. At normal temperatures, these elements exist as molecules containing two atoms joined together. So, hydrogen is best represented as H_2 not H, oxygen as O_2 not O, nitrogen as N_2 and chlorine as Cl_2 (Figure 1.13).

How are elements arranged in the Periodic Table?

In the Periodic Table (Figure 1.14) the 100 or so elements are arranged in order of their atomic number (the number of protons in an atom of the element). You will learn more about this in Chemistry 2 (Section 5.5) .

The **Periodic Table** shows the elements arranged in order of their atomic number.

So, the first element in the table is hydrogen (atomic number = 1), then helium (atomic number = 2), then lithium (atomic number = 3) and so on.

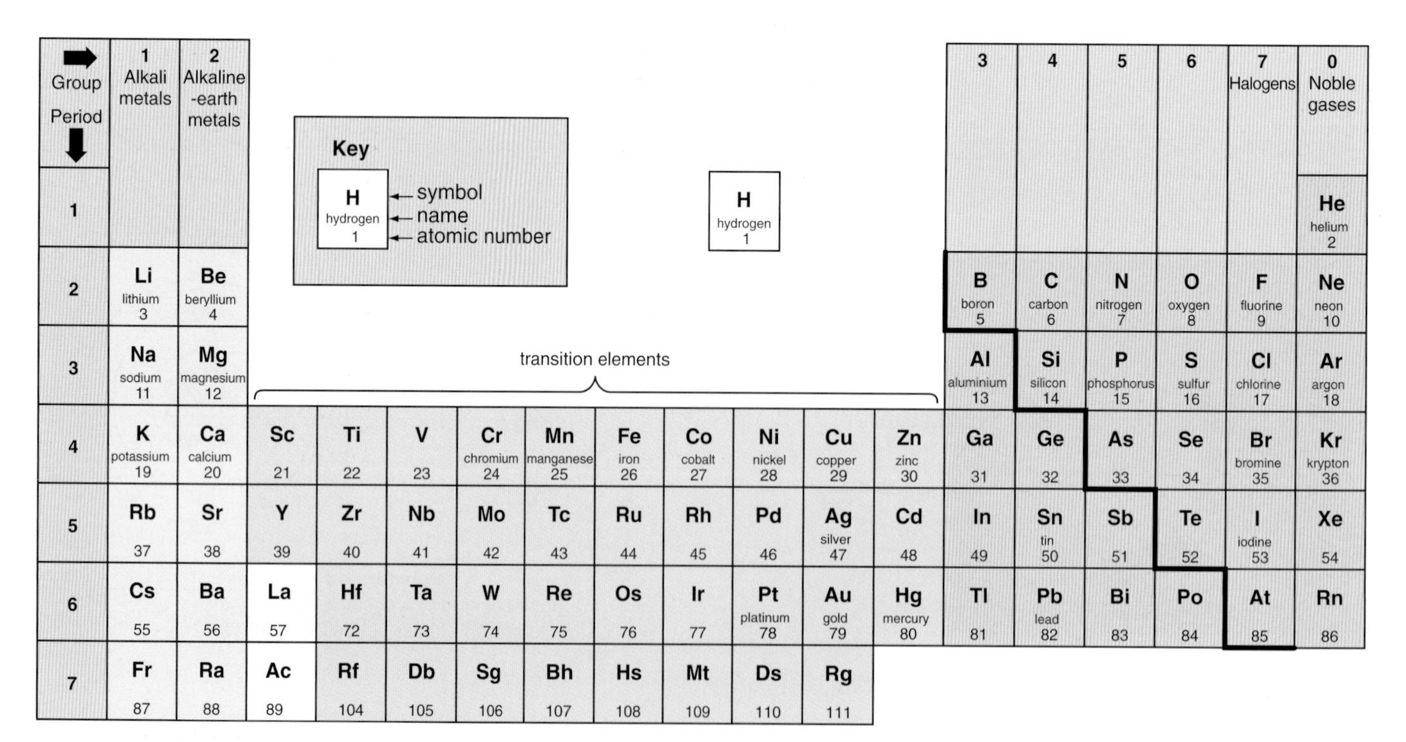

Key: H (symbol), hydrogen (name), 1 (atomic number)

Group → / Period ↓	1 Alkali metals	2 Alkaline-earth metals	transition elements										3	4	5	6	7 Halogens	0 Noble gases
1										H hydrogen 1								He helium 2
2	Li lithium 3	Be beryllium 4											B boron 5	C carbon 6	N nitrogen 7	O oxygen 8	F fluorine 9	Ne neon 10
3	Na sodium 11	Mg magnesium 12											Al aluminium 13	Si silicon 14	P phosphorus 15	S sulfur 16	Cl chlorine 17	Ar argon 18
4	K potassium 19	Ca calcium 20	Sc 21	Ti 22	V 23	Cr chromium 24	Mn manganese 25	Fe iron 26	Co cobalt 27	Ni nickel 28	Cu copper 29	Zn zinc 30	Ga 31	Ge 32	As 33	Se 34	Br bromine 35	Kr krypton 36
5	Rb 37	Sr 38	Y 39	Zr 40	Nb 41	Mo 42	Tc 43	Ru 44	Rh 45	Pd 46	Ag silver 47	Cd 48	In 49	Sn tin 50	Sb 51	Te 52	I iodine 53	Xe 54
6	Cs 55	Ba 56	La 57	Hf 72	Ta 73	W 74	Re 75	Os 76	Ir 77	Pt platinum 78	Au gold 79	Hg mercury 80	Tl 81	Pb lead 82	Bi 83	Po 84	At 85	Rn 86
7	Fr 87	Ra 88	Ac 89	Rf 104	Db 105	Sg 106	Bh 107	Hs 108	Mt 109	Ds 110	Rg 111							

Figure 1.14 The Periodic Table (elements 58–71 and 90–103 have been omitted)

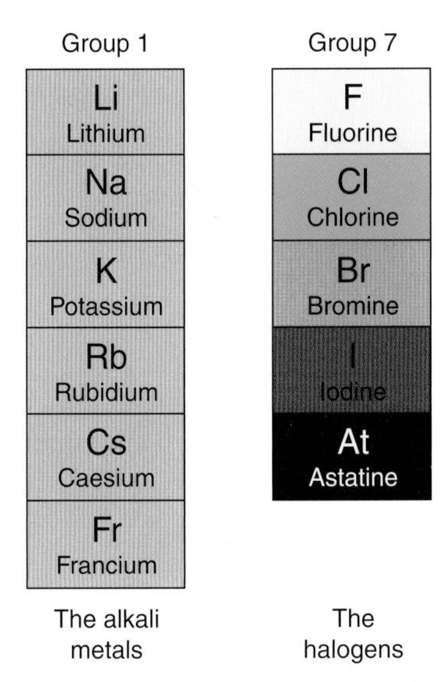

Figure 1.15 Group 1 and Group 7 of the Periodic Table

The Periodic Table is also set out so that elements with similar properties are in the same vertical column.

- The vertical columns are called **groups**. Group 1 contains the metals lithium, sodium and potassium, which have very similar properties, and Group 7 contains the non-metals chlorine and bromine which are also very similar (Figure 1.15).
- Some of the groups have names as well as numbers. These are shown below the group numbers across the top of Figure 1.14.
- The horizontal rows in the Periodic Table are called periods. So, period 1 contains just two elements – hydrogen and helium, and period 2 has eight elements from lithium (atomic number 3) to neon (atomic number 10).
- Metals are clearly separated from non-metals. The 20 or so non-metals are packed into the top right-hand corner, above the thick stepped line in Figure 1.14.
- In each group of the Periodic Table the elements have similar properties, but there is a gradual change in properties from the top to the bottom of the group.

The vertical columns in the Periodic Table are called **groups**. Elements in the same group have similar properties.

⓬ Look at Figure 1.15.
a) Which of the elements, sodium or potassium, is the most reactive?
b) Which element in Group 1 do you think is i) the most reactive; ii) the least reactive?
c) Do the elements in Group 1 get more or less reactive as you go down the group from Li to Fr?

⓭ Draw a large outline of the Periodic Table similar to Figure 1.14. On your outline, indicate where you would find:
a) metals;
b) elements with atomic numbers 14 to 17;
c) the noble gases;
d) the transition metals;
e) the most reactive metal;
f) one magnetic element;
g) two elements used in expensive jewellery;
h) an element used to disinfect water supplies.

1.4 Using symbols and formulae to write balanced equations

Figure 1.16 shows sparks from a sparkler. The sparks are tiny bits of burning magnesium. Let's use symbols and formulae to write an equation for this reaction.

Figure 1.16 The sparks from a sparkler are tiny bits of burning magnesium.

When magnesium burns, it reacts with oxygen to form magnesium oxide.

$$\text{magnesium} + \text{oxygen} \rightarrow \text{magnesium oxide}$$

Chemists usually write symbols and formulae rather than names in equations. So, in the word equation, we should write Mg for magnesium, O_2 for oxygen and MgO for magnesium oxide.

$$Mg + O_2 \rightarrow MgO$$

This is more helpful than the word equation, but it doesn't balance. There are two oxygen atoms in O_2 on the left, but only one oxygen atom in MgO on the right. So MgO must be doubled to give

$$Mg + O_2 \rightarrow 2MgO$$

Unfortunately, the equation still doesn't balance. We now have one Mg atom on the left, but two Mg atoms in 2MgO on the right. This is easily corrected by writing 2Mg on the left to give:

$$2Mg + O_2 \rightarrow 2MgO$$

The numbers of different atoms are now the same on both sides of the arrow. This is a **balanced chemical equation**.

A **balanced chemical equation** shows the atoms involved in a chemical reaction. No atoms are lost or made in a chemical reaction, so the numbers of each kind of atom are the same on both sides of a balanced equation.

Figure 1.17 shows a picture equation for this reaction. Picture equations help us to understand how the atoms are rearranged in reactions.

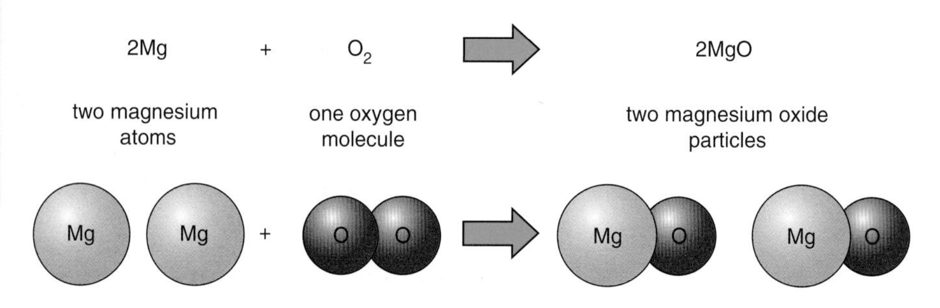

Figure 1.17 A picture equation for the reaction between magnesium and oxygen

Picture equations show that no atoms are lost during a chemical reaction. The atoms in the reactants are all still there at the end of the reaction but they are arranged differently to make different substances. As all the atoms are present at the start and finish of the reaction, the mass of the products must equal the mass of the reactants.

This is summarised in the law of conservation of mass, which says:

In any chemical change, the total mass of the products equals the total mass of the reactants.

No atoms are lost or made during chemical reactions so we can write balanced equations by making sure there are the same number of atoms of each element on both sides of the arrow.

The following example shows the three steps to follow.

Step 1 Write a word equation.

$$\text{hydrogen} + \text{oxygen} \rightarrow \text{water}$$

Step 2 Write symbols or formulae for the reactants and products.

$$H_2 + O_2 \rightarrow H_2O$$

Remember that hydrogen, oxygen, nitrogen and chlorine exist as molecules and are written as H_2, O_2, N_2 and Cl_2. All other elements are shown as single atoms (C for carbon, Fe for iron).

Step 3 Balance the equation by making the number of atoms of each element the same on both sides.

$$2H_2 + O_2 \rightarrow 2H_2O$$

Figure 1.18 shows a picture equation for this reaction.

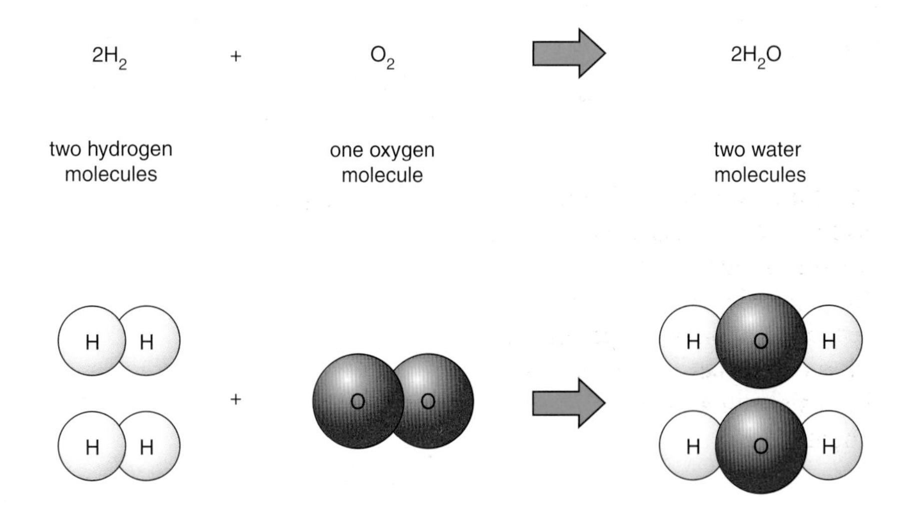

Figure 1.18 A picture equation for the reaction of hydrogen with oxygen to form water

Remember that you must never change a formula to make an equation balance. The formula for water is always H_2O and never HO or HO_2. Similarly, the formula of magnesium oxide is always MgO and never MgO_2 or Mg_2O.

You can only balance an equation by putting numbers in front of symbols or formulae, for example, 2Mg, 2MgO, $2H_2$ and $2H_2O$.

Figure 1.19 In a barbecue, charcoal (carbon) burns in oxygen in the air to form carbon dioxide.

14 a) Write a word equation for the reaction of charcoal (carbon) burning in oxygen to form carbon dioxide.
b) Why is it important that equations balance?
c) Why is the formula for water always H_2O and never HO or HO_2?

Balanced equations are more useful than word equations because they show:

- the symbols and formulae of the reactants and products;
- the relative numbers of atoms and molecules of the reactants and products;
- the rearrangement of atoms from reactants to products.

State symbols

State symbols are used in equations to show the state of a substance. (s) after a formula or symbol indicates that the substance is a solid. (l) is used for a liquid, (g) for a gas and (aq) for an aqueous solution (where a substance is dissolved in water). For example,

$$2Mg(s) + O_2(g) \rightarrow 2MgO(s)$$

Figure 1.20 When natural gas burns on a hob, methane (CH_4), reacts with oxygen in the air to form carbon dioxide and water.

15 Copy and balance the equation for the reaction in Figure 1.20.

$$CH_4 + \_\_ O_2 \rightarrow CO_2 + \_\_ H_2O$$

16 Propane, C_3H_8, is used in large red cylinders as a portable fuel in caravans and in some homes. The word equation and balanced chemical equation for burning propane are shown below.

propane + oxygen → carbon dioxide + water

$$C_3H_8(g) + 5O_2(g) \rightarrow 3CO_2(g) + 4H_2O(g)$$

Write down six things that the balanced equation tells you and which you could not know from the word equation. This should help you to appreciate why chemists like to write balanced equations for chemical reactions.

17 a) The following equations are not balanced. Write out the equations and balance them.

i) $N_2 + H_2 \rightarrow NH_3$

ii) $Na + H_2O \rightarrow NaOH + H_2$

iii) $C_2H_6 + O_2 \rightarrow CO_2 + H_2O$

b) Write balanced equations with state symbols for the following word equations.

i) calcium + oxygen → calcium oxide (CaO)

ii) hydrogen + chlorine → hydrogen chloride

iii) zinc + hydrochloric acid (HCl) → zinc chloride ($ZnCl_2$) + hydrogen

1.5 How do rocks provide building materials?

Figure 1.21 Blocks of limestone have been used to build castles, cathedrals and houses for hundreds of years.

Rocks can be quarried or mined to provide essential building materials such as stone for building homes, factories and offices. One of the most important rocks and naturally occurring resources is limestone. Limestone is mainly calcium carbonate, $CaCO_3$. It contains calcium ions, Ca^{2+} combined with carbonate ions, CO_3^{2-}. Each carbonate ion has one carbon atom bonded to three oxygen atoms with an overall charge of 2−. Limestone is quarried and used as building stone. The stone can be broken into smaller pieces and used as aggregate and chippings in concrete. Large amounts of limestone aggregate are used for making roads every year.

Figure 1.22 Mining engineers like those in these photos must survey quarry sites with great care before setting explosives to dislodge the rock and quarry it safely.

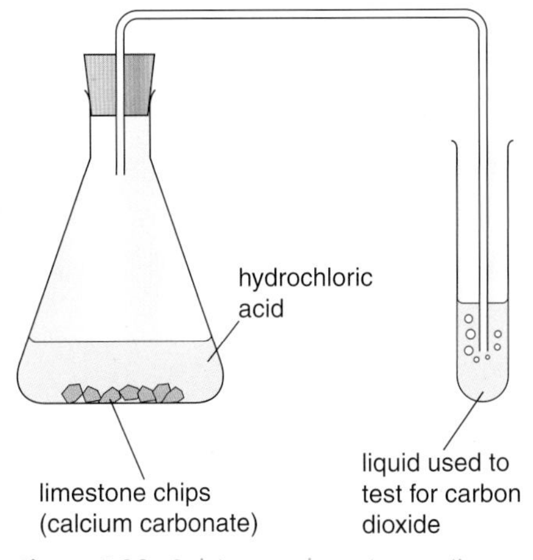

Figure 1.23 Calcium carbonate reacting with hydrochloric acid, with a test for the carbon dioxide produced

Using limestone to neutralise acidity

Limestone chippings are crushed to produce powdered limestone. This is used to neutralise acidity in soils and lakes. In this reaction, the limestone (calcium carbonate) reacts with acid to produce a calcium compound, water and carbon dioxide, which causes the mixture to 'fizz'. For example, with hydrochloric acid:

calcium carbonate + hydrochloric acid → calcium chloride + carbon dioxide + water

$$CaCO_3(s) + 2HCl(aq) \rightarrow CaCl_2(aq) + CO_2(g) + H_2O(l)$$

Figure 1.23 shows calcium carbonate and hydrochloric acid reacting with a test for the carbon dioxide produced.

Decomposing limestone to make quicklime and slaked lime

Limestone (calcium carbonate) decomposes when it is heated strongly. The products are calcium oxide (commonly called quicklime) and carbon dioxide.

calcium carbonate → calcium oxide + carbon dioxide

$$CaCO_3(s) \rightarrow CaO(s) + CO_2(g)$$

18 a) Name the liquid used to test for carbon dioxide.
b) What happens to this liquid as carbon dioxide bubbles into it?
c) Why is the soil in limestone areas neutral or slightly alkaline?

Calcium hydroxide is slightly soluble in water. The solution is called lime water, which can be used to test for carbon dioxide. When carbon dioxide is bubbled into lime water, calcium carbonate forms as a milky precipitate.

This is an example of thermal decomposition – using heat to break down a compound.

Carbonates of other metals decompose on heating in a similar way to calcium carbonate. The lower a metal is in the reactivity series, the more easily its carbonate decomposes.

Calcium oxide (quicklime) reacts vigorously with water to form calcium hydroxide (often called slaked lime). The reaction is strongly exothermic, which means that heat is produced.

$$\text{calcium oxide} + \text{water} \rightarrow \text{calcium hydroxide}$$
$$CaO(s) + H_2O(l) \rightarrow Ca(OH)_2(s)$$

Both quicklime and slaked lime are useful substances. They are used in industry as cheap alkalis to neutralise acidity. Calcium hydroxide (slaked lime) is used by water companies to neutralise acid in water supplies and to make bleaching powder. Farmers and gardeners also use calcium hydroxide like powdered limestone on acid soils to neutralise acidity.

Figure 1.24 summarises the reactions of calcium carbonate, calcium oxide and calcium hydroxide.

19 Limestone is insoluble in water, but slaked lime (calcium hydroxide) is slightly soluble. Why does this make slaked lime better than limestone for use on acid soils?

20 Copy Figure 1.24 and fill in the blank spaces.

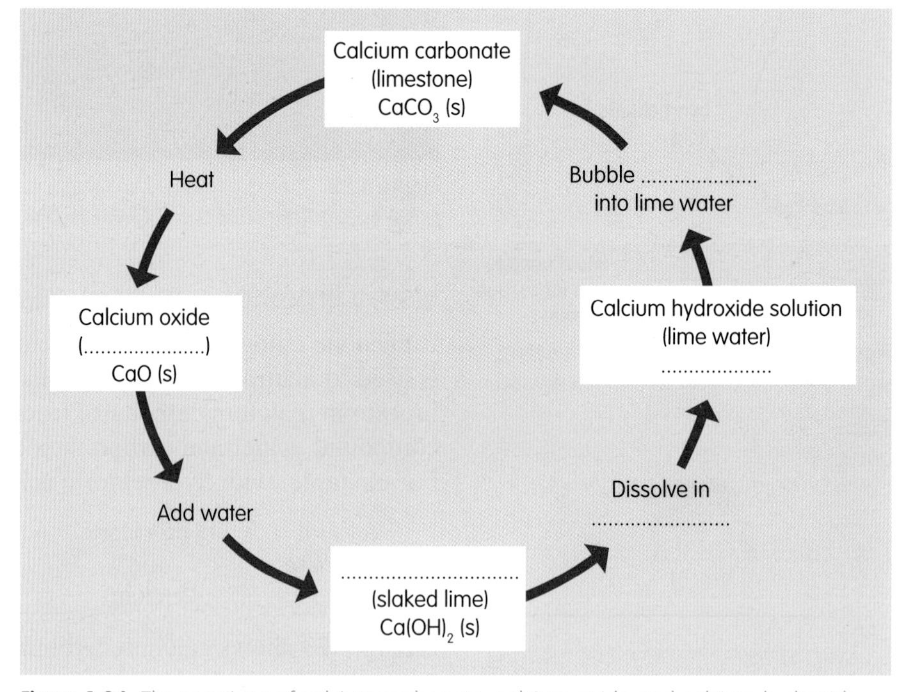

Figure 1.24 The reactions of calcium carbonate, calcium oxide and calcium hydroxide

How is limestone used to manufacture other useful materials?

In addition to quicklime and slaked lime, limestone also provides a starting point for the manufacture of cement, concrete and glass. It is also used in the extraction of iron from iron ore (see Section 1.7). All these uses make limestone an extremely valuable resource for the chemical and building industries (Figure 1.25).

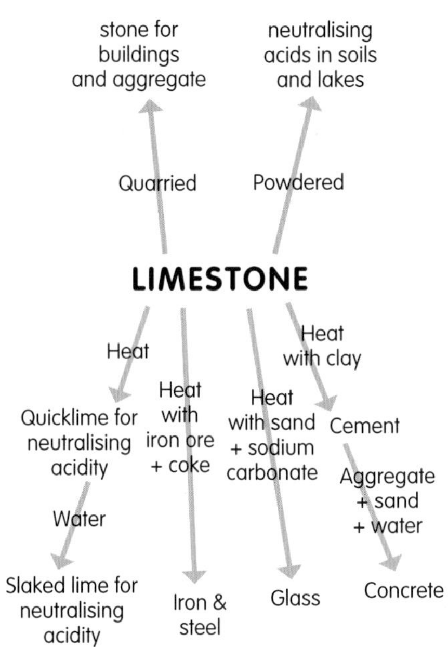

Figure 1.25 Important uses and products of limestone

Cement is made by heating limestone with clay in a kiln. When cement is used, it is normally mixed with two or three times as much sand as well as water. This mixture is called mortar. Mortar reacts slowly and sets to form a very hard material. Bricklayers use mortar to hold bricks firmly together.

Concrete is made by mixing mortar with aggregate (small pieces of broken rock). As the mortar sets around the aggregate, it produces a hard, stone-like building material. The mixture will even set under water at room temperature. Isn't that strange and extraordinary?

Glass is usually made by heating a mixture of metal oxides or metal carbonates with pure sand (silicon dioxide, SiO_2) in a furnace. At the high temperatures in the furnace, carbonates decompose to oxides and bubbles of carbon dioxide escape from the mixture. If the mixture is heated further, a runny liquid forms. This liquid is allowed to cool until it is thick enough to be moulded or blown into different shapes. On further cooling, the glass sets solid.

Ordinary glass for bottles and windows is made by heating a mixture of limestone (calcium carbonate), soda (sodium carbonate) and sand. This is sometimes called soda glass.

Figure 1.26 Cutglass dishes and ornaments are made of lead glass, which is harder and shinier than ordinary glass.

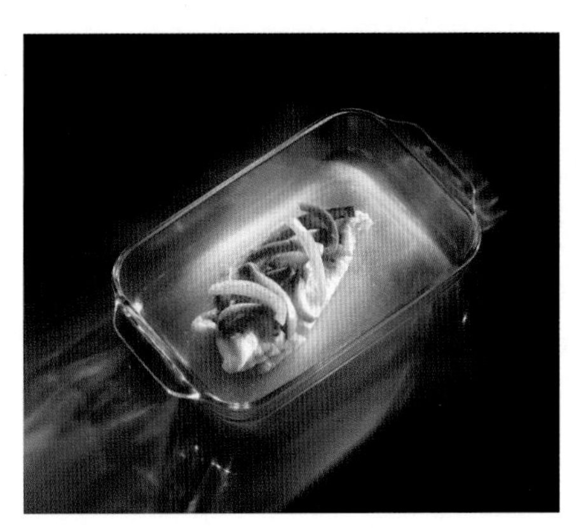

Figure 1.27 Glass ovenware and laboratory glassware are made of borosilicate glass (Pyrex®), which is heat resistant.

Figure 1.28 Blue glass is made by adding cobalt oxide to the usual constituents for making glass.

21 Various types of special glass are produced for particular purposes. Some of these are shown in Figures 1.26, 1.27 and 1.28.

a) What substances do you think are added to the usual constituents in order to make
   i) lead glass;
   ii) Pyrex®?
b) Which element produces the colour in blue glass?
c) Which other elements might produce colour in glass?
d) What benefits do these photos show in developments in the use of glass?
e) What are the risks and drawbacks of using glass?

Activity – Quarrying or countryside?

There are important environmental, social and economic issues involved in the quarrying and mining of rocks and ores. These issues are well illustrated in the UK by the quarrying of limestone and the production of building materials from it. Limestone occurs in some of the most beautiful areas of Britain: the Yorkshire Dales, the Peak District in Derbyshire, the Chilterns in Buckinghamshire and the Sussex Downs. The quarrying of limestone can spoil the countryside and create environmental problems.

On the other hand, limestone is a very important raw material for industry. Every year about 90 million tonnes of limestone are quarried in Britain. The limestone industry provides useful products for society; it also creates jobs and increases our wealth as a country.

So, how do we balance the benefits of quarrying with the problems it causes?

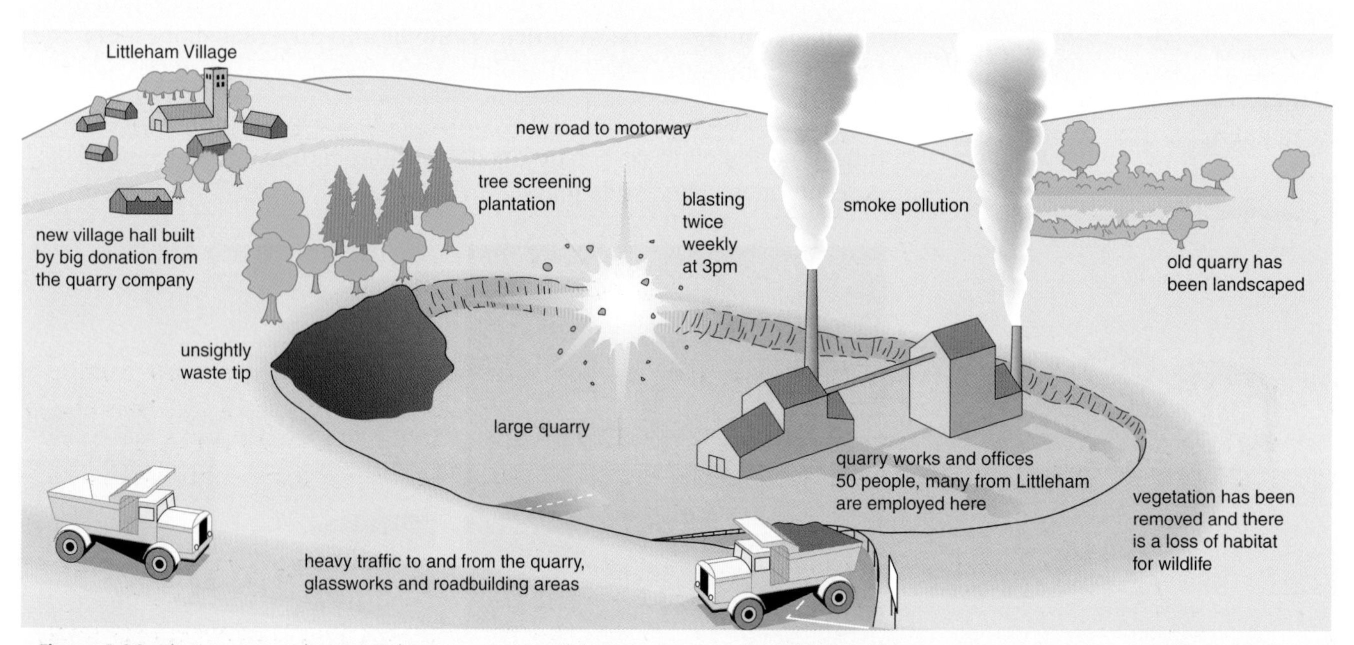

Figure 1.29 The quarry and surrounding area near Littleham

Look carefully at Figure 1.29, which shows a quarry near Littleham.

1. Suppose you live in Littleham. Make separate lists showing three advantages and three disadvantages of the nearby quarry for Littleham.
2. Suppose you are the Chairman of the Littleham Residents' Association. Write a letter to the Chief Executive of Limestone UK, the quarry operators, expressing the complaints you have had from members of the Residents' Association.
3. Suppose you are the Chief Executive of Limestone UK. Write a reply to the Chairman of the Littleham Residents' Association describing your efforts to reduce problems caused by the quarry and the improvements and advantages it has created for the area and people of Littleham.
4. Science can help us in many ways, but there are some questions that science cannot answer at all. These tend to be questions where beliefs are important, where views are personal or where we cannot obtain reliable evidence. Science cannot tell us whether the quarry at Littleham has provided the residents with a richer or a poorer lifestyle. Why not?

How do rocks provide metals?

Besides building materials, rocks also provide important metal **ores**. These ores contain enough metal or metal compounds to make it economic to extract the metal (Table 1.2).

Figure 1.30 Attractive crystals of gold on quartz

Name of ore	Name and formula of metal compound in the ore	Metal obtained
Bauxite	Aluminium oxide, Al_2O_3	Aluminium
Iron ore (haematite)	Red iron oxide, Fe_2O_3	Iron
Copper pyrites	Copper sulfide, CuS, and iron sulfide, FeS	Copper

Table 1.2 The ores from which we extract some important metals

An **ore** is a rock or mineral from which a metal can be extracted.

From ores to metals

Extracting (getting) metals from their ores usually involves four stages (Figure 1.31):

1. Mining (digging up) the ore.
2. Concentrating (separating) the ore.
3. Reducing (converting) the ore to the metal.
4. Purifying the metal.

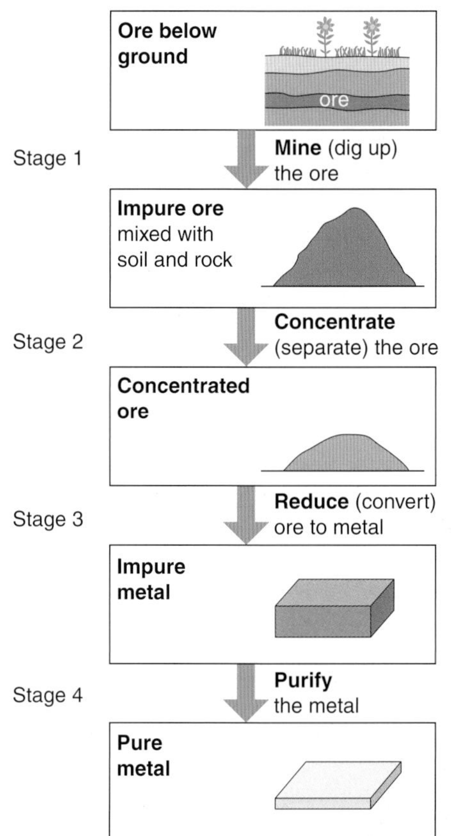

Figure 1.31 The four stages in extracting a metal from its ore

Mining metal ores involves quarrying, tunnelling or open-cast mining. After mining, the ore must be separated from impurities such as soil and waste rock. This is called concentrating the ore. First the rock is crushed. Then, the ore is separated from the waste by a process that relies on their different densities.

The ores of some metals are in very limited supply. Others are more plentiful, but even the richest ores are impure. Iron ore (haematite) is over 80% pure Fe_2O_3 in many parts of the world, but copper ores rarely contain more than 1% of the pure copper compound.

A few metals, such as gold, are so unreactive that they are found in the Earth as the metals themselves. So, extracting these metals does not involve a chemical process. In the case of metals like gold, the concentrated ore is the impure metal itself. So, these metals can be obtained by simply mining, concentrating and purifying the metal.

Most metals are too reactive to exist on their own in the Earth. Their ores are compounds – usually metal oxides or substances such as sulfides and carbonates that can easily be changed into oxides. The metal can then be obtained from the metal oxide by removing oxygen. This loss of oxygen by the metal oxide is an example of reduction. Reduction is studied in more detail in Section 1.7.

There are two main methods of reducing metal compounds depending on the position of the metal in the reactivity series (Figure 1.32).

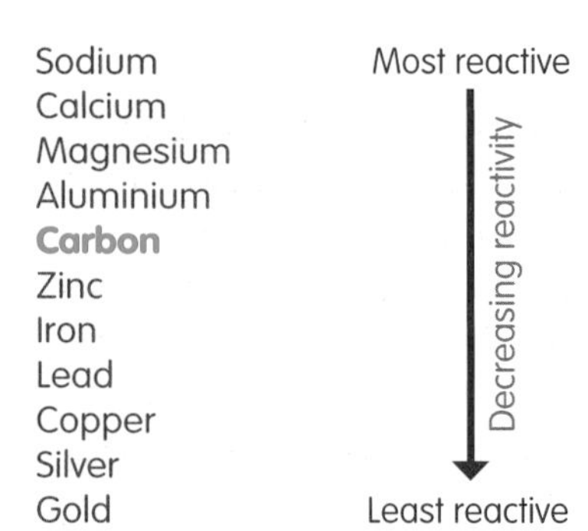

Figure 1.32 The reactivity series showing the position of carbon

Notice that carbon has been included in Figure 1.32 even though it is not a metal. Any element higher in the reactivity series can displace an element lower down from its compounds.

Figure 1.32 shows that metals, like zinc and iron, in the middle of the reactivity series are less reactive than carbon. This means that they can be extracted by reduction of their oxides with carbon (coke) or carbon monoxide.

Figure 1.33 Derelict tin mine buildings in Cornwall. Tinstone (tin oxide, SnO_2) was once mined in Cornwall and reduced to tin using carbon (coke) in furnaces near the mines. Decisions about whether to mine metal ores are often made on economic grounds. If the price of tin increased greatly in world markets, tin mining might well return to Cornwall

22 a) Copy and complete the following word equation for the formation of tin.
tin oxide + carbon → ?
b) Now write a balanced chemical equation for the process.
c) Why do you think that tin is no longer produced in Cornwall?

23 a) Name two metals, other than zinc or iron, that can be obtained by reducing their oxides with carbon (coke) or carbon monoxide.
b) Name two metals, other than sodium and aluminium, that are obtained by electrolysis of their molten compounds.

Metals like sodium and aluminium are above carbon in the reactivity series, so their compounds cannot be reduced to the metals using carbon. They are extracted by electrolysis. This involves decomposing the molten oxide or chloride to the metal using electricity.

Counting the cost of extracting metals

Extracting metals from their ores involves turning huge quantities of raw materials into useful and much more valuable metals. The metals can be used to manufacture a vast range of desirable products – vehicles, tools, pans, cutlery, cans, jewellery, pipes and girders.

The extraction of metals followed by the production and sale of valuable metal products creates jobs for many people. This improves their standard of living and adds to the wealth of a nation. But we, as a society, don't get these benefits for nothing. There are social, environmental and economic costs.

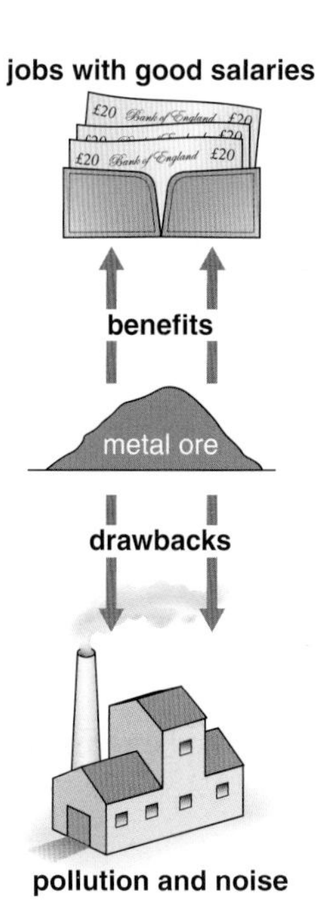

Figure 1.34 One of the benefits and one of the drawbacks of exploiting metal ores

Social costs

People who work in the mining and metals industries are exposed to health and safety risks. Some of the processes involve chemicals which are toxic (poisonous) and workers can suffer damage to their hearing from loud factory noise. It is important to remember, however, that it is possible to work safely with hazardous chemicals and in noisy factories by taking suitable precautions. Adverse effects may extend beyond the mines and factories to people living in the area. But this will not happen if appropriate health and safety regulations are followed.

Large industrial operations like mining and quarrying often involve rapid changes in the number of people employed and living in an area. This can put great strain on social services like schools and hospitals.

Environmental costs

Where there is mining, quarrying and the use of heavy machinery, wildlife habitats and farmland will be destroyed. The extraction of ores and transport of materials also create noise and pollution. The pollution comes in various forms: air pollution from factory chimneys and vehicle exhausts, dust from blasting and unsightly tips of waste materials.

Economic costs

Large industrial operations also incur huge economic costs due to the use of expensive machinery and materials, particularly fuels. Fuels are needed to operate machinery, to heat buildings and maintain chemical processes, as well as to transport workers and materials.

When you next buy a can of Coke® or a piece of jewellery remember that although these items may add to your enjoyment, their production has social, environmental and economic costs.

24 Look carefully through the last sub-section headed 'Counting the cost of extracting metals'.
a) Make a list of other benefits and drawbacks similar to the two already shown in Figure 1.34.
b) Sketch out Figure 1.34 and add further artwork and labels to your own diagram to show benefits and drawbacks.

How is iron extracted from iron ore?

The main raw material for making iron is iron ore (haematite). Haematite is impure iron oxide, Fe_2O_3. The iron ore is usually obtained by open-cast mining.

Iron is extracted from the iron ore in a **blast furnace**. A blast furnace is a large tower about 15 metres tall. Figure 1.35 on the next page shows a diagram of a blast furnace with an explanation of the chemical processes alongside.

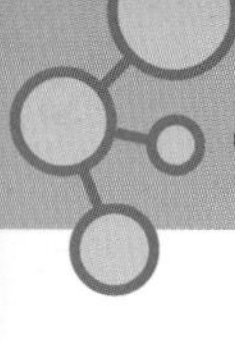

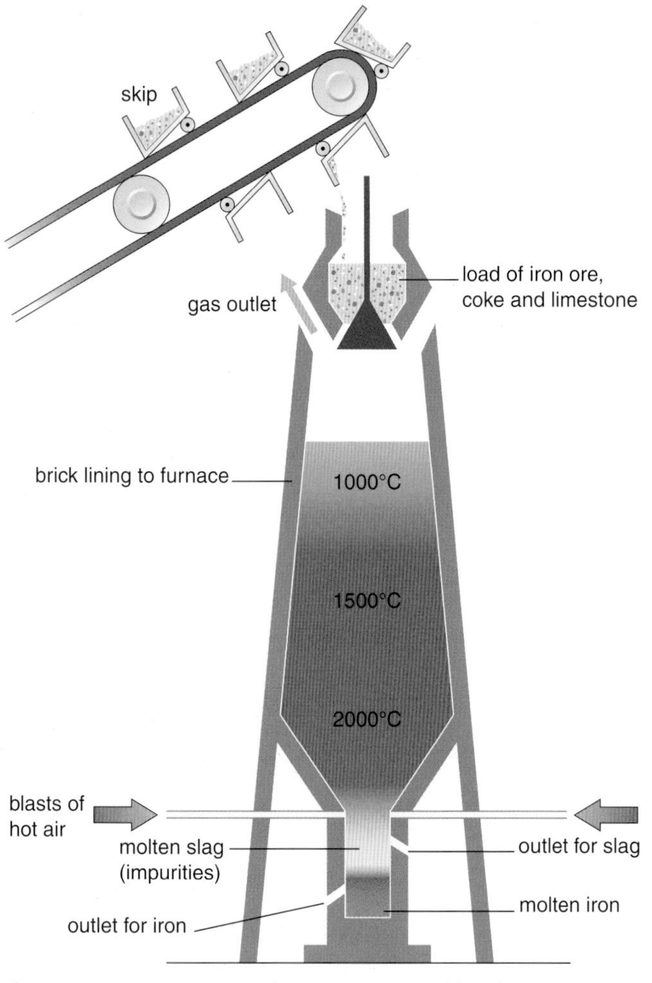

Figure 1.35 Extracting iron from iron ore in a blast furnace

1. Solid raw materials (iron ore, coke (carbon) and limestone) are added at the top of the furnace.
2. Blasts of hot air (which give the furnace its name) are blown in near the bottom of the furnace.
3. Oxygen in the blasts of air causes the coke (carbon) to burn, forming carbon dioxide and releasing energy (heat).

carbon + oxygen → carbon dioxide

$$C + O_2 \rightarrow CO_2$$

4. At the high temperatures in the furnace, carbon dioxide reacts with more coke (carbon) to form carbon monoxide.

carbon dioxide + carbon → carbon monoxide

$$CO_2 + C \rightarrow 2CO$$

5. The carbon monoxide reacts with the iron ore (red iron oxide) producing carbon dioxide and molten iron.

red iron oxide + carbon monoxide → iron + carbon dioxide

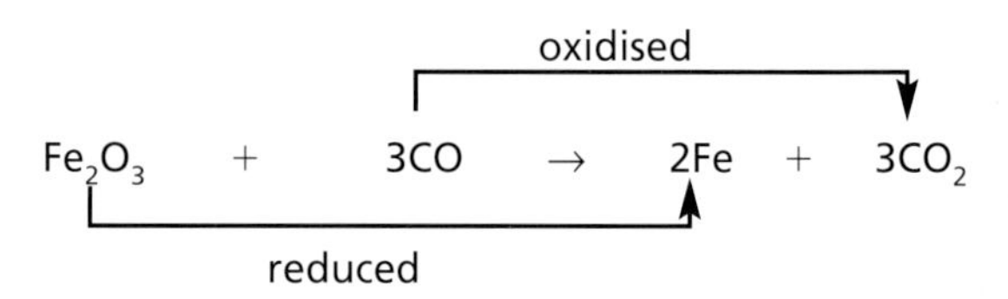

6. Molten iron runs to the bottom of the furnace and is tapped off from time to time.

---

25 The reaction between carbon dioxide and coke (carbon) to form carbon monoxide is a redox reaction.

a) What is a redox reaction?

b) In this redox reaction, which substance is:

i) oxidised;

ii) reduced;

iii) the oxidising agent;

iv) the reducing agent?

In the reaction between Fe_2O_3 and CO, carbon monoxide gains oxygen forming carbon dioxide. This gain of oxygen is called **oxidation** and the carbon monoxide is said to be oxidised. At the same time, iron oxide loses oxygen. This loss of oxygen is called **reduction** and the iron oxide is said to be reduced. Iron oxide, which supplies oxygen, is described as the oxidising agent and carbon monoxide, which takes oxygen, is described as the reducing agent.

Oxidation and reduction always occur together. If one substance gains oxygen and is oxidised, another substance must lose oxygen and be reduced. We call the combined process **redox** (**red**uction + **ox**idation).

It may seem confusing, but notice that during redox reactions:

- the oxidising agent (in this case iron oxide) is reduced;
- the reducing agent (in this case carbon monoxide) is oxidised.

Why is limestone used in the furnace?

The main impurity in iron ore is sand (impure silicon dioxide, SiO_2). This is removed by limestone.

Oxidation is the addition of oxygen to a substance.

Reduction is the removal of oxygen from a substance.

Reduction and oxidation reactions always occur together, as a combined **redox** process.

Figure 1.36 Molten iron being poured from a furnace

At the high temperatures in the furnace, limestone decomposes forming calcium oxide and carbon dioxide. The calcium oxide, which is basic, reacts with sand (SiO_2), which is acidic, to form 'slag', calcium silicate.

calcium oxide + silicon dioxide → calcium silicate

$$CaO + SiO_2 \rightarrow CaSiO_3$$

The molten 'slag' falls to the bottom of the furnace and floats on the molten iron. This can be tapped off at a different level from the molten iron. The 'slag' is used in road making and cement manufacture.

Why is iron converted to steel?

Iron from the blast furnace contains about 96% iron. The main impurity in this iron is carbon. This makes it brittle, so it has only limited uses. Removing all the impurities from the iron would produce pure iron. This is too malleable (easily shaped) and too soft for most uses. Most hot molten iron from the blast furnace goes straight to a steel making furnace. Here it is converted into steels with the ideal strength and hardness. Steels are alloys – mixtures of iron with carbon and often other metals.

Steel is made by blowing oxygen under pressure onto the hot, molten impure iron. The oxygen converts excess carbon to carbon dioxide which escapes as a gas.

Using alloys

Alloys can be designed and manufactured to have properties for specific uses. Some are designed for hardness, some for resistance to corrosion and others have special magnetic or electrical properties.

Alloys are usually made by melting the main metal and then dissolving the other elements in it.

The most important alloys are steels. The composition, properties and uses of various steels are shown in Table 1.3.

Type of steel	Composition	Properties	Uses
Low-carbon steel (mild steel)	99.8% iron 0.2% carbon	Easily pressed into shapes	Car bodies
High-carbon steel	98.0% iron 1.7% carbon 0.3% manganese	Hard but brittle	Tools
Stainless steel	73.7% iron 0.3% carbon 18.0% chromium 8.0% nickel	Hard and rresistant to corrosion	Cutlery, pans

Table 1.3 The composition, properties and uses of various steels

Figure 1.37 Spectacle frames made of 'smart alloys' will revert to their original shape in a second, after being bent and twisted

Most metals in everyday use are alloys. Like iron, pure copper, pure aluminium and pure gold are too soft for most uses. So, they are mixed with small amounts of other metals to make them harder. During the last 40 years, aluminium alloys have been used more and more. These include duralumin, which contains 4% copper. Aluminium alloys are light, strong and corrosion resistant. They are used for aircraft bodywork, overhead electricity cables and lightweight tubing.

In recent years, smart alloys, sometimes called 'memory metals', have been developed. These can return to their original shape after being deformed. Smart alloys are excellent for use in spectacle frames and in the braces fitted by dentists. The smart alloy braces are made so that, after fitting, they will return to their original shape and pull the teeth into better alignment.

The main elements in important alloys are transition metals in the central block of the Periodic Table. These transition metals include iron, copper, chromium, nickel and titanium.

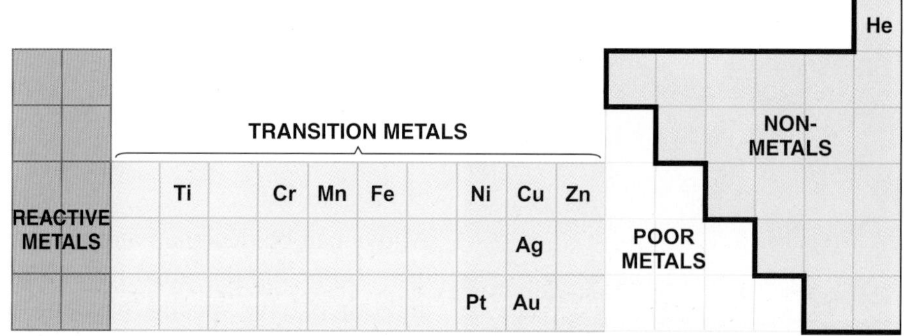

Figure 1.38 The position of the transition metals in the Periodic Table

Like other metals, transition metals are good conductors of heat and electricity. They can also support heavy loads and can be bent or hammered into shape. So, they are useful as structural materials and for making things that must conduct heat or electricity easily. Steel is one of our most important structural materials. It is used in girders, joists and bridges. Copper has properties that make it useful for electrical wiring and in pipes for plumbing.

Aluminium and titanium are two other useful metals because of their low density and resistance to corrosion. Both metals occur in ores as their oxides, but these oxides cannot be extracted by reduction with carbon. Current methods of extracting the two metals are expensive because large amounts of energy are needed and there are several stages in each process. These costs have limited the use of titanium.

Figure 1.39 A hip joint made of titanium alloy

26 This question is about new ways to extract copper. Answer the questions as you read the passage.

Copper ores contain copper sulfide (CuS). At one time, all copper was extracted from these sulfide ores by first converting the copper sulfide to copper oxide and then reducing this to copper by heating with carbon.

During the last 20 years, the supply of copper-rich ores has become very limited. This has led chemists to look for new ways of extracting copper from low-grade ores.

a) What is the main environmental problem of mining low-grade ores that contain vast amounts of worthless rock?

Fortunately, there are helpful bacteria in most copper ores. These bacteria use oxygen in the air to oxidise insoluble copper sulfide to soluble copper sulfate ($CuSO_4$).

b) Write a word equation for this reaction.

The bacteria found in the ore can tolerate:

- acidic conditions;
- heat generated by the reaction;
- the copper compounds which are poisonous to most organisms.

To extract the copper a heap of crushed rock is simply sprayed with very dilute sulfuric acid and dilute copper sulfate solution trickles from the bottom of the pile.

c) What conditions help this process to go faster?

d) What conditions in this process might have a damaging effect on wildlife and the environment?

Finally, copper metal is extracted from the copper sulfate solution by electrolysis.

Explaining the properties of alloys

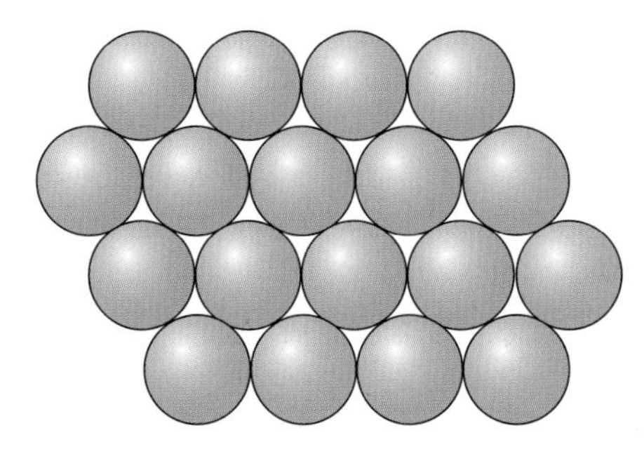

Figure 1.40 Close packing of atoms in a metal

X-ray analysis shows that the atoms in most metals are packed as close together as possible. This arrangement is called close packing. Figure 1.40 shows a few close-packed atoms in one layer of a metal.

The bonds between atoms in a metal are strong, but they are not rigid. When a force is applied to a metal, the layers of atoms can slide over each other, allowing the metal to be soft and malleable. This movement of atoms in a metal is called slip. After slipping, the atoms settle into position again and the close-packed structure is restored. Figure 1.41 shows the positions of atoms before and after slip. This is what happens when a metal is hammered or pressed into different shapes.

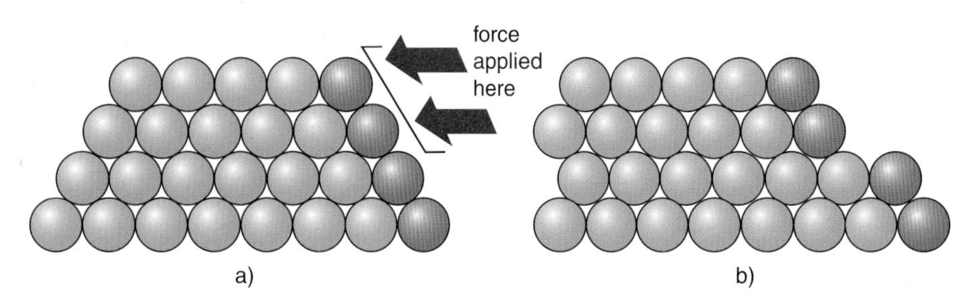

Figure 1.41 The positions of atoms in a metal a) before and b) after 'slip' has occurred

A variable is something that can vary. There are different kinds of variables.

Continuous variables can have any numerical value, such as the length of a piece of string or the mass of a stone.

Discrete variables are restricted to whole numbers, for example the number of atoms in a molecule.

Ordered variables have a clear order of size or mass or length, such as small, medium and large pairs of socks.

Categoric variables are different types of something. For example, if you were investigating the effect of acid on different metals, the type of metal would be a categoric variable.

In an investigation, the variable that you change or select is called the **independent variable**.

The variable that you measure for each and every change of the independent variable is called the **dependent variable**.

A **control variable** is one which, in addition to the independent variable, may affect the outcome of an investigation. All control variables should be kept constant.

In a **fair test**, only the independent variable is allowed to affect the dependent variable.

In steel, the regular arrangement of iron atoms is disrupted by adding smaller carbon atoms. The different-sized carbon atoms distort the layers of iron atoms in the structure of the pure metal. This makes it more difficult for the layers to slide over each other, so the steel alloy is harder and much less malleable (Figure 1.42).

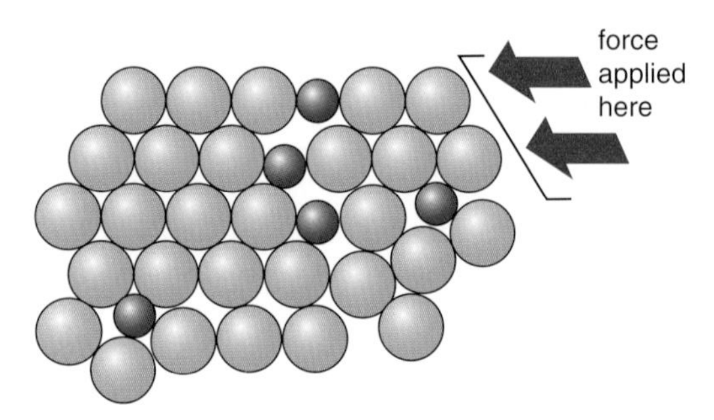

Figure 1.42 Slip cannot occur so easily in the steel alloy because the atoms of different sizes cannot slide over each other.

Activity – Recycling metals

Do you and your family recycle metals? Scrap metals such as drinks cans, aluminium foil and old domestic machines can be melted down and used again. Local authorities are now expected to achieve certain targets for recycling waste materials such as metals, glass, paper and plastics. In many areas, there are strong reminders to encourage the recycling of these materials.

1. Extracting metals usually involves four stages (see Figure 1.31). Which of these stages are avoided if metals are recycled?
2. Now look carefully at 'Counting the cost of extracting metals' in Section 1.6. Use this section and Figure 1.29 in the activity 'Quarrying or countryside' to list the advantages of recycling metals.
3. Recycling is not always easy. Scrap metal has to be collected and then transported to where it can be re-used. The metal has to be separated from other material and sometimes one metal has to be separated from another.
   a) Bearing in mind that the metal has to be melted down during the recycling process, how do you think metals are separated from paper?
   b) The two metals recycled in the largest amounts are aluminium and iron (steel). How do you think these are separated?
4. Practically all the gold we use is recycled, but only about half the aluminium. Why do you think there is this difference?
5. Find out about the plans and targets for recycling metals in your area and write a few sentences about them.
6. What further initiatives could be taken to improve the recycling of metals in your area?

27 a) Why are the furnaces used to make iron called *blast* furnaces?
b) Why are blast furnaces usually built near coal fields?
c) Why is most iron from the blast furnace converted to steel?
d) Suggest three reasons why steel (iron) is used in greater quantities than any other metal.

28 Explain the following terms: *close packing*; *slip*; *malleable*.

29 The strengths of two metal wires can be compared by measuring the force needed to break each wire using the apparatus in Figure 1.43.

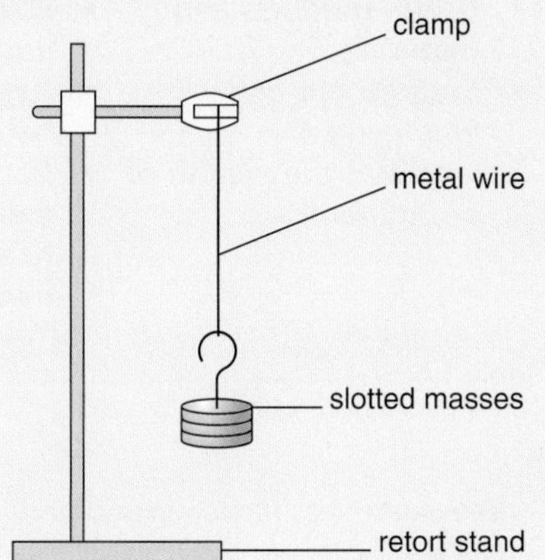

Figure 1.43 Comparing the strengths of metal wires

a) Describe briefly how you would carry out the experiment.
b) What measurements would you record?
c) Copy and complete the following sentences using words from the box below.

dependent	fair	independent
measured	same	selected

In this experiment the __________ variable is the type of metal wire used. This is the variable that is changed or __________ by the investigator.
The __________ variable in the experiment is the force needed to break each wire. This is the variable that is __________ when the independent variable changes.
In order to make the experiment __________, it is important that only the independent variable affects the dependent variable. All other possible variables must be kept the __________.

d) State two variables that you would control to ensure the wires are tested fairly.

Summary

- ✓ Chemistry is the study of materials and substances. Chemists and chemical engineers study materials and try to change **raw materials**, like iron ore and limestone, into useful **manufactured materials**, like steel, cement and glass.
- ✓ All substances are made of **atoms**. Atoms have a small central **nucleus** around which there are **electrons**.
- ✓ **Elements** are the simplest substances. They cannot be broken down any further. Each element contains only one kind of atom.
- ✓ In a chemical **compound** the atoms of two or more different elements are joined together by chemical bonds.
- ✓ **Mixtures** are two or more substances that are mixed together but not combined chemically. An **alloy** is a mixture of a metal with one or more other elements.
- ✓ When atoms combine, they can:
  – either share electrons to form **molecules**;
  – or give and take one or more electrons to form **ions**.
- ✓ An atom is the smallest particle of an element.
- ✓ A **molecule** is a particle containing two or more atoms joined by chemical bonds.
- ✓ An **ion** is a charged particle formed from an atom by the loss or gain of one or more electrons.

✓ Elements and atoms of elements are represented by **symbols**. Compounds and molecules of a compound are represented by **formulae**. A formula shows the number of atoms of the different elements that are joined together in one molecule of a compound.

✓ The **Periodic Table** shows all the known elements in order of atomic number. In each **group** of the Periodic Table the elements have similar properties.

✓ **Balanced chemical equations** use symbols and formulae to summarise the reactants and products in a reaction.

✓ Limestone is mainly calcium carbonate, $CaCO_3$. It is one of the most important and useful naturally occurring building materials. It provides us with quicklime and slaked lime and is a starting point for the manufacture of cement, concrete and glass.

✓ Extracting metals from their **ores** involves four stages:
- mining the ore;
- concentrating the ore;
- reducing the ore to the metal;
- purifying the metal.

✓ There are two main methods of extracting metals from their ores:
- electrolysis of fused (molten) compounds for metals above carbon in the reactivity series;
- reduction of oxides with coke (carbon), or carbon monoxide, for metals below carbon in the reactivity series.

✓ The quarrying and mining of rocks and ores, followed by the extraction of metals, brings both benefits and drawbacks (Table 1.4).

Benefits	Drawbacks
• Useful products and materials • Creates jobs and employment • Increases the wealth of a community	• Pollution from smoke, dust and noise • Destroys wildlife habitats • Damages the environment with spoil heaps, quarries and mines

Table 1.4 The benefits and drawbacks of quarrying, mining and metal extraction

✓ **Redox** involves reduction and oxidation.
Oxidation occurs when a substance gains oxygen.
Reduction occurs when a substance loses oxygen.

✓ The recycling of metals is important because:
- it helps to conserve limited resources of the metals;
- it reduces damage to the environment;
- it saves the cost of extracting the metals.

EXAMQUESTIONS

1 At one time, gutters and drainpipes were made of iron. Today they are made of plastics like PVC (polyvinyl chloride).

a) What properties of iron made it useful for gutters and drainpipes? *(2 marks)*

b) What were the problems of using iron? *(2 marks)*

c) Why has iron been replaced by plastics? *(2 marks)*

2 a) Limestone is used to make some important products. Name two important products from limestone listed below. *(2 marks)*

cement diesel glass petrol plastic

b) In an experiment a student heated a piece of limestone very strongly as shown in Figure 1.44.

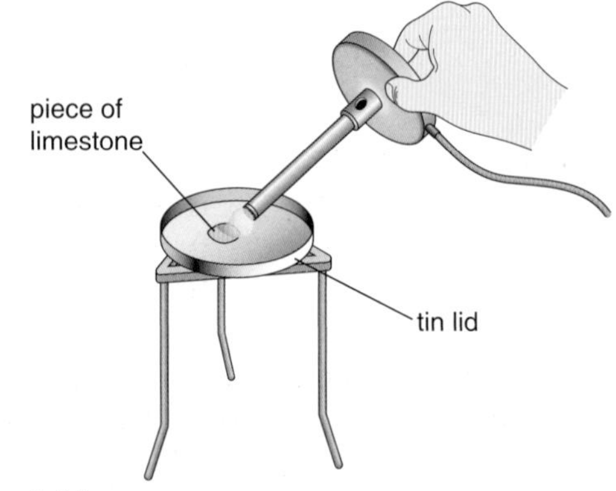

Figure 1.44

i) State one safety precaution that the student should take during this experiment. *(1 mark)*

ii) When limestone is heated, it forms two products: a white powder and carbon dioxide. What is the chemical name of the white powder? *(1 mark)*

c) The student did a second experiment using 3.00 g of limestone. The limestone was weighed before and after being heated. The student then repeated this experiment using a new sample of 3.00 g of limestone. The results are shown in Table 1.5.

	Experiment 1	Experiment 2
Mass of limestone before heating	3.00 g	3.00 g
Mass of limestone after heating	1.68 g	1.72 g
Mass lost	1.32 g	1.28 g

Table 1.5

i) What is the average mass lost for the two experiments? *(1 mark)*

ii) Why is it important to repeat this experiment? *(1 mark)*

iii) Why is the mass lost not the same for the two experiments? *(1 mark)*

iv) Why is a balance which measures to the nearest 0.1 g not suitable for this experiment? *(1 mark)*

3 Car bodies are made from low-carbon steel.

a) Explain in terms of atoms, why pure iron is useless for making car bodies. *(2 marks)*

b) Steel used for making car bodies is an alloy containing 99.8% iron and 0.2% carbon. Explain in terms of atoms how such a small percentage of carbon makes the steel suitable for car bodies. *(2 marks)*

4 Look at Figure 1.45.

a) Which process produces the strongest alloy – chill casting (rapid cooling) or sand casting (slow cooling) of the liquid alloy? *(1 mark)*

b) Are chill casting and sand casting examples of discrete variables, ordered variables or categoric variables? *(1 mark)*

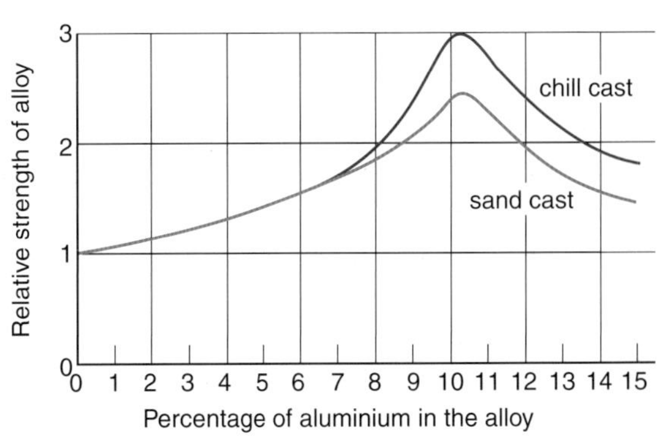

Figure 1.45 The effect of aluminium on the strength of copper alloys

c) What percentage of aluminium produces the strongest alloy? *(1 mark)*

d) How many times stronger is this alloy than pure copper? *(1 mark)*

e) What percentage of aluminium produces a sand-cast alloy twice as strong as pure copper? *(1 mark)*

f) Why do you think the strength of the alloy increases at first and then decreases as more aluminium is added? *(4 marks)*

5 a) A word equation for making a simple glass of calcium silicate ($CaSiO_3$) is:

limestone + sand → calcium silicate + carbon dioxide

Write a balanced chemical equation for the reaction. *(2 marks)*

b) Glass is used for bottles and ovenware.

i) List four properties which make glass so suitable for these uses. *(4 marks)*

ii) List two drawbacks of using glass for bottles and ovenware. *(2 marks)*

6 a) Many industrial processes involve the removal of minerals by quarrying. All quarrying has some effect on the environment and on people's lives. Make six important points about the social, economic, health or environmental effects of quarrying. *(6 marks)*

b) Aluminium is more expensive than iron. Why then is aluminium and not iron used for the central core of power cables between pylons in the National Grid? *(2 marks)*

Chapter 2
How does crude oil provide useful materials?

At the end of this chapter you should:

- ✓ understand that crude oil (petroleum) is a mixture of alkanes that can be separated into useful products;
- ✓ know how the process of oil refining works;
- ✓ understand that burning fossil fuels such as crude oil causes environmental problems, including acid rain and climate change;
- ✓ understand how cracking produces different hydrocarbons, which are useful as fuels;
- ✓ know that cracking produces alkenes, which can be used to manufacture a range of products;
- ✓ understand that ethene can react with steam to make ethanol;
- ✓ know that alkenes can be polymerised to make new materials known as polymers;
- ✓ know that polymers have a range of useful properties;
- ✓ understand the different uses of polymers;
- ✓ know that some polymers can be recycled easily;
- ✓ understand that most polymers are not biodegradable;
- ✓ be able to evaluate the effects on the environment of burning hydrocarbon fuels;
- ✓ know about the development of better fuels;
- ✓ be able to evaluate the uses of crude oil as a fuel and as a chemical to manufacture other materials;
- ✓ know about the development of polymer materials;
- ✓ be able to evaluate the manufacture of ethanol from renewable and non-renewable sources.

Figure 2.1 In the last 50 years, the 'garage' has changed its function. Gone are the mechanics and the repair of cars. The garage is now where you get your milk, lottery ticket, newspaper, sweets, phone top-up card, and fuel for the family car. Garages are an essential part of our everyday living and not just for the 'petrol heads'. The oil industry touches our lives many times a day. Often there are benefits, sometimes there are drawbacks.

Crude oil

Crude oil is a naturally occurring material found underground. It is a mixture of very similar compounds, called **hydrocarbons**.

A **hydrocarbon** is a compound that contains hydrogen and carbon atoms only.

Crude oil forms over thousands, if not millions, of years as dead organisms slowly break down. The hydrocarbons in crude oil vary a great deal in molecular size. These different sized molecules can be separated by fractional distillation.

Activity – Mixtures

1. Crude oil is sticky stuff. It often gets mixed with other materials. How could you separate the following mixtures:
   a) crude oil and seawater, to get clean seawater;
   b) crude oil and seawater, to recover the crude oil;
   c) crude oil and sand, to get clean sand (you may need to use another chemical)?
2. a) How would you clean crude oil off sea-bird feathers?
   b) What adverse effects might this have on the bird?

Figure 2.2 The bird may not appreciate this treatment!

Distillation

Mixtures of liquids can be separated by **distillation**.

Different substances have different boiling points. When a liquid boils, the liquid particles have to be separated from each other to make a gas. Three factors affect the boiling point of a liquid:

- atmospheric pressure – substances boil at a lower temperature under reduced pressure;
- the mass of the molecules in the liquid;
- the strength of the forces between the liquid particles.

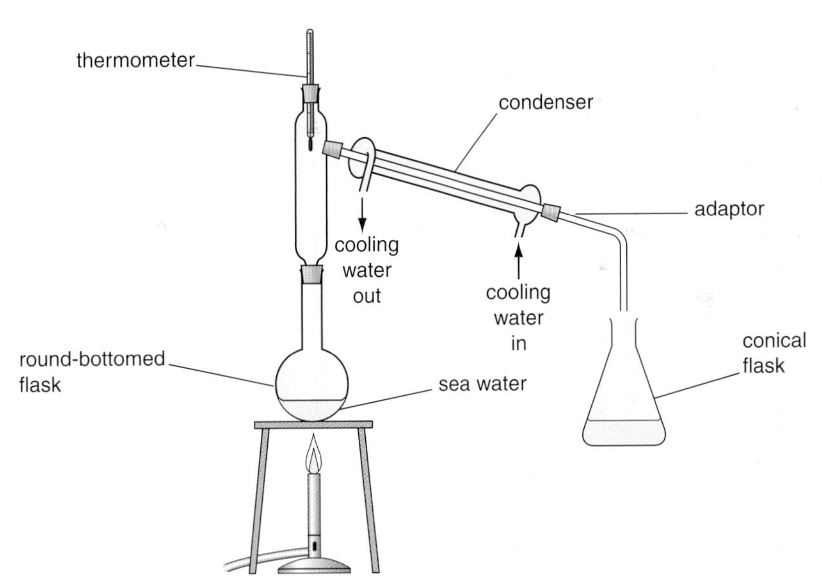

Figure 2.3 A liquid can be separated from a solution using this apparatus. Water or air can be used to cool the condenser. In this case, the condenser is cooled by water.

When a simple mixture like salt water (sodium chloride solution) is boiled, only water molecules are turned into vapour. The sodium ions and chloride ions in the water do not vaporise. They remain in the boiling liquid. The vapour that boils off is pure water. This can be cooled and condensed back to a liquid.

Various distillation methods are used to produce pure water from salt water. One of these is shown in Figure 2.4.

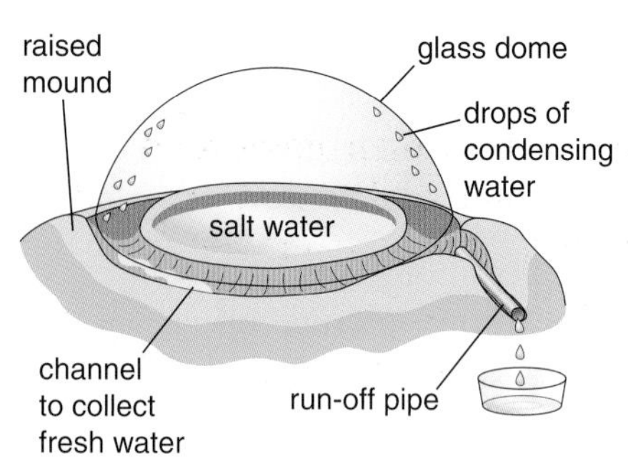

Figure 2.4 Turning salt water into pure water – sunshine is often used as an energy source.

Fractional distillation

Ordinary distillation cannot separate liquids with boiling points that are close together. This is done by fractional distillation.

For example, when a mixture of ethanol (boiling point 78 °C) and water (boiling point 100 °C) is heated, both the ethanol and the water molecules will vaporise together. When the vapour is condensed, it is still a mixture but there is more ethanol than water in this condensed mixture because ethanol boils at a lower temperature. The ethanol molecules 'escape' from the boiling liquid more easily.

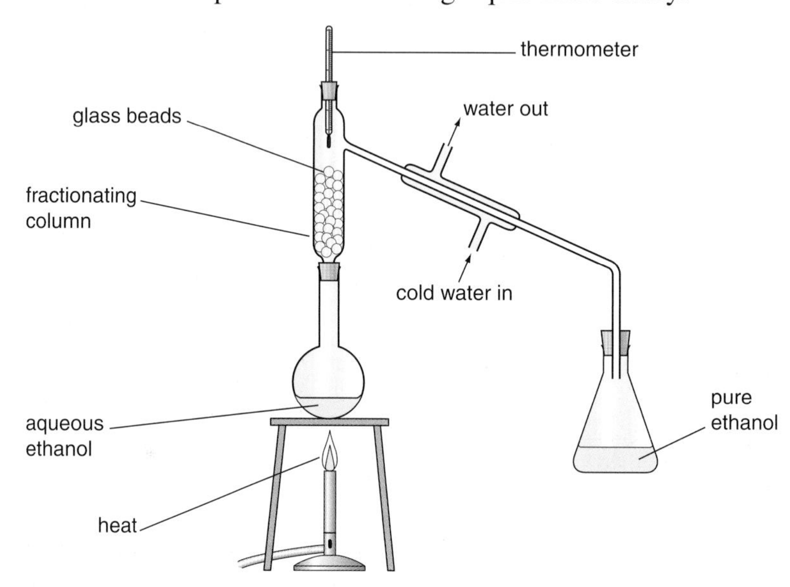

Figure 2.5 During fractional distillation, the vapours condense and evaporate several times in the vertical column.

When a liquid boils, or vaporises, its particles gain enough energy to escape from the surface and become a gas (vapour). **Distillation** occurs when this vapour is removed and condensed back to liquid in a separate container.

During fractional distillation, hot vapour condenses on cooler surfaces inside the column. The condensed mixture then evaporates again as more hot vapour rises up the column. Each time this happens, more of the lower boiling point liquid evaporates, and the higher boiling point liquid is left behind.

The higher boiling point liquid runs back down the column, and into the boiling mixture.

After the vapour has evaporated and condensed several times on its way up, only the lower boiling point liquid reaches the top of the column. This flows out of the column and is collected.

The way that fractional distillation is used in an oil refinery is described in Section 2.3.

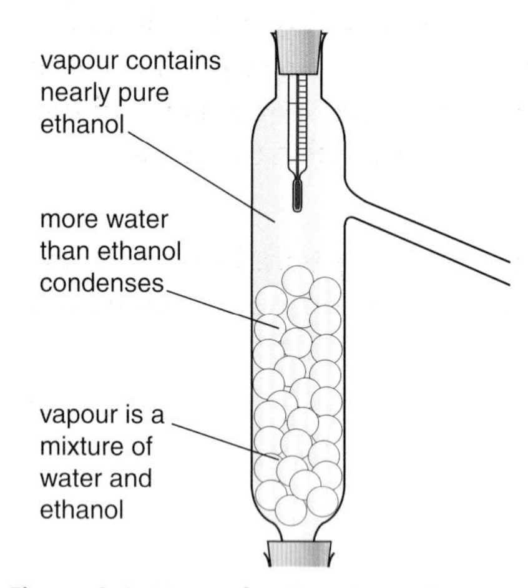

Figure 2.6 How a fractionating column works

Alkanes

The simplest hydrocarbons are alkanes. When carbon atoms form compounds, they can form four chemical bonds with other atoms, but hydrogen atoms can form only one chemical bond. So, the simplest alkane of all is methane, CH_4 (Figure 2.7).

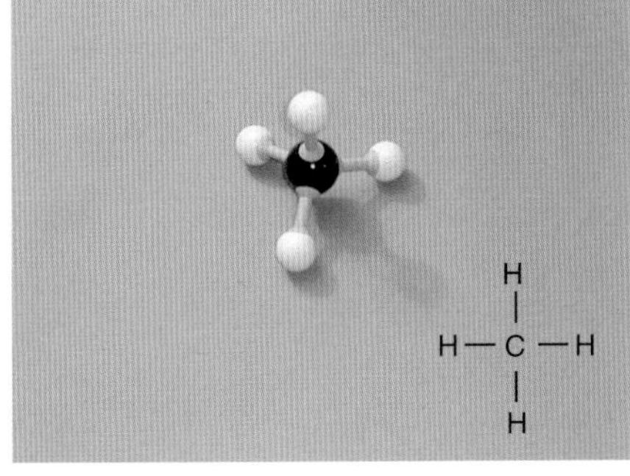

Figure 2.7 A molecular model of methane and its structural formula

The hydrocarbon with two carbon atoms is shown in Figure 2.8. It is called ethane.

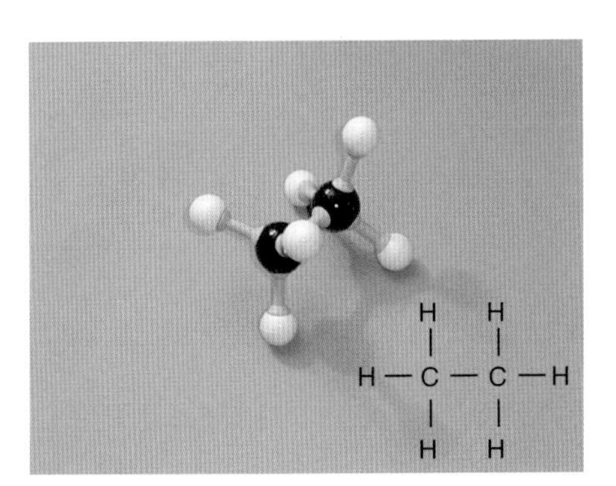

Figure 2.8 A molecular model of ethane and its structural formula

Can you see that ethane is like methane with an extra $-CH_2-$ unit? If you add another $-CH_2-$ unit you get a heavier molecule called propane (Figure 2.9a). Another $-CH_2-$ unit makes butane (Figure 2.9b).

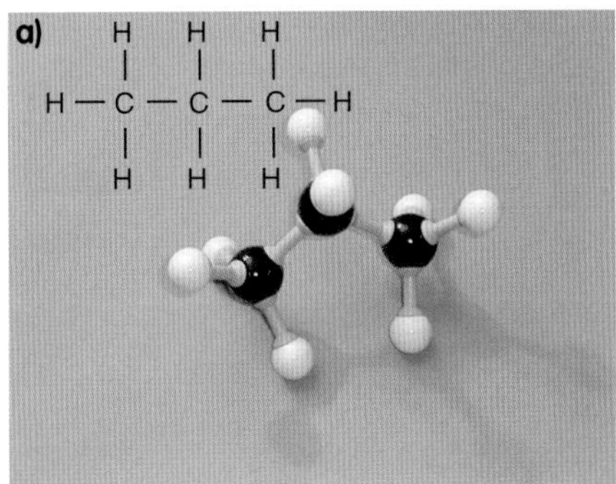

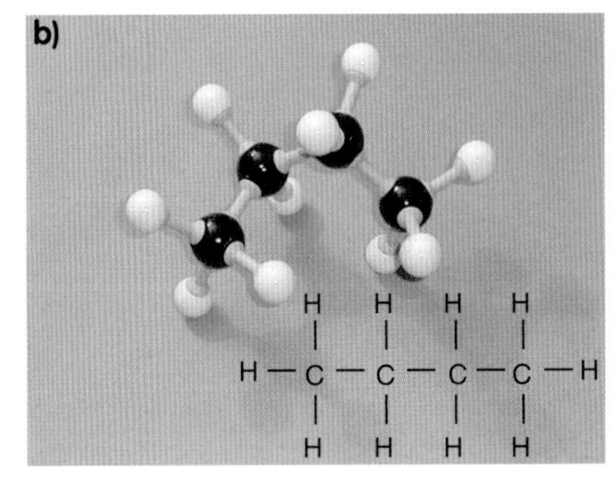

Figure 2.9 Molecular models and structural formulae of a) propane and b) butane

It is easy to see that this is becoming a series of similar molecules. This series of hydrocarbons are called the **alkanes**. Any alkane will have a chemical formula that can be written as $C_n H_{2n+2}$, where 'n' is a whole number. $C_n H_{2n+2}$ is known as the general formula for alkanes. All the alkanes have similar properties because of their similar molecular formulae.

The **alkanes** are a series of hydrocarbons with the general formula C_nH_{2n+2}. They are used as fuels.

The most important property of alkanes is that they burn in air to form carbon dioxide and water vapour. This is why the alkanes are all used as fuels (Table 2.1).

Alkane	Use
Methane	Kitchen gas for cookers
Propane	Camping (calor) gas for cookers
Butane	Bottled gas for heating
Octane	Petrol for cars
Dodecane	Diesel for vans

Table 2.1 Alkanes as fuels

Activity – Is there a link between the size of the molecules in an alkane and its boiling point?

Alkane	Formula	Boiling pt in °C
Methane	CH_4	−161
Ethane	C_2H_6	−89
Propane	C_3H_8	−42
Butane	C_4H_{10}	0
Pentane	C_5H_{12}	36
Hexane	C_6H_{14}	69
Heptane		98
Octane	C_8H_{18}	125
Nonane		151
Decane	$C_{10}H_{22}$	174

Table 2.2 Boiling points of the alkanes

1. Write down the missing formulae of heptane and nonane.
2. Plot a graph of the boiling points for all the alkanes in the table against the number of carbon atoms in the molecules.
3. a) What pattern does the graph show?
   b) Why is there a pattern?
4. Estimate the boiling point of the alkane $C_{12}H_{26}$.
5. The boiling point of a pure substance is a very 'reliable' piece of data. This means that if you repeat the measurement you get the same result time after time. Is the pattern in the graph reliable? Explain your answer.

Activity – How much energy is transferred when 1 g of hydrocarbon is burned?

Plan an investigation to find out how much energy is transferred when 1 g of a hydrocarbon is completely burned in air (oxygen). You could use:

- camping gas as a source of butane, C_4H_{10};
- paraffin oil as an alkane with about 12 carbon atoms in each molecule;
- engine oil as a long chain hydrocarbon.

If you have sufficient time, you can also compare how easily and how cleanly these alkanes burn.

You should always use chemicals from an education supplier, rather than commercial products, for this type of investigation. Rooms should be well ventilated for these experiments and you must always wear eye protection.

In a **saturated hydrocarbon** all the carbon atoms are joined by single bonds.

Hydrocarbons that contain one or more carbon–carbon double bonds are called **unsaturated hydrocarbons**.

Saturated and unsaturated hydrocarbons

Ethane (C_2H_6) and propane (C_3H_8) (Figure 2.10, page 33) are saturated hydrocarbons.

In alkanes, all the carbon atoms are joined together with single carbon–carbon bonds. These bonds (see Section 1.2) are made as the carbon atoms share their electrons with each of the carbon atoms next to them. This results in a very stable skeleton.

Alkanes, in which all the carbon atoms are joined by single bonds, are called **saturated hydrocarbons**.

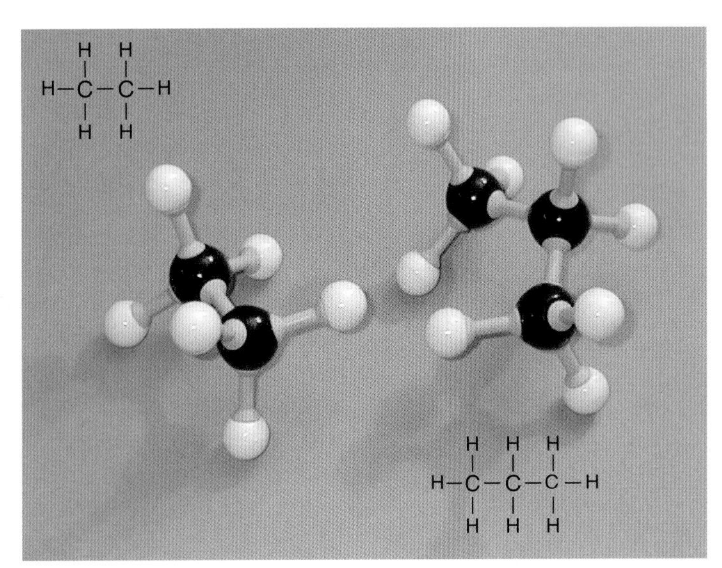

Figure 2.10 Molecular models and structural formulae of ethane and propane. Remember the plastic model is only a representation of the molecule. It is not a 'real' picture of the structure.

Saturated hydrocarbons burn in air, but they don't react with many other chemicals.

Ethene (C_2H_4) and propene (C_3H_6) (Figure 2.11) are very similar to ethane and propane, except that they contain a double bond between two of the carbon atoms in their molecules. These hydrocarbons containing a double carbon–carbon bond are called **unsaturated hydrocarbons**. Unsaturated hydrocarbons with a carbon–carbon double bond are also called alkenes (see Section 2.6).

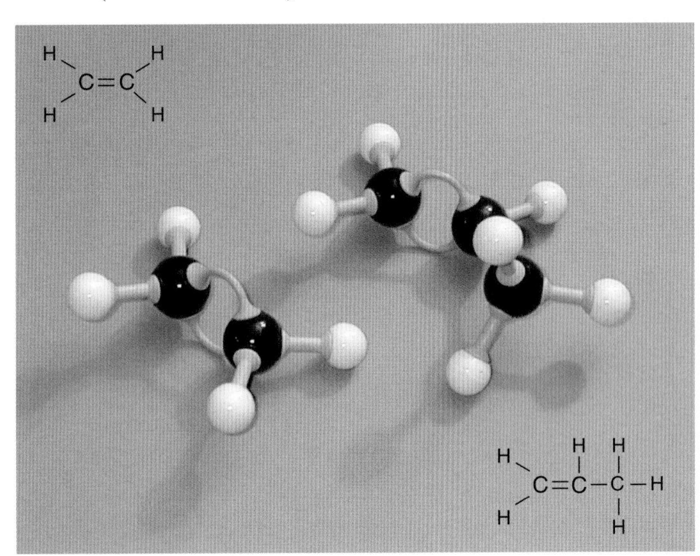

Figure 2.11 Ethene and propene – unsaturated hydrocarbons

The carbon–carbon double bond is a very reactive area in these molecules. It can take part in lots of reactions. Unsaturated hydrocarbons are therefore much more reactive than saturated hydrocarbons.

1. How many chemical bonds can a carbon atom make with other atoms in a molecule?
2. Predict the chemical formulae of the alkenes with 15 carbon atoms and 20 carbon atoms.
3. What is meant by a saturated hydrocarbon?
4. What is meant by an unsaturated hydrocarbon?
5. 'Black and white molecular models are useful for thinking about chemical reactions even though they are models and not real molecules.' Explain whether you agree or disagree with this statement.

2.3 Oil refinery

Figure 2.12 The fractionating tower in an oil refinery works like a fractionating column in the laboratory, but on a much larger, industrial scale.

An oil refinery is a large chemical works which processes crude oil and separates it into useful products. As it comes out of the ground, crude oil is a useless mixture of hydrocarbons.

The evaporation and condensation of substances inside the fractionating tower sorts molecules of the crude oil mixture into separate groups. The biggest molecules remain on the lower levels of the tower where the temperature is highest. The smaller molecules don't condense so easily. They travel up the tower to higher levels where they condense at lower temperatures. The products that are taken from the fractionating tower at different levels are called **fractions**. The different 'fractions' are not pure substances. They are just mixtures of substances with similar boiling points and similar-sized molecules.

A **fraction** from an oil refinery (for example, naptha, kerosene) contains a mixture of compounds with molecules of similar size and boiling point.

6. How is the petroleum gas fraction used in the refinery? See Table 2.3 on page 35.
7. What is the naphtha fraction used for?
8. Why can you not get pure octane from the fractionating tower?
9. Look carefully at Table 2.3.
   a) What is the trend in the colour and viscosity of the fractions as you move down the fractionating tower?
   b) Explain the trend in viscosity.
   c) How does the trend in colour and viscosity relate to the size of the molecules in the fractions?
10. How does the trend in flammability (how easily it burns) relate to the size of the hydrocarbon molecule in a fraction?

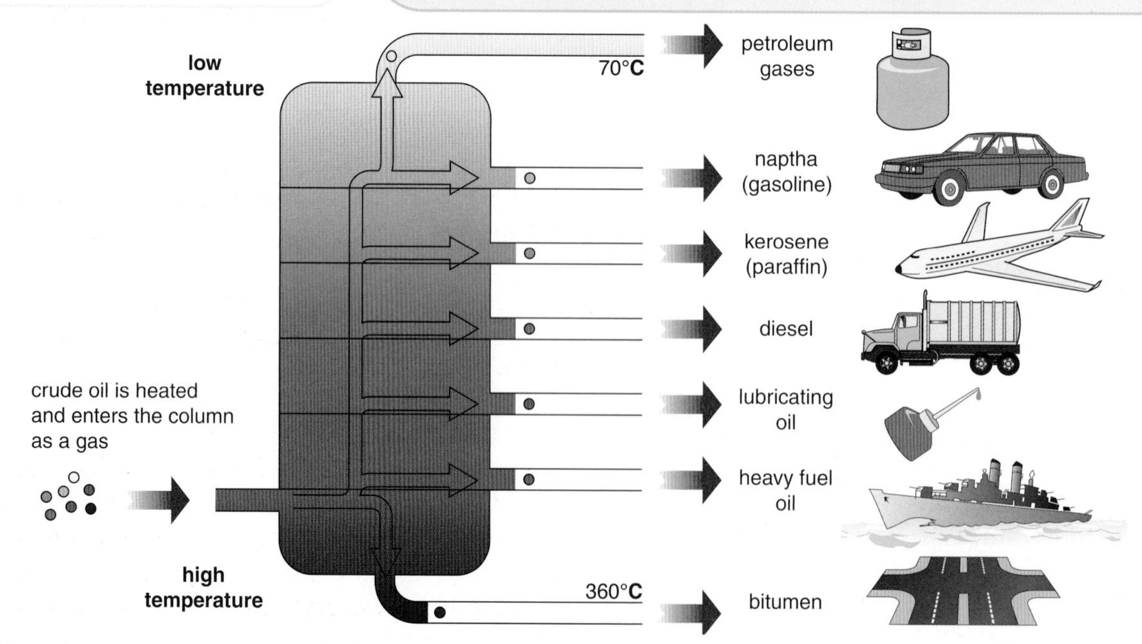

Figure 2.13 A fractionating tower in an oil refinery

Fraction	Number of carbon atoms in the molecule	Description	Boiling pt in °C	Flammability	Uses
Natural gas	1–4	Colourless gas	Less than 40	Explodes if mixed with air and lit	Used as a fuel in the refinery Bottled and sold as LPG
Naphtha	5–10	Yellowish liquid flows very easily	25–175	Evaporates easily, vapour mixed with air is explosive	Petrol Also used for making other chemicals
Kerosene	10–14	Yellowish liquid flows like water	150–260	Will burn when heated	Aircraft fuel
Light gas oil	14–20	Yellow liquid thicker than water	235–360	Needs soaking onto a wick or other material to burn	Diesel fuel
Heavy gas oil	20–50	Yellow brown liquid	330–380	Only burns when soaked onto a wick – very smoky	Used in the catalytic cracker (see Section 2.5)
Lubricants (motor car engine oils)	50–60	Thick brown liquid like syrup	340–575	Needs to be hot and soaked onto a wick before it burns	Grease for lubrication Used in the catalytic cracker
Fuel oil	60–80	Thick brown sticky liquid	above 490	Needs to be hot and soaked onto a wick before it will burn	Fuel oil for power stations and ships
Bitumen	more than 80	Black semi solid	above 580	Hardly burns at all, unless very hot	Road and roof surfaces

Table 2.3 Fractions from a oil refinery

Activity – Rocville

Rocville is an imaginary town in a rural area on the west coast of Britain. At one time, Rocville attracted tourists, but cheap flights and cheap accommodation in more exotic destinations has led to reduced numbers of visitors.

A big oil company wants to purchase a deep-water harbour just to the north of Rocville in order to build a new oil refinery terminal there.

Different people from the surrounding area have different views about the proposal. Here are some of their views.

A Rocville town councillor

'The town cannot provide employment for all our young people so they are leaving to find work elsewhere. We need to bring industry to the area, and an oil refinery would be a good option. It will make our town more prosperous and attract other businesses.'

A Rocville fishing boat owner

'The deep-water estuary is one of our best fishing grounds. My family have been fishing these waters for two hundred years. The big tankers will pollute the water and kill the fish. Using seawater as a coolant in the refinery will destroy the balance of life in the sea.'

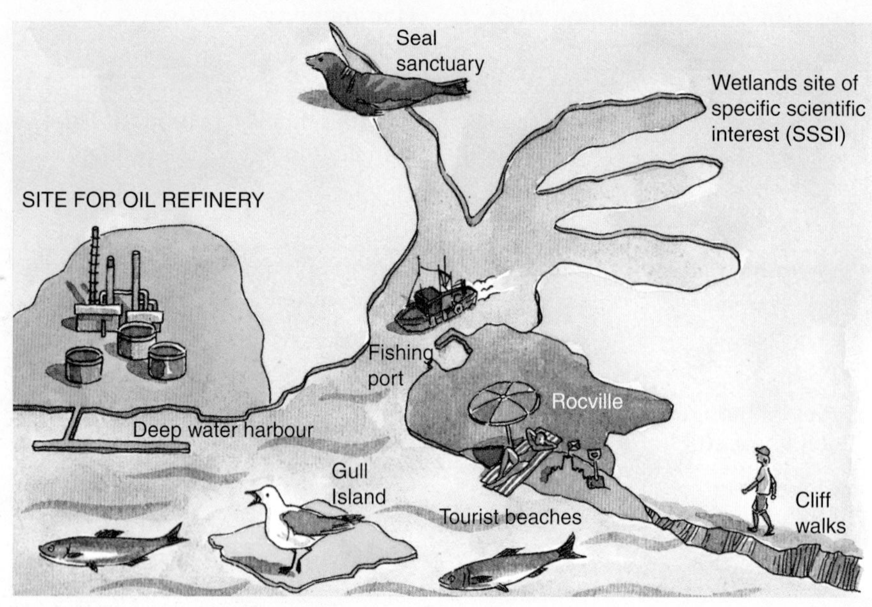

Figure 2.14 Rocville and the surrounding area

The environmental warden for the coast
'Crude oil is a major problem if it contaminates coastal waters. There is a seal sanctuary on our coastline and rare birds nest along the cliffs. The offshore oyster beds and salt marshes that support rare species will be destroyed.'

An executive from the oil company
'This site is ideal for us. The risk of environmental damage from a spill is relatively low. It is true that there is a decline in local air quality near a refinery, but we will bring people and employment to the area.'

An unemployed 17-year-old from Rocville
'My life is rubbish. The holiday business only has jobs for the summer months. All I want is a regular job and money. I can put up with the smells of an oil refinery if I get that.'

A young mother with a 2-month-old baby
'I've read stories in the paper that lots of the chemicals in crude oil can cause cancer. I visited a refinery with my school. It smelt really foul.'

A retired person living in Rocville
'We came here after I retired from the saw mill. My wife was a classroom assistant. We want a quiet life and healthy sea air. I have bad asthma after working around sawdust and could not stay here with the fumes from the refinery – it would be the death of me.'

Think about the views of other people in the area who would be affected by a new oil refinery.

1. Write a short statement that might be made by a young chef working in a Rocville restaurant. He doesn't want to see the Rocville holiday trade vanish, but is keen for the town to prosper.
2. A local landowner has phoned the Rocville radio station about the proposal. Write what she might say on air. She'd like to make money from selling some land, but she's also worried about her farm workers leaving for higher wages elsewhere.
3. Work in groups of three or four for this question.
   You are the council planning officers for Rocville. Write a briefing paper to advise the local council on the proposal. In your briefing paper:
   a) collect together all the arguments in favour of the refinery;
   b) collect together all the arguments against the refinery;
   c) add a short introduction at the beginning of your report, giving some background information;
   d) make a decision about the refinery as a group and write this as your recommendation at the end of the paper.
   If you have access to ICT, add pictures and other graphics. You could make the briefing paper into a presentation and present it to your class.

Air pollution

Acid rain

Figure 2.15 Over time, acid rain pollution can destroy beautiful monuments.

The worst acid rain ever recorded was in 1983 at the Inverpolly Forest, Scotland. The pH value recorded for the acid rain was 1.87. The rain was more acid than strong vinegar and would turn Universal Indicator red. Unpolluted rainwater is naturally acid and always has been. Rainwater reacts with carbon dioxide in the air to form a weak acid. The acid in unpolluted rain is carbonic acid.

During the last 200 years, the problem of acid rain has got much worse. Pollutants from industrial and domestic processes have made rainwater more acidic. Acid rain attacks marble and decorative stonework. Worse still, it kills vegetation and runs off into rivers and lakes where it kills the water life.

When fossil fuels are burnt, the sulfur atoms in the fuels eventually end up as sulfuric acid. When petrol and diesel burn at high pressure in vehicle engines, oxides of nitrogen are formed. These react with rainwater to make nitric acid. Sulfuric acid and nitric acid are much stronger acids than carbonic acid. They give the acid rain a much lower pH.

11. What is Universal Indicator solution used for?
12. What colour is Universal Indicator solution in a) pure water; b) strong acid?
13. What acid is naturally present in unpolluted rain?
14. What acids are present in acid rain?
15. Write word equations for the formation of the acids which pollute rainwater.
16. Catalytic converters in car exhausts can convert oxides of nitrogen into nitrogen gas. Explain why this reduces acid rain.
17. Explain how sulfur dioxide scrubbers in factories and low sulfur diesel fuels for lorries help to solve the problem of acid rain.
18. Design a poster to explain the effects of acid rain to a non-reader.

Climate change and global warming

The Earth's surface is getting warmer. Scientists predict increasing extreme weather and say there is growing evidence that human activity, particularly burning fuels, is to blame.

What is the 'greenhouse effect'?

The 'greenhouse effect' is caused by a layer of gases which trap heat from the Sun in the Earth's atmosphere. Without them, the planet would be too cold to sustain life as we know it. These gases include carbon dioxide which is released by the burning of fossil fuels.

What is the evidence for global warming?

Temperature records show that average global temperature increased by about 0.6 °C in the twentieth century. Sea levels have risen by 10–20 cm and Arctic sea-ice has thinned by 40% in recent decades.

If nothing is done to reduce 'greenhouse gas' emissions, scientists predict a global temperature increase of about 5 °C by 2100. There will be more rainfall overall but droughts in inland areas will increase. More flooding is expected from storms and rising sea levels. Poorer countries, which are least equipped to deal with rapid change, will suffer most.

Global dimming happens when particles in the atmosphere prevent sunlight reaching the Earth. Soot and ash particles from burning fossil fuels, ocean spray and aircraft contrails all add to particles in the air.

Other factors may slow down the warming – such as global dimming or plants taking up more carbon dioxide as their growth rate is increased by warmer conditions. Scientists are not sure how the complex balance between the positive and negative effects will play out.

What about the sceptics?

Most global warming sceptics do not deny that the world is getting warmer. But they do doubt that human activity is the cause. Some say the changes, now being witnessed, are not extraordinary because similar, rapid changes have occurred at other times in the Earth's history.

19 Why are tiny particles in the air particularly harmful to those with breathing problems?

20 a) What evidence is there that global warming is just a natural process?
b) What evidence is there that contradicts this view?

21 Which of the following three predictions do you think will actually happen? Write a paragraph to explain your answer.
a) We'll worry and blame ourselves for climate change for thousands of years.
b) Fossil fuels will run out and renewable energy will save us.
c) The oceans will evaporate as the Earth heats up and humans will all die.

Activity – Acid rain

This table shows the emissions of sulfur dioxide from large combustion plants and the total sulfur dioxide emissions in thousands of tonnes per year.

Year	1970	1980	1990	1995	2000	2001	2002	2003
Emissions from large combustion plants	3717	3449	2935	1756	899	822	745	734
Total emissions	6456	4841	3711	2354	1194	1118	1002	979

1 a) Plot a graph of the emissions from large combustion plants, along the *y*-axis, against the year, along the *x*-axis. Take care with the *x*-axis scale; make sure you space out the years correctly. Plot the *y*-axis scale from 0 to 4000. Remember that this scale is thousands of tonnes of sulfur dioxide per year.
b) Draw an appropriate line through the points on the graph.

2 An industry spokesman commented on this data. He said: 'The acid rain problem of the 1970s and 80s is over. There is no longer anything to worry about. The industry has put its house in order. The sulfur dioxide emissions from large combustion plants are a thing of the past. Our acid gas emissions from burning fossil fuels are now insignificant and not a pollution hazard.'
a) Do you agree with the industry spokesman?
b) Write a paragraph to explain why you agree or disagree with what he said.

Activity – $PM_{10}s$

Major sources	1970	1980	1990	2000	2001	2002	2003
Road transport	42	51	61	40	39	39	38
Residential heating	209	93	47	27	29	27	20
Energy and other industries	86	83	74	25	21	12	13
Total	**486**	**332**	**287**	**168**	**169**	**150**	**141**

Table 2.4 Emission of $PM_{10}s$ in thousands of tonnes per year

Particles that are less than 10 micrometres in size ($PM_{10}s$) are produced when fossil fuels burn. The particles are too small to see. They are also too small to settle to the ground and remain suspended in the air.
Table 2.4 gives some data about PM_{10} emissions.

1. Display the information about total emissions in a suitable way so that it can be used in a presentation.
2. Draw charts to compare the PM_{10} emissions from the different sources in 1970, 1990 and 2003.
3. The decline in residential heating emissions over the period of 1970–2003 can be explained by the reduction in the use of coal for domestic heating. What forms of heating have replaced coal, and why has this led to a reduction in emissions?
4. Look at the trends in the data. In your opinion, should transport and energy industries be congratulated on the progress they have made to reduce this source of air pollution?

Catalytic cracking

Light fractions	Use
Gas Naphtha / Petrol Kerosene Diesel fuel	All can be used directly as fuels without any processing
Heavy fractions	
Light & heavy gas oil Lubricants Fuel oil Bitumen	These cannot be used as fuels except in large engines such as those on a ship, where the oil is heated up before burning

Table 2.5 Comparison of the flammability of light and heavy fractions from crude oil

Figure 2.16 Heavier fractions from crude oil are not much use as a fuel.

Cracking is the name given to the breaking up of large hydrocarbon molecules into smaller and more useful bits. You can do this in the laboratory at school using high temperatures and a catalyst, but in the oil industry they also use high pressures because it is more efficient.

Catalytic cracking is the process of breaking up large hydrocarbon molecules into smaller, more useful, molecules using high temperatures, high pressures and a catalyst.

The **catalytic cracking** process is controlled by careful choice of conditions to produce mixtures of smaller hydrocarbons. Some of these smaller hydrocarbons are alkenes with carbon–carbon double bonds.

For example, during cracking, the hydrocarbons are vaporised and passed over the catalyst at 500 °C. The temperature and pressure used for the cracking process are selected to give a high percentage of hydrocarbons with 5–10 carbon atoms. These are the hydrocarbons needed for petrol.

Figure 2.17 When a large molecule like $C_{18}H_{38}$ in the left hand picture is cracked, you can end up with several fragments like those in the right hand picture. Octane (C_8H_{18}) is one of the molecules found in petrol. Ethene (C_2H_4) and propene (C_3H_6) are useful starting substances for the manufacture of plastics and other chemicals.

22. What is meant by a catalyst?
23. Why are the heavier fractions from crude oil unsuitable for use as fuels?
24. 'Cracking produces more useful molecules.' Explain this statement.
25. Draw a flow diagram to summarise the catalytic cracking process.
26. Catalytic cracking is a thermal decomposition reaction. Why is heating needed to start the reaction?
27. Balance the following equation for the complete combustion of octane, C_8H_{18}.

    $\_\_\_\_ C_8H_{18} + \_\_\_\_ O_2 \rightarrow \_\_\_\_ CO_2 + \_\_\_\_ H_2O$
28. Copy and complete this equation for the incomplete combustion of $C_{11}H_{24}$ in a limited supply of air.

    $C_{11}H_{24} + \_\_\_\_ O_2 \rightarrow \_\_\_\_ C + \_\_\_\_ H_2O$

Figure 2.18 Which of these pasta shapes can become the most tangled?

Viscosity measures how easily a liquid can flow. Thick liquids have a high viscosity. Runny liquids have a low viscosity.

Viscosity

Some hydrocarbons have long, flexible molecules. The longer the molecules, the more tangled they become with other molecules (Figure 2.18). This makes the liquid thicker (more viscous) and less easy to pour. Motor oils have a range of **viscosities** displayed on their containers. They are written as numbers with a 'W' after them.

2.6 Alkenes

In Section 2.2, we noticed that the alkanes formed a series of similar hydrocarbons. The **alkenes** also form a series of similar compounds. Each alkene has one carbon–carbon double bond in its molecule. So, the simplest alkene is **ethene** (C_2H_4). If you add one $-CH_2-$ link to this molecule, it becomes propene (C_3H_6), the next molecule in the series. Add more $-CH_2-$ links and the next three members of the series are butene (C_4H_8), pentene (C_5H_{10}), and hexene (C_6H_{12}). This makes a series of similar compounds, with a general formula of C_nH_{2n} for the alkenes.

The **alkenes** are a series of hydrocarbons with a single carbon–carbon double bond with the general formula C_nH_{2n}. The alkenes can be used to make polymers.

Ethene is the simplest alkene, formula C_2H_4. It can be used to make polythene and ethanol.

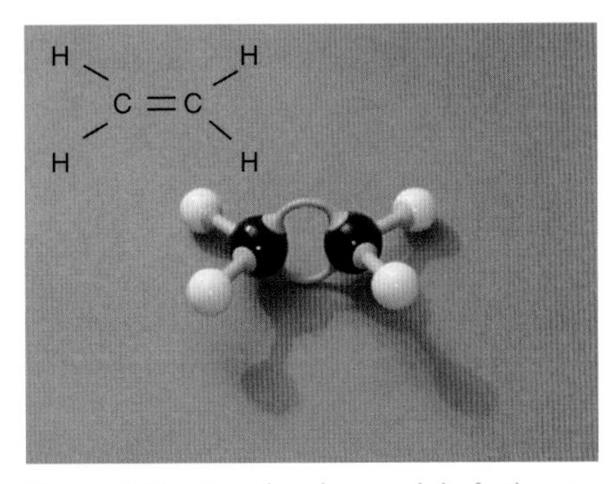

Figure 2.19 A molecular model of ethene and its structural formula

Figure 2.20 A molecular model of propene and its structural formula

The reactivity of alkenes

The carbon–carbon double bond in alkenes is a target for reactions. One of the two bonds can easily be broken, allowing the carbon atoms to link with other atoms. For example, propene will act as in Figures 2.21 to 2.23 in the presence of a reactive substance such as bromine (Br_2).

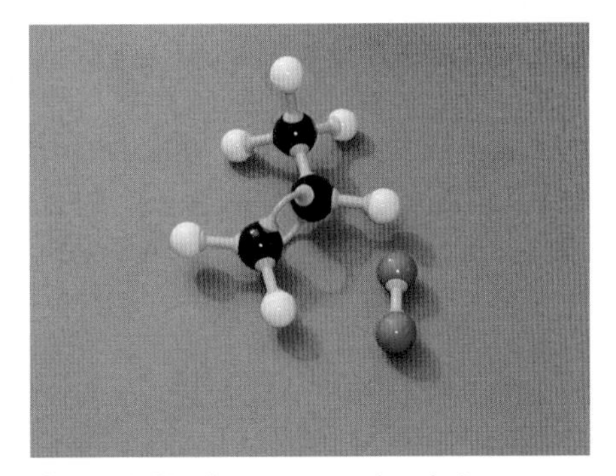

Figure 2.21 The green molecule is a molecule of bromine that can react with propene.

Figure 2.22 When propene and bromine water react, one of the bonds in the double bond of the propene molecule opens up. Each bromine atom then bonds with a carbon atom.

When one of the bonds in the double bond of the propene molecule has opened up, the two atoms in the bromine molecule are added to the propene molecule.

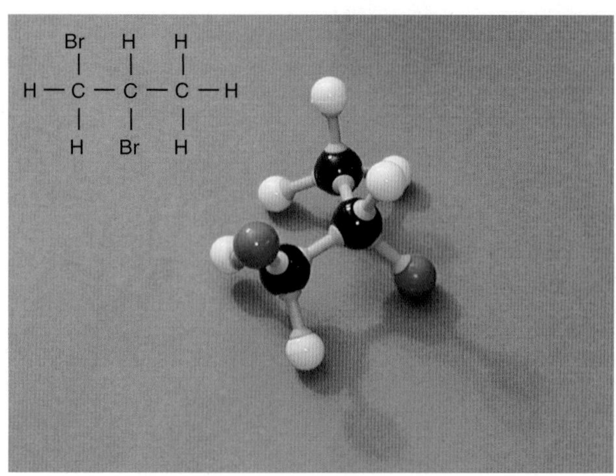

Figure 2.23 The product of the reaction has two bromine atoms added to the alkene. The product of the reaction is a saturated compound because it contains only carbon–carbon single bonds.

29 Draw one possible structure for decene ($C_{10}H_{20}$).

30 Why is a carbon–carbon double bond more reactive than a single bond?

C_3H_6	+	Br_2	→	$CH_2Br–CHBr–CH_3$
propene		bromine solution		dibromopropane
(colourless gas)		(orange/yellow)		(colourless liquid)

2.7 Spirit fuel

NEWS: Brazil has started to export 'alcohol motor fuel' based on ethanol spirit made from sugar cane

Brazilian energy giant Petrobras made its first shipment of fuel alcohol this week. The first shipment of alcohol left on Monday from Rio de Janeiro and was destined for Venezuela. A monthly shipment of around 25 000 cubic metres of the fuel is planned.

Figure 2.24 Here, some cotton wool has been soaked in ethanol and set alight. Ethanol is clean, safe and carbon neutral.

Ethanol has a lot of benefits as a fuel.

- It burns with a clean, smokeless, hot flame.
- It is a liquid so it can be stored and transported easily.
- It is not oily and can be washed away with water.
- It mixes with water and does not pollute it.
- It is present naturally in the environment, for example, in rotting (fermenting) fruit.
- It is 'carbon neutral' when produced from plant products. The same amount of carbon dioxide is released into the air when it burns as that which came out of the air when the plant crop was growing.
- Cars do not need modification to run on it.

There are two methods of producing ethanol: either from ethene or from plants. Let's compare the two methods.

```
H         H              H   H
 \       /               |   |
  C = C     + H2O →  H — C — C — H
 /       \               |   |
H         H              H   OH
```

Figure 2.25 The hydration of ethene. The reaction adds a water molecule to an ethene molecule to make ethanol.

Making ethanol from ethene

It is possible to make ethanol by adding the atoms from a water molecule to the double bond in an ethene molecule. This reaction is an example of hydration because it involves adding water.

Steam and hot ethene are mixed under pressure at 300 °C in a reactor. Even with all the molecules squashed together and colliding with one another, the reaction is very slow. We can reduce the energy required for the reaction to start using a catalyst. The catalyst used is solid silicon dioxide coated with phosphoric acid.

In this process

- Two volumes of ethanol vapour are mixed with one volume of steam at 300 °C.
- This is compressed to 65–70 times atmospheric pressure, then it is passed over the catalyst in the reaction chamber.
- The gases are cooled and ethanol and water condense out.
- Much of the ethene is unchanged – this is recycled and returns to the start of the process.
- The ethanol and water mixture is separated by fractional distillation.
- The water is recycled through the process.

In spite of the extreme conditions in the reactor, only 5% of the ethene is converted into ethanol. By removing ethanol from the mixture and recycling the ethene, it is possible to achieve an overall conversion of 95%.

Making alcohol by fermentation of plant products

Starchy material such as barley is used as the starting material for this process. Sugar cane is used in hot countries. Starch is a complex carbohydrate. The starch must first be broken down into sugars. The starchy material is heated with hot water to extract the starch and then turned into sugars using natural enzymes in the plant material.

Yeast is then added and the mixture is kept warm at 35 °C for several days until fermentation is complete. Enzymes in the yeast convert the sugars into ethanol and carbon dioxide. This is the fermentation process. The yeast is killed when the ethanol concentration reaches 15%.

$$C_6H_{12}O_6 \xrightarrow[\text{enzyme}]{\text{yeast}} 2C_2H_5OH + 2CO_2$$

$$\text{sugar} \xrightarrow[\text{enzyme}]{\text{yeast}} \text{ethanol} + \text{carbon dioxide}$$

Finally, ethanol is separated from the watery mixture by fractional distillation.

31. Make a list of the benefits of ethanol fuel over petrol. Put the most important benefits at the top.
32. Explain the term 'carbon neutral'.
33. Comment on the following statement: 'The source of ethene, to make ethanol, is crude oil, so ethanol made from ethene is just another fossil fuel.'
34. Ethene does not react easily with water.
    a) Why is the reaction carried out at a high temperature?
    b) Why is the reaction carried out at high pressure?
    c) Why is a catalyst needed?
35. Why must the ethene be passed through the reactor again and again?
36. What is the starting material for making ethanol by fermentation:
    a) in a country like the UK;
    b) in a hot tropical country?
37. Draw a flow chart to summarise the production of ethanol by fermentation. Start the flow chart with barley grain in the fields and end with ethanol.

Activity – Spirit fuel factory

Ethanol has a part to play as a motor fuel in our future, but at present not enough ethanol is produced to satisfy even half our needs.

1. Compare the two methods described in Section 2.7 for producing spirit fuel (ethanol). List the benefits and drawbacks of each method. Write an evaluation of each method and decide which is better.
2. Devise a plan for setting up a small factory to produce ethanol fuel for the petrol stations in a small town.
   Write down a) the starting materials you would need; b) the processes you would carry out; c) how you would store the ethanol fuel before delivery to the petrol stations.

2.8 Polymers

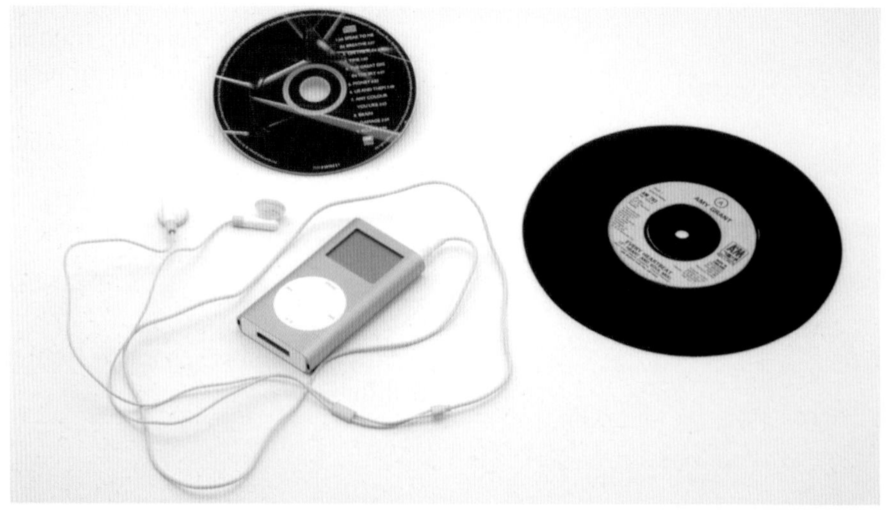

Figure 2.26 High-quality music on plastic!

Our music collections are important to all of us. 'Records' are what they used to be called – that now sounds really old-fashioned. About 20 years

Plastics are manufactured materials consisting of molecules with long chains of carbon atoms.

A **monomer** is a small molecule capable of forming bonds, or links, with other small molecules to form a larger molecule called a polymer.

A **polymer** is a large molecule made from lots of monomers joined together. Polymers can be made from thousands of monomer units. They can be natural materials such as cellulose or manufactured materials such as polystyrene.

ago, records started to be replaced by CDs. Records and CDs are made of 'vinyl'. Vinyl is a manufactured **plastic**. Without plastics, we wouldn't be able to enjoy recorded music.

Plastics

Plastics are materials called **polymers**, made from small molecules called **monomers**.

Polythene

Polythene is the material used for supermarket carrier bags and for most types of flexible packaging. The problem with polythene is that it is not biodegradable. This means that it is not broken down by bacteria, like paper and wood.

Figure 2.27 Bags, and other items made of polythene, litter our streets. They get into rivers, and the sea, after being thrown away. They can remain there for years and years.

Polythene is made from ethene. Under high pressure, ethene molecules are so squashed together that they react and take up less space. The double bonds open up and link the molecules together.

Thousands of molecules can be linked in this way creating a very long molecule of polythene.

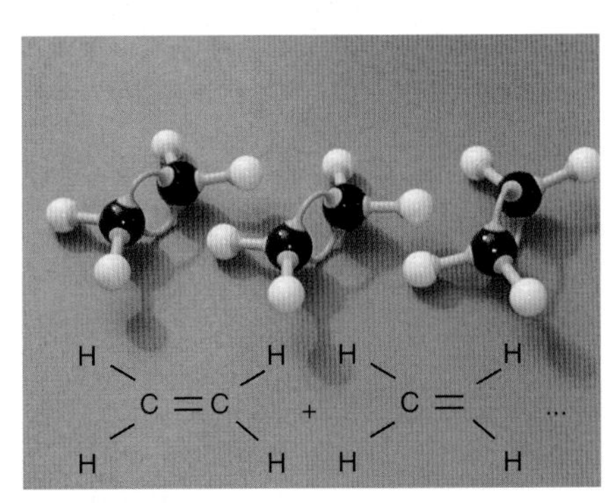

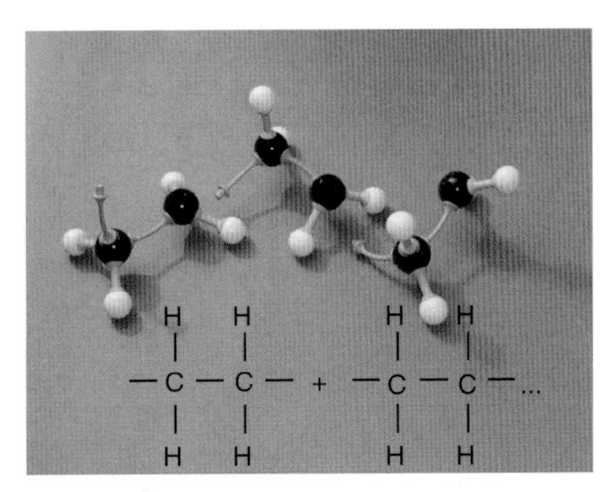

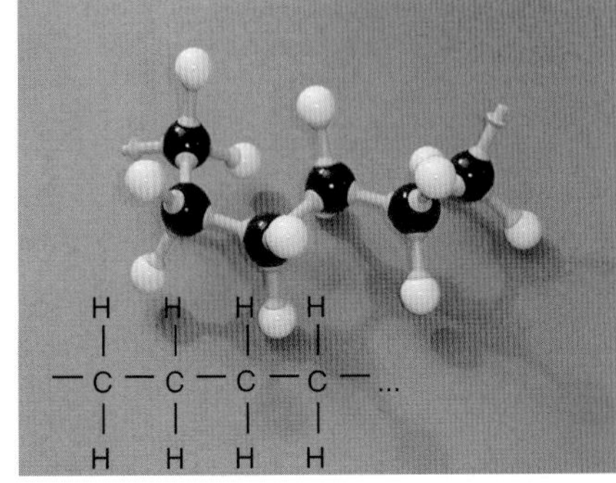

Figure 2.28 We can use plastic models to help us understand how polythene molecules are formed from ethene.

Choosing the monomer to give the best plastic

Polythene (or polyethene) is made of long, thin and very flexible but strong molecules.

Polypropene (polypropylene) is a more rigid plastic that keeps its shape better than polythene. It is used for washing up bowls and plastic toolboxes. Although it is stiff, thin pieces are still flexible, so it can be used to make its own hinge. Other monomers give different properties to the polymers they form.

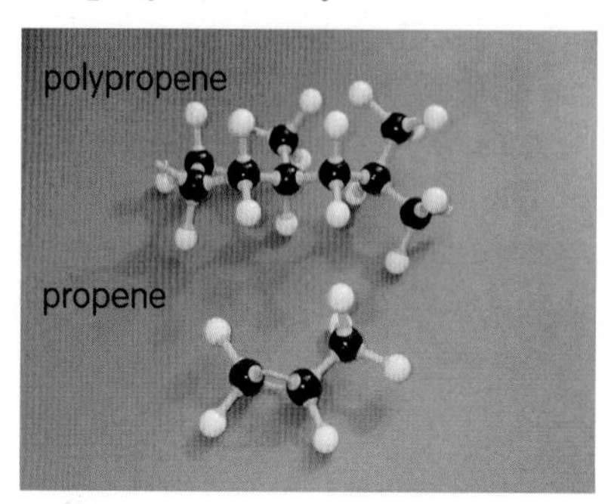

Figure 2.29 Molecular models of propene and part of polypropene

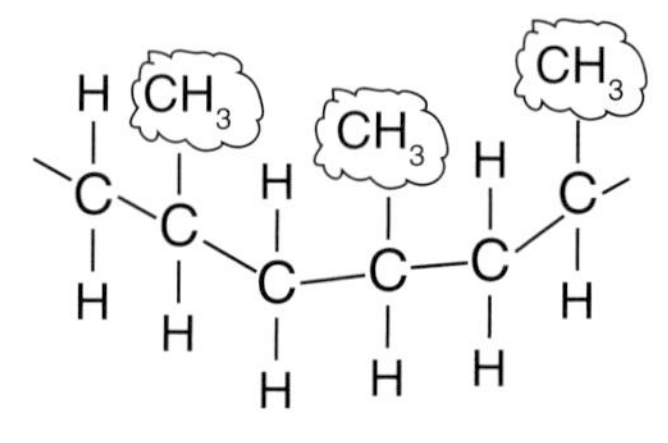

Figure 2.30 Polypropene has better rigidity than polyethene. The CH_3 groups make it less flexible.

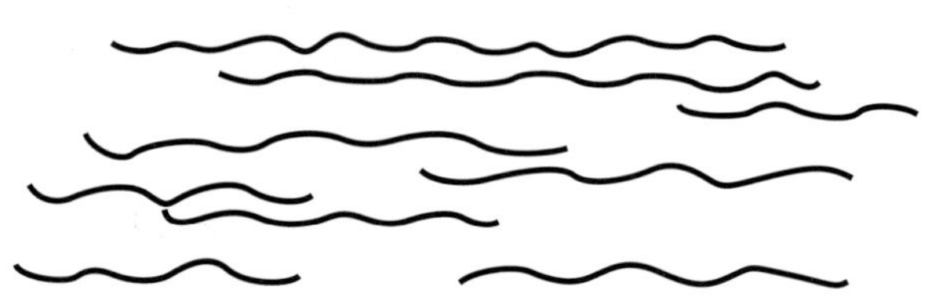

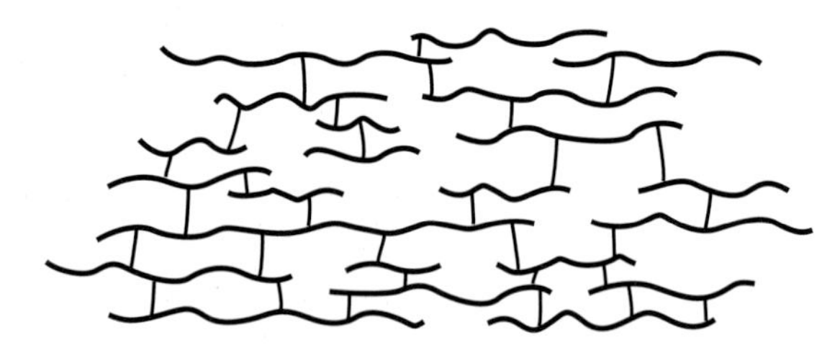

Figure 2.31 Polythene and polypropene are made of long flexible molecules. In rigid polymers, chemical bonds have formed between neighbouring molecules. These bonds join different parts of the polymer together in a three dimensional network. The crosslinks lock the molecules together and stop them from being flexible.

38. Explain the difference between a monomer and a polymer.
39. Draw a diagram of a short section of a polypropene molecule containing three monomer units.
40. What properties of PVC ('vinyl') make it suitable for CDs and records?
41. Explain what happens when a polymer forms 'crosslinks'.

2.9 The benefits and problems with plastics

Plastics can meet very specific technical requirements to fulfil a whole range of very different uses. They can be used for items as different as hygienic, transparent boxes to hard, rubber-like balls.

There are many benefits of plastics:
- They weigh less than other materials.
- They are extremely long lasting.
- They are resistant to chemicals, water and physical damage.
- They have excellent insulation properties.
- They are inexpensive to produce.

There are also many problems with plastics. Most plastics will not rot away quickly like other organic matter. This is partly because there are no bacteria that naturally live on the plastics we have created. With more and more plastics being thrown away, the space required to dispose of the waste is a concern. A massive 85% of plastic waste is sent to landfill sites. Only 8% is incinerated and only 7% is recycled. Re-using plastic is even better than recycling. It reduces waste and uses less energy. Multi-trip plastic packaging has become widespread, replacing less durable and single-trip alternatives. For example, supermarkets now use returnable plastic crates for transport and display purposes. They usually last up to 20 years.

Making degradable plastics

There are now biodegradable carrier bags. These are made from plastic which rots under certain conditions or after a certain length of time. There are two types of degradable plastic:
- biodegradable plastics, which contain a small percentage of non-oil-based material;
- photodegradable plastics, which break down when exposed to sunlight.

In Sweden, McDonalds have used biodegradable cutlery since 2002. This means that all their catering waste can be turned into compost. Carriers for packs of drinks cans are also produced which photodegrade in six weeks.

There are three major concerns related to degradable plastics.
- They will only degrade if exposed to the right conditions. So, a photodegradable plastic will not degrade if it is buried in a landfill site.
- Some produce the greenhouse gas methane as they degrade.
- They may lead to an increase in plastic waste, if people believe that discarded plastics will simply disappear.

Activity – Contents of your dustbin

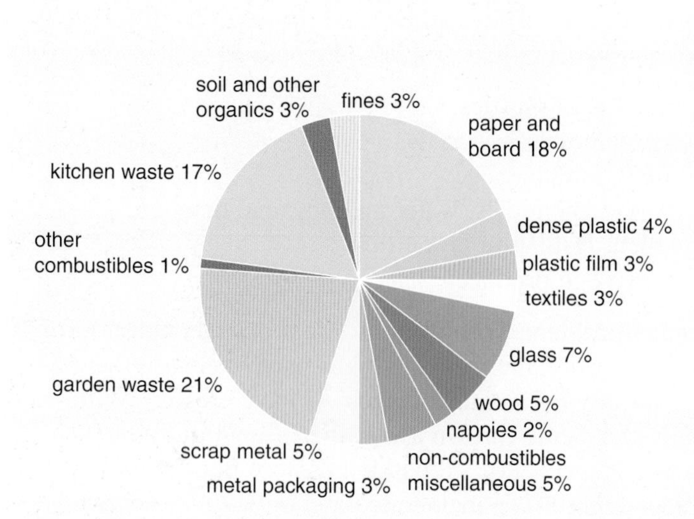

Figure 2.32 The contents of a dustbin

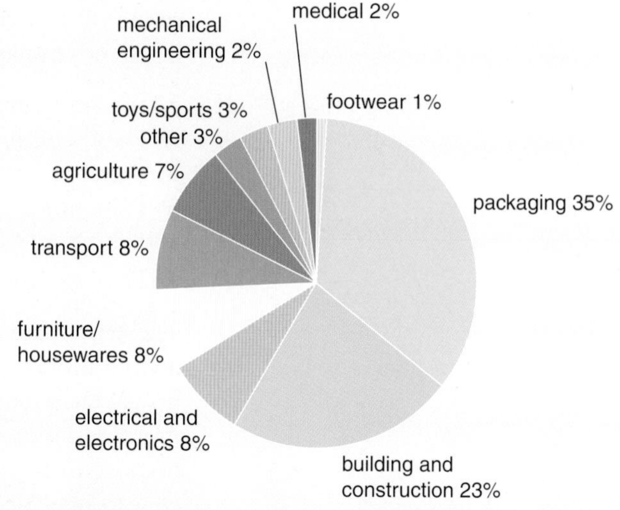

Figure 2.33 The uses of plastics in the UK

Read this statement and then answer the questions below.

'Plastic can be thought of as oil on its way to being burnt. Instead of just being burnt as a fuel, the initial material from crude oil gets a life as a plastic bag, before it's burnt. When waste plastic is burnt, the energy produced can be used to generate electricity or in neighbourhood heating schemes. The large-scale burning of waste and use of the energy produced should be the priority, rather than getting every busy person to sort household waste and send it to the right place. Life in the twenty-first century is just too busy and interesting for people to do that.'

1. What percentage of your rubbish is plastic?
2. a) What percentage of total plastic production goes towards making items for 'household' use'?
   b) What percentage of plastic production goes towards 'job related' use?
3. Describe three ways in which recycled polythene helps the environment.
4. Are the opinions in the statement above valid? Say whether you agree or disagree with what is said and give your reasons.

Producing smart polymer materials

The rate of innovation in polymers is so fast that by the time a new advance gets into the papers, it may already be out of date. Many companies invest huge sums of money in developing new plastics.

Shape memory polymers

Doctors put a straightened plastic thread into a main artery. Once in place, the shape memory returns the plastic to a coiled spring. This widens the artery, improving the patient's blood flow.

Figure 2.34 Shape memory polymers in the casing of mobile phones are activated by heating. The phone then falls apart into its components, so that the expensive materials can be sorted and recycled. This also makes it easier to dispose of any parts containing toxic chemicals.

These polymers can also be used in surgical sutures (stitches to hold flesh together). After insertion, the sutures regain their memory shape and slowly tighten, making a better repair.

Dentistry

Smart polymers can also be used in dental braces. They can be moulded into the shape of the mouth, then hardened by exposure to a special ultraviolet lamp. (The uses of smart memory metals are described in Section 1.7.)

Wound dressings

New water-based polymers can be applied to a wound. These are smart hydrogel materials. They will react to the state of the wound, sometimes hydrating it, sometimes drawing water from it. They are particularly useful for messy wounds, where there is a lot of discharge and pus. The smart hydrogel absorbs this and keeps the wound healing steadily.

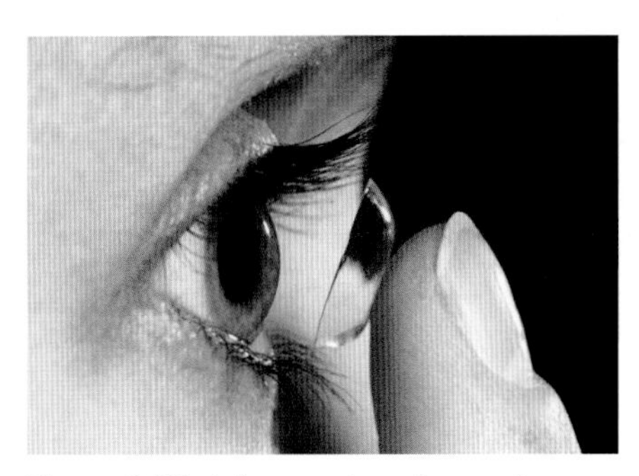

Figure 2.35 Advances in polymers have allowed us to make throw away contact lenses.

Contact lenses

Ordinary contact lenses are easily scratched and can be difficult to keep clean. The wearer can also find them uncomfortable. One-day contact lenses, that you wear once and then throw away, are the answer. This is another advance brought about by silicone hydrogel polymers.

Figure 2.36 Supergel for cleaning oil slicks is a hydrogel polymer. It can absorb up to half its own weight of oil. A liquid form of the gel can be sprayed on an oil slick. The gel absorbs the oil and thickens. The resulting mixture is strong and solid enough to be rolled up and lifted out of the water.

Thermochromic materials

Thermochromic materials are polymers that change colour at different temperatures. They are often used on T-shirt and coffee mug designs. When they cool down, the colour changes back again.

Figure 2.37 Thermochromic materials on a coffee mug

Conducting polymers

These are really amazing materials. When a small electrical current is passed through them they contract, stretch or twist like a muscle fibre. They can be incredibly small fibres that nanorobots of the future will use as moving parts rather than electric motors.

Slime

Polyvinyl alcohol consists of long chain molecules that are free to move around in solution. When sodium tetraborate (borax) solution is added, links are made between the chains. As you add more cross-linking borax solution, the 'slime' you produce behaves more like a solid (Figure 2.38).

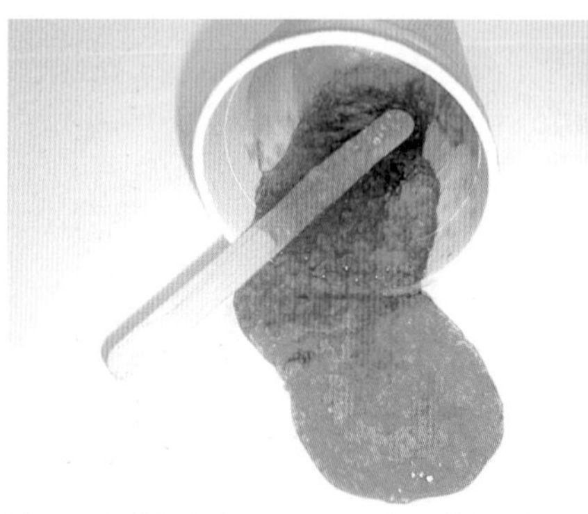

Figure 2.38 Polymers: two liquids make a semi-solid.

Activity – A novel use for plastic

Choose one of the novel uses of plastics presented on pages 48–50. Devise a storyboard and script for a TV advert for this product. The ad will last for 20 seconds. There should be approximately four scenes and 50 words spoken.

Summary

- ✓ Crude oil is a mixture of hydrocarbons that can be separated into **fractions** of similar molecular size by fractional **distillation**.
- ✓ **Hydrocarbons** are compounds containing only carbon and hydrogen.
- ✓ The **alkanes** are a series of similar **saturated** hydrocarbons.
- ✓ The heavier fractions in crude petroleum can be turned into useful materials by **catalytic cracking**.
- ✓ The **alkenes** are a series of **unsaturated** hydrocarbons with one carbon–carbon double bond per molecule. They are more reactive than alkanes.
- ✓ **Polymers** are large molecules made by joining together smaller molecules called **monomers**.
- ✓ **Ethene** can be polymerised to form polyethene. The double bond in ethene molecules can be opened up to link the molecules together.
- ✓ Different polymers based on alkenes can be produced with a wide range of properties and uses.
- ✓ The **viscosity** (runniness) of a hydrocarbon fraction depends on how long and tangled the molecules are. As the viscosity increases, the fractions get stickier and less easy to pour.
- ✓ New polymers with novel uses are continually being developed.
- ✓ Burning fossil fuels can lead to a range of atmospheric pollution problems.
- ✓ **Plastics** create pollution problems if they are not biodegradable.

EXAMQUESTIONS

1 a) Octane is an alkane with eight carbon atoms per molecule. What is its formula? *(1 mark)*

b) Write the formula of an alkene molecule with four carbon atoms. *(1 mark)*

c) Draw the structural formula of an alkene with four carbon atoms. *(1 mark)*

2 How does an alkene decolorise bromine water? *(2 marks)*

3 a) What structural feature of an alkene allows it to undergo polymerisation reactions? *(1 mark)*

b) Draw a short section of a poly(ethene) molecule showing six carbon atoms. *(1 mark)*

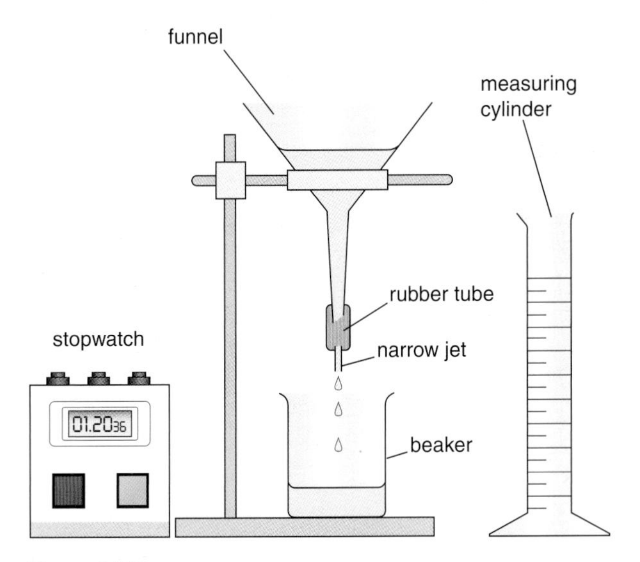

Figure 2.39

EXAMQUESTIONS

4 Some fractions of crude oil were tested for viscosity (runniness) using the apparatus shown in Figure 2.39.

a) Explain how you would use this apparatus to compare the viscosities of fuels. *(2 marks)*

b) Name one factor that must be constant to make this a fair test. *(1 mark)*

The results of four tests are shown below.

Fraction	Time in minutes
Petrol	2.0
Kerosene	2.3
Diesel	3.0
Lubricating oil	6.0

c) Copy the grid in Figure 2.40 and plot these results on to it. *(2 marks)*

d) Turn the plot of the results into a bar chart or a line graph, whichever you think is more suitable. *(1 mark)*

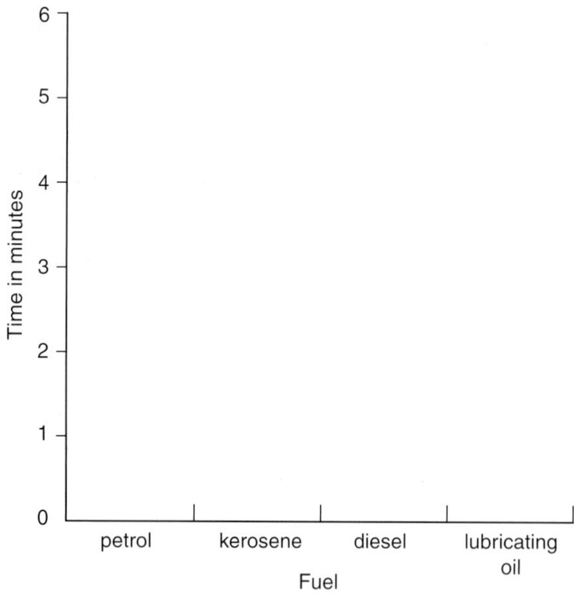

Figure 2.40

e) Explain the pattern in the results. Use what you know about the different molecular sizes in the fractions to write your answer. *(2 marks)*

5 Latex is a naturally occurring polymer. Latex is quite soft, like a pencil eraser. In order to manufacture the rubber for tyres, sulfur is added to latex. The sulfur makes cross-links between the polymer molecules in the latex.

a) Draw diagrams to show latex molecules before the sulfur is added and then the tyre rubber with sulfur crosslinks between the latex molecules. *(2 marks)*

b) How does sulfur change the properties of latex? *(2 marks)*

6 Crude oil is separated by fractional distillation using a column as in Figure 2.41.

a) Explain fully how fractional distillation works. *(2 marks)*

b) Explain why naphtha burns more easily than diesel oil. *(1 mark)*

c) Naphtha contains a saturated hydrocarbon with the formula C_7H_{16}. Draw the structural formula of this compound. *(1 mark)*

d) Lubricating oil is used in a catalytic cracker. Describe what a catalytic cracker does. *(1 mark)*

e) Draw three molecules that you might obtain from cracking one molecule of $C_{15}H_{32}$. *(1 mark)*

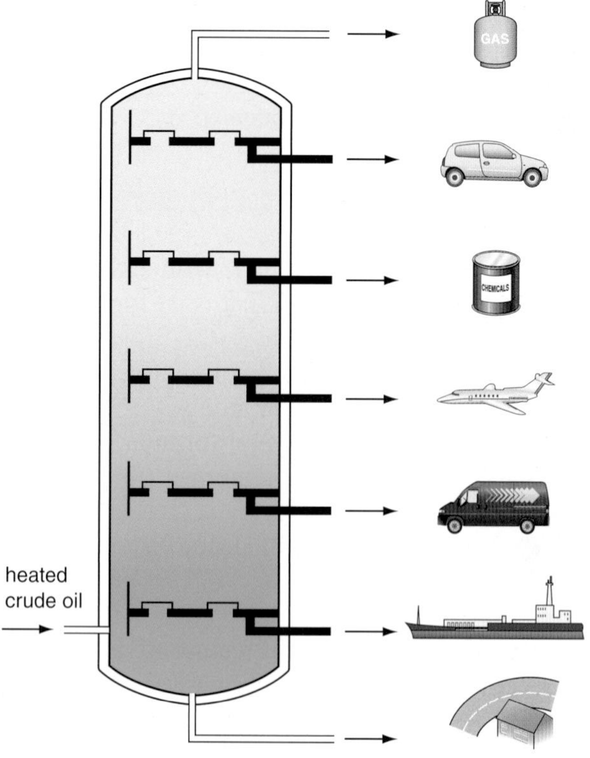

Figure 2.41

Chapter 3
How can plant oils be used?

At the end of this chapter you should:

- ✓ know how plants provide us with plant oils;
- ✓ appreciate that the natural products from plants can be changed chemically to make substances that are more useful in our everyday lives;
- ✓ know about the extraction of plant oils to produce processed foods and biodiesel fuel;
- ✓ appreciate the use, benefits and drawbacks of ingredients and additives in processed foods.

Figure 3.1 Plants use raw materials in their environment. These include air, water and dissolved minerals in the ground. Animals and humans then make use of the materials that plants have made.

3.1 Plant products

Figure 3.2 Plants capture energy from the Sun that ends up in our food.

Many plants contain oils. These plant oils (often called vegetable oils) are used in food and for other products, such as cosmetics and biodiesel fuel. Plant oils have been used for thousands of years for cooking and lamp oil. Years ago, the oil was extracted by pulping the plant, heating this pulp and then squeezing it in a press until the oil flowed out.

These pulping methods are still used, but more modern methods have been developed where the oil is extracted by dissolving it in a solvent such as hexane. The solvent is then distilled off to leave the oil behind.

3.2 Oil in the kitchen

Plant oil	What part of the plant does the oil come from?
Olive oil	Fruit
Rape oil	Seeds
Peanut oil	Nut (food store for seeds)
Avocado oil	Fruit
Jojoba oil	Seed
Palm oil	Fruit

Table 3.1 Different types of plant oil

A mixture in which oil and water are bound together is called an **emulsion**.

Oil and water do not mix. Shake a container of oil and water for as long as you like, but as soon as you stop the oil and water will separate. The water will sink to the bottom and the oil will float to the top because it is less dense. Oil and water can be made to mix, but to do this you need an emulsifier – a substance that links together oil molecules and water molecules. Emulsifier molecules have a 'water loving' end and a 'water hating' ('oil loving') end. They bind the oil and water together forming a mixture of the two, which is called an **emulsion**.

Emulsifiers allow small droplets of one liquid to remain suspended inside the other. Egg yolk, mustard and sugars can act as emulsifiers.

1. What raw materials do plants take in from their environment? Explain how plants obtain each of these materials.
2. Name three products obtained from plant oil.
3. a) What are the food groups that are essential for a healthy diet?
   b) Are there any food groups we cannot obtain from plants?
4. Name three sources of plant oil.
5. Old fashioned oil lamps burned plant oil. Where did the energy of the oil originally come from?
6. Plant oils are energy stores. Explain why plant oils are mainly found in fruit (containing seeds) and in the seeds themselves.

Figure 3.3 Plant oils provide lots of energy and nutrients. They are important as foods. But beware they are very high in energy content. Eating too much plant oil-based food will make you fat.

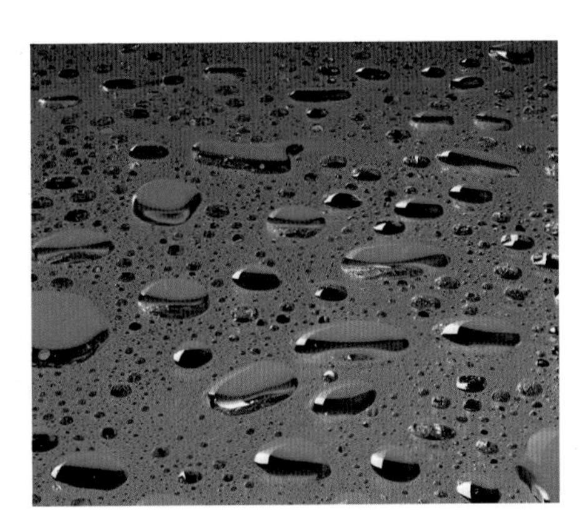

Figure 3.4 Oil and water do not mix without an emulsifier.

Figure 3.5 Soap is a common emulsifier. When you wash your hands, the soap allows droplets of oily substances to become suspended in water.

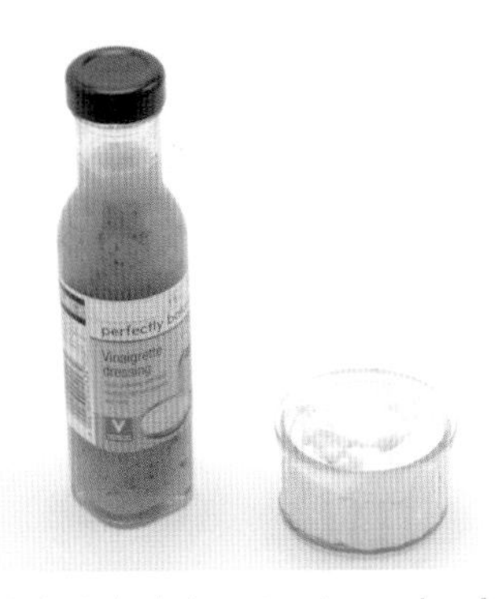

Figure 3.6 Salad dressing is made of oil and water. The thick mixture coats the salad.

Mayonnaise and vinaigrette

To make an emulsion the oil must be broken up into tiny droplets and then prevented from coming together. Vigorous shaking or whisking breaks the oil into droplets. The emulsifier then prevents the droplets from pooling together again.

Emulsions can be temporary, like vinaigrette, or permanent, like mayonnaise. The thicker an emulsion, the less likely it is to separate, because the droplets move more slowly through the thick mixture.

Fresh milk is an emulsion of fat droplets in water. With time, the creamy fat separates from the water and floats to the top. Emulsifiers are used in many types of food including sauces, salad dressing, ice cream and cakes. In ice cream, tiny crystals of flavoured ice are trapped in droplets of cream or vegetable fat. Tiny bubbles of air are also whisked into the mixture.

Figure 3.7 Mayonnaise was invented in 1756 by the French chef of the Duc de Richelieu. After the Duc beat the British at Port Mahon, Minorca, his chef created a victory feast that included a sauce normally made of cream and eggs. As there was no cream in the kitchen, the chef used olive oil instead and a new culinary creation was born. The chef named the new sauce 'Mahonnaise' in honour of the Duc's victory.

Figure 3.8 There are many uses for emulsions, depending on their specific properties. Ice cream has tiny ice crystals suspended in oily droplets with a creamy smooth texture and delicious appearance! Gravy and salad dressings are thick so they stick to food. In fact, nearly all the sauces a chef makes are emulsions.

Figure 3.9 Frying adds to the flavour of food.

Frying in oil

Frying is a fast cooking method. Plant oils can be heated up to 230 °C before they start smoking, but water can only be heated to 100 °C. Direct contact between the food and the oil transfers energy rapidly to the food, causing chemical changes. You should always fry food with the oil above 100 °C. This seals in the flavour and also prevents oil soaking into the food as steam escapes from the food to stop this happening. Frying adds flavour. The high temperature changes proteins and sugars on the surface so they turn brown and taste good. You should always take care when helping to prepare fried food. In the UK alone there are 3000 injuries from chip pan fires every year!

7. Draw step-by-step diagrams to show how olive oil and lemon juice (a watery solution) can be used to make mayonnaise. Egg white is used as the emulsifier. The oil must be added a little at a time.
8. Why are emulsions used as sauces and salad dressings?
9. Why do chips cook faster than boiling potatoes?
10. Why do chips taste different from boiled potatoes?

Food additives

Processed foods may contain added chemicals called **additives** to improve their appearance, taste or shelf-life.

Adding chemicals to food isn't a recent invention. Saltpetre was used in the Middle Ages to preserve meat. Saltpetre is the common name for potassium nitrate. Nowadays, nitrate, the active ingredient in saltpetre, is used. It prevents meat from becoming contaminated with the bacteria that cause food poisoning.

Additives are used to make food look and taste more attractive, and to prevent it from turning mouldy or stale. But, there is a concern that food additives make unhealthy, processed foods cheaper and more attractive than healthy, fresh foods.

Why are additives given E-numbers?

The European Union (EU) requires food additives to be labelled clearly in the list of ingredients, either by name or by E-numbers.

If an additive has an E-number, it means that it has passed safety tests.

Figure 3.10 Lots of foods contain additives.

What types of additives are there?

There are many different types of food additives which are grouped together by what they do to food. Additives include:

- antioxidants, which keep food fresh;
- colorants;

- emulsifiers, stabilisers and thickeners;
- flavourings;
- preservatives;
- sweeteners.

The use of additives can be beneficial.

- Antioxidants stop butter being oxidised and going rancid if it is left out of the fridge for too long.
- Some additives prevent bread going mouldy.
- Preservatives kill the bacteria that produce deadly poisons in meat.
- A teaspoonful of high-intensity sweetener has the same sweetening effect as a kilogram of sugar and yet it is calorie-free and virtually harmless to teeth.

Substances present in fresh tomatoes

Flavouring:
- E621 monosodium glutamate

Colorants:
- E160a carotene
- E160d lycopene
- E101 riboflavin (vitamin B6)

Antioxidant:
- E300 ascorbic acid (vitamin C)

Acids:
- E330 citric acid
- E296 malic acid
- Oxalic acid

Figure 3.11 Most additives occur naturally in food, they are not manufactured. All the substances listed on the left with E-numbers are present in a fresh home-grown tomato. They have not been added, they are in the tomato naturally. Sometimes these substances are added to other foods to improve them.

Additives can also have drawbacks.

- Aspartame, a sweetener, is widely used in soft drinks. When aspartame was fed to rats, it resulted in low levels of tryptophan in the brain. Low tryptophan levels are linked with aggressive and violent behaviour.
- Tartrazine (E102), the orange colour in some soft drinks, has been linked to asthma, rhinitis and hyperactivity.
- Sunset Yellow (E110), used in biscuits, has been found to damage the kidneys.

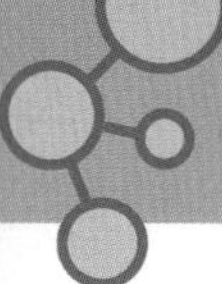

Figure 3.12 You've seen simple colour chromatography – analytical chromatography follows the same principles.

Analysing food additives

Chemists use liquid chromatograms to test foods for additives. Their machines use the same chromatography principles as separating ink colours with water on a filter paper chromatogram. In the machine, the paper is replaced by a column of granules and, instead of water, the liquid is a mixture of solvents. After passing through the chromatography machine, the substances are analysed by a mass spectrometer. This machine helps to identify the separated substances.

11. Explain simply how food can be tested for artificial colours.
12. a) What are E-numbers?
    b) Why are they used?
13. Give two benefits of using food additives.
14. Give two drawbacks of using food additives.
15. Why are food additives blamed for making unhealthy food more attractive?

3.4 The marge story

Have you noticed the pictures of sunflowers on margarine tubs? But, what's the connection between sunflowers and margarine?

Sunflower oil is extracted from sunflower seeds. The seeds are crushed between heavy rollers to squeeze out the oil and this is used to make margarine. Sometimes the seeds are heated to obtain a higher yield of the oil, but this can lead to lower quality, more acidic oils.

Olive oil is also used to make 'marge'. There are several grades of olive oil. Extra virgin olive oil is produced by pressing the flesh of ripe olives at room temperature. This gives the highest possible quality of olive oil. Ripe olives yield about 200 cm^3 of oil per kg of fruit. Commercial growers can get more oil out of the olives, but this has a higher acidity and is used for making soap.

Butter is made from the cream in milk. This is an animal fat. Unfortunately, animal fats contain cholesterol which causes heart disease. Margarine is a butter substitute. You can remove a lot of the cholesterol from your diet by eating margarine instead of butter.

Liquid plant oils, like sunflower oil and olive oil, can be changed into solid margarine by hydrogenation. This means that hydrogen is added to the oil molecules.

Figure 3.13 Sunflowers are a common crop in Southern Europe.

Plant oils contain **unsaturated** molecules with carbon–carbon double bonds. Molecules with these double bonds tend to be liquids (oils) at room temperature. If you add hydrogen to the double bond, the liquid oil turns into a much more solid fatty material which is easy to spread on bread (Figure 3.14).

The liquid oil is heated to 60 °C with a powdered nickel catalyst and hydrogen gas is bubbled through it. When the oil is filtered and cooled it turns into a solid.

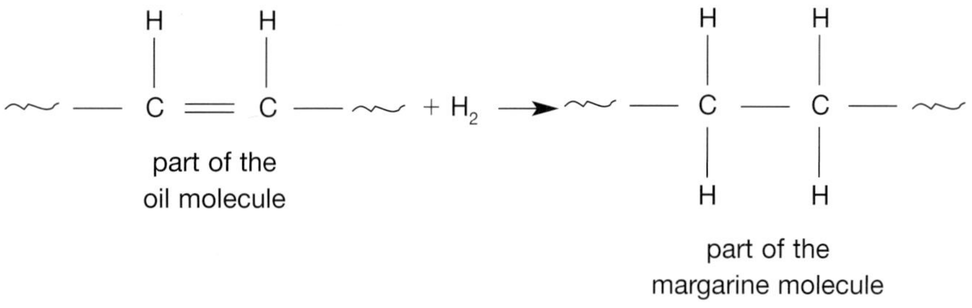

$$\sim\!\!-\underset{}{\overset{\displaystyle H}{\overset{|}{C}}}=\overset{\displaystyle H}{\overset{|}{C}}-\!\!\sim \; + H_2 \longrightarrow \sim\!\!-\underset{\displaystyle H}{\underset{|}{\overset{\displaystyle H}{\overset{|}{C}}}}-\underset{\displaystyle H}{\underset{|}{\overset{\displaystyle H}{\overset{|}{C}}}}-\!\!\sim$$

part of the oil molecule

part of the margarine molecule

Figure 3.14 Hydrogenation of a plant oil

Plant oils belong to a family of chemicals called triglycerides (Figure 3.15). They are not the same as crude oil, which contains hydrocarbons. The shape of a plant oil molecule indicates how useful it is.

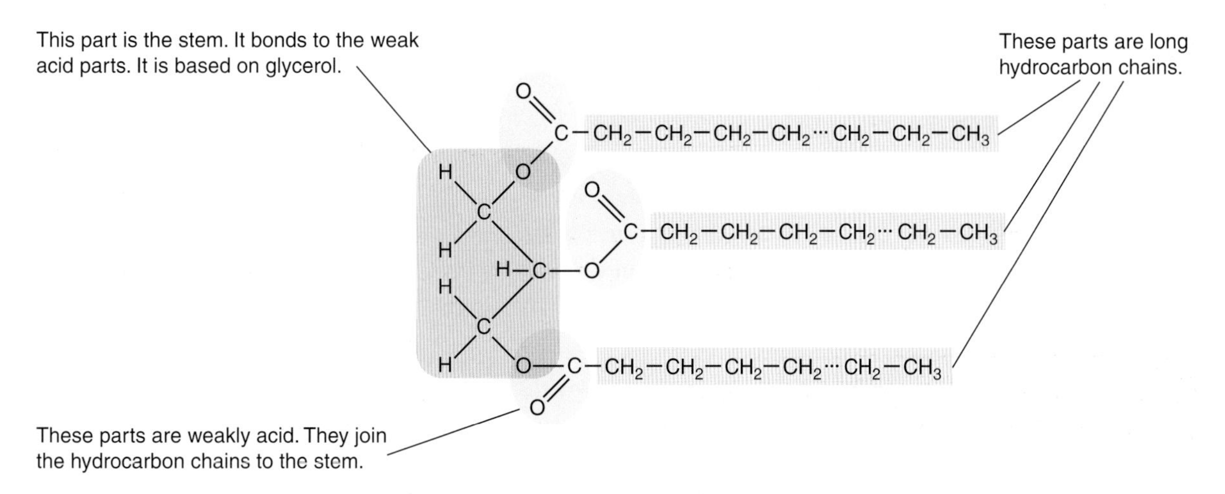

Figure 3.15 The structure of a typical plant oil molecule

16. One kilogram of olives produces 200 cm^3 of oil. Olive oil has a density of 0.9 g per cm^3. What percentage of the olives is oil?
17. Why could eating a lot of olives make you fat?
18. Why is extra virgin olive oil the best olive oil for cooking?
19. Draw a flow chart to summarise the process of changing sunflower oil into margarine.
20. What is a triglyceride?

3.5 Polyunsaturated fat

An **unsaturated** fat is a fat which contains one or more carbon–carbon double bonds in its molecules. Such fat molecules are mono-unsaturated if they contain one double bond and polyunsaturated if they contain more than one double bond. See also page 33.

In a **saturated** fat all the carbon atoms are joined by single bonds.

An unsaturated fat has chains like this:

A saturated fat has chains like this:

Figure 3.16 The structure of molecules of saturated and unsaturated fats

Saturated fats are present in meat and dairy products such as milk, cheese and cream. When margarine or cooking fat is made from corn oil, soyabean oil or other plant oils, hydrogen atoms are added producing saturated fat molecules and consequently the oil becomes more solid.

When we reduce the amount of saturated fat in our diet, it reduces both our blood cholesterol level and our chances of developing heart disease. It is therefore important to read food labels and check how much saturated fat is in our food.

Unsaturated fats are usually liquid at room temperature. They are found in most plant oils. Choosing foods containing polyunsaturated fats can reduce your risk of heart disease. But, fat is fattening, and unsaturated fats pile on the kilojoules (calories) of energy, just like saturated fats. So, eating too much unsaturated fat will increase your weight just like saturated fat.

Testing an oil or fat to see if it is unsaturated

1. Place a few drops of oil or melted fat in a test tube.
2. Put the test tube in a 250 cm^3 beaker of hot water (Figure 3.17).
3. Add three drops of bromine water, or 2% iodine solution.
4. Stir briefly.
5. Time how long it takes for the red or orange colour to fade away.

Bromine and iodine react with any unsaturated carbon–carbon double bonds in the oil or fat molecule. This reaction forms a colourless compound. If the colour fades away in a few seconds the fat or oil is high in unsaturated fat.

Bromine water is an irritant and eye protection should be worn during this test.

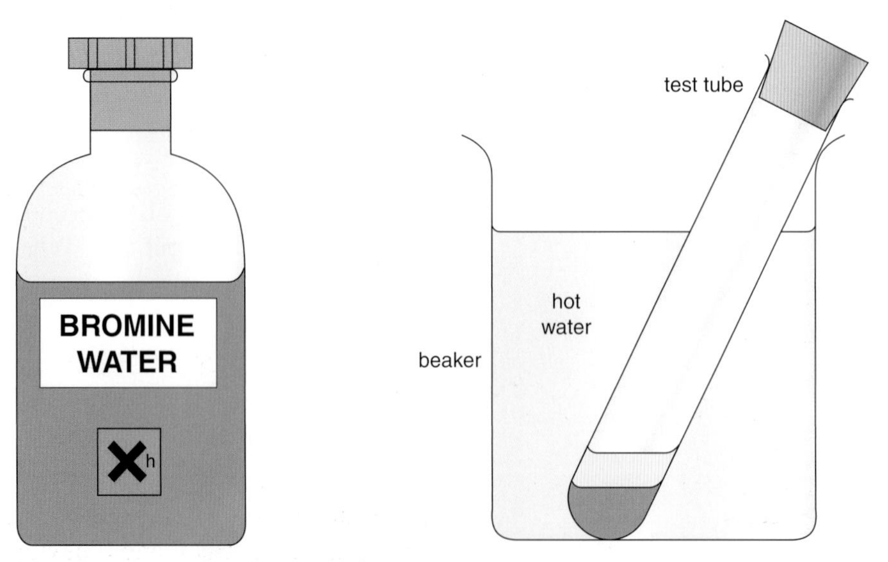

Figure 3.17 Testing for unsaturated fats

Here are some tips to help you reduce fat in your diet:

- eat less fatty foods;
- eat more low-fat foods such as fruit, vegetables, bread, rice, pasta and cereals;
- use less fat, less oil, less butter and margarine in cooking;
- use skimmed milk and low-fat cheeses instead of whole milk and cheese;
- choose margarine instead of butter;
- trim the fat off meat and remove the skin from poultry;
- read and compare food labels to find foods that have less total fat.

Figure 3.18 Yellow flowers of oilseed rape. The name 'rape' is derived from the Old English word for turnip, *rapum*.

21 Why is saturated fat bad for you?

22 Why is eating too much unsaturated fat also bad for you?

23 Draw an unsaturated fat molecule and write a caption to explain why it is unsaturated.

24 How would you test olive oil to see if it was an unsaturated fat?

3.6 Do we need so much oilseed rape?

Oilseed rape is a very useful crop. Its seeds are 42% oil and the material (meal) left after removing the oil is 42% protein and very good as animal food.

Rape seeds are tiny round black seeds which germinate rapidly. After a period of growth, the rape plant produces bright yellow flowers, smelling faintly of honey.

Oilseed rape is widely cultivated throughout the world for the production of animal feed, plant oil and biodiesel. Forty million tonnes of seeds are produced worldwide every year.

Oilseed rape is the third leading source of plant oil in the world after soyabean oil and palm oil. It is also the world's second leading source of protein meal for animals.

After storage, the seeds are crushed between rollers, to extract the oil. There is so much oil in the seeds that other extraction methods are unnecessary.

25 Describe an oilseed rape plant. What would you see and smell?

26 Name three sources of plant oil.

27 How is oil extracted from rape seeds?

28 a) What use is made of the material left after the oil is extracted?
b) Why is this material so useful?

GM controversy

Some seed companies sell genetically modified (GM) rape seeds that are resistant to the effects of certain weedkillers. This allows farmers to spray weedkiller all over their fields without affecting the growth of their GM oilseed rape crop.

In recent years, these companies have prosecuted farmers found to be growing their seeds without paying for them. The farmers say that GM

29. Name a non-food use of rapeseed oil.
30. What are the drawbacks of cultivating rape seed?
31. On balance, do you think the cultivation of oilseed rape should be increased or decreased? Explain your decision.

pollen was carried to their fields and combined with the non-GM rapeseed which they had planted.

Other farmers find unwanted GM plants in their fields that are not killed by weedkiller.

The extensive use of weedkillers when farming these GM crops leads to a significant loss of biodiversity, as wildflowers ('weeds') are killed, leaving wildlife, such as bees and butterflies which are dependent on the wildflowers, unable to survive.

Biodiesel

Rapeseed oil can be turned into diesel fuel by chemically pulling the plant oil molecules apart. This fuel is 'carbon neutral' because the carbon dioxide from the burning fuel was removed from the air by the rapeseed plants when they were growing.

Petroleum-based fuel is running out and becoming very expensive. This has led to the suggestion that locally produced motor fuel could become one of the fastest-growing industries in the future. Small-scale factories could spring up in every town and city to produce biodiesel. You could become a multi-millionaire working in this industry!

Alternatives to petrol and diesel

Figure 3.19 H_2 Eyes Cool is hi-tech. She sells photovoltaic cells that turn water into hydrogen fuel and converts your car to run on it. Sunshine, rainwater and some technology are all you need. She's light and bright.

Activity – What happens when petrol runs out?

In a few years petrol will be running out.

Locally produced motor fuel has become the fastest growing industry and small-scale factories are springing up in every town and city producing alternatives to petrol and diesel. You have to decide which fuel you will support.

Work in groups for this activity.

H_2 Eyes Cool is selling hydrogen and fuel cells

- The raw materials for this form of energy are sunlight and moisture in the air.
- Photovoltaic (solar) cells are becoming common and more efficient. They split up water forming hydrogen for the fuel cells and oxygen, which is vented to the air.
- There will never be a shortage of fuel, as long as the Sun shines.
- Photovoltaic cells are expensive.
- The cells will only make hydrogen fuel in daylight hours.
- Hydrogen fuel is clean – the waste materials from the process are oxygen and water.
- Storing hydrogen is difficult, but the problem has been solved using carbon nanotubes. They can store 65% of their mass as hydrogen.

- Hydrogen cars don't have an engine that burns fuel, so noisy exhausts will be a thing of the past.
- Hydrogen cars have an electric motor. The hydrogen powers a fuel cell to make electricity. The oxygen you need comes from the air.
- Hydrogen car technology is expensive.

Figure 3.20 Betty Biodiesel makes diesel motor fuel from recycled cooking oil and plant oil from local farmers. She's keen and green.

Betty Biodiesel

- You don't have to modify your ordinary diesel engine to burn biodiesel.
- Biodiesel can be mixed with ordinary diesel to make it last longer.
- Biodiesel contains 80% less CO_2 and does not produce SO_2, so acid rain will be reduced.
- Biodiesel leaves 90% less unburnt hydrocarbons to make ozone and smog.
- Biodiesel produces lubrication, so engines last longer.
- Biodiesel has been used successfully in Europe for 20 years.
- Biodiesel is safe to handle and transport. It does not burst into flames.
- Biodiesel is made from crops such as soy, sunflowers and rape seed.
- Biodiesel is as biodegradable as sugar and ten times less toxic than salt.
- Biodiesel exhaust fumes smell of popcorn.

Figure 3.21 Sugar Dude runs a fermenting plant. She makes ethanol fuel for motor vehicles. She's cooking up a clean future.

Sugar Dude – Ethanol fuel

- Ethanol is made by the natural process of fruit rotting.
- Crops such as maize, barley and potatoes can be used instead of fruit.
- The fuel is 'carbon neutral' because the original crops took in carbon dioxide.
- Ethanol burns very cleanly producing no polluting gases such as sulfur dioxide.
- Ethanol is not a hydrocarbon, so there can be no unburnt hydrocarbons to cause smog.
- Ethanol mixes with water and does not pollute it.
- Making ethanol fuel is a two stage process – fermentation and distillation.
- Ethanol is very flammable and bursts into flames easily.
- Pure carbon dioxide is produced by fermentation. This can be collected and used to make fizzy drinks.
- After fermentation the 'mash' can be made into animal feed.
- Ethanol keeps your engine clear and clean.

1. Compare each of the alternative fuels with petrol and diesel. Focus on the disadvantages of the 'new fuel'.
2. Make a table to compare the benefits and drawbacks of each fuel.
3. Appoint one person to make a short (3-minute) presentation for each fuel.

4. Decide as a group which fuel is best. You will need to decide on criteria such as cost, cleanliness, availability and supply.

5. Devise a plan for setting up a small factory to produce your chosen fuel for the petrol stations in a small town.

Summary

- ✓ Plants can provide oil from seeds and fruits.
- ✓ Mixtures of oil and water are called **emulsions** and are widely used in cooking.
- ✓ Frying in oil cooks food faster than boiling and changes the flavour.
- ✓ **Additives** are chemicals used to improve food, but they can have drawbacks.
- ✓ Plant oils can be made into margarine, an alternative to butter.
- ✓ **Unsaturated** fats from plant oils are more healthy than **saturated** animal fats. The most healthy diet is low in all fats.
- ✓ Plant oil crops can be made into a motor fuel called biodiesel.

EXAMQUESTIONS

1. a) Where do plants get their energy from? *(1 mark)*

   b) Copy and complete these sentences.
   A plant collects energy using __________ .
   This makes __________ in the leaves of the plant. Plants also store energy as __________ in their seeds. This energy store is to help the seed __________ . *(4 marks)*

2. Food with additives looks better, tastes stronger and lasts longer on the shelf. Who benefits most from this – the manufacturer, the retailer or the customer who eats the food? Explain your answer. *(3 marks)*

3. Write a 50-word report to be read on a news programme, warning people about saturated fats.

   The message should be amusing to catch the attention of the audience. *(4 marks)*

4. This question is about homemade mayonnaise.

Homemade mayonnaise

Ingredients

1 egg yolk
150 cm^3 olive oil
10 cm^3 grainy mustard
50 cm^3 fresh lemon juice
salt and freshly ground black pepper to taste

Method

1 Place the egg yolk in a large bowl.
2 Gradually whisk the oil into the egg yolk.
3 When the egg and oil are well combined, whisk in the mustard and lemon juice.
4 Season the mayonnaise.

   a) What plant oil is used to make the mayonnaise above? *(1 mark)*

   b) What is the watery solution in the mayonnaise? *(1 mark)*

   c) Which ingredients act as emulsifiers? *(1 mark)*

   d) How do the emulsifiers prevent the oil and water from separating out? *(1 mark)*

Chapter 4
What changes have occurred in the Earth and its atmosphere?

At the end of this chapter you should:

- ✓ know that the Earth has a layered structure comprising a crust, a mantle and a core;
- ✓ appreciate that the Earth's crust and the upper part of the mantle are cracked into a number of very large tectonic plates;
- ✓ understand how the slow movement of these tectonic plates can cause earthquakes, volcanic eruptions and changes in the landscape;
- ✓ understand why scientists cannot predict accurately when earthquakes and volcanic eruptions will occur;
- ✓ know about the composition of the Earth's atmosphere;
- ✓ be able to explain the changes that have occurred and are continuing to occur in the Earth's atmosphere.

Figure 4.1 The Earth from space – a blue and white planet. Can you see continents, oceans and swirls of white clouds which form part of the atmosphere? In this chapter we shall be studying the Earth and its atmosphere.

4.1 The Earth and its atmosphere

❶ Write the following steps in the order they occur when oxygen is separated from argon and nitrogen in the atmosphere:
A allow liquid air to warm up
B compress and cool air
C nitrogen boils off first, then argon, then oxygen
D oxygen liquefies first, then argon, then nitrogen.

In the last three chapters, we have seen that the Earth and its atmosphere provide the raw materials for everything we need:

- the Earth itself provides crude oil, rocks for building materials and metal ores;
- the seas provide salt (sodium chloride) from which we obtain sodium, chlorine and sodium hydroxide;
- the atmosphere provides important gases including oxygen and nitrogen;
- living things, animals and particularly plants provide foods and raw materials for clothing and medicines.

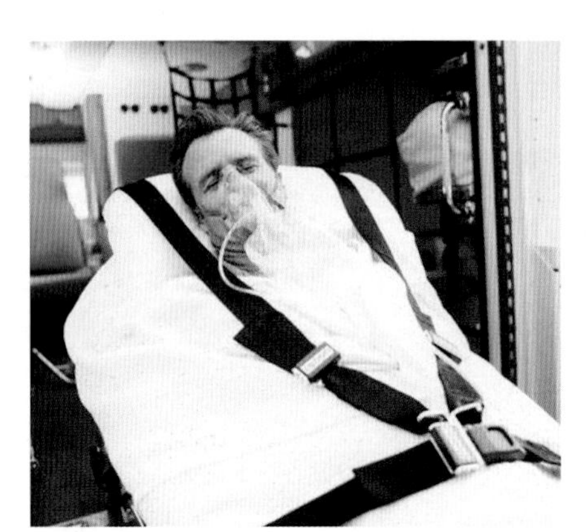

Figure 4.2 A variety of people require extra supplies of oxgyen to survive. These include hospital patients, victims of accidents, divers and mountaineers. Pure oxygen can be separated from other gases in the atmosphere by fractional distillation of liquid air.

4.2 Layers of the Earth

The Earth is a huge ball of rock, iron and nickel. It has a radius of about 6400 km. When the Earth was first formed, about 4500 million years ago, it was a mass of molten rock and metals. Over millions of years, the Earth cooled down. At the same time, heavier materials sank towards the centre of the Earth, less dense materials rose to the surface and this resulted in a layered structure (Figure 4.3).

The outer layer of the Earth is called the **crust**. It is made of rock.

The **mantle** is a layer of denser rock between the crust and the core.

The Earth has a hot central **core** made of iron and nickel.

There are three distinct layers in the Earth:

- a thin outer **crust** of less dense material on the surface of the Earth about 50 km thick. In places where the crust is thicker, its surface is above sea level.
- a **mantle** of moderately dense rock about 3000 km thick that extends almost halfway to the centre of the Earth. Temperatures in the mantle range from 1500 to 4000 °C. The hottest rocks in the mantle form a very viscous liquid called magma, which changes shape and flows very slowly.
- a **core** of very dense iron and nickel with a radius of 3500 km (about half the Earth's radius).

Most of the evidence for the Earth's layered structure comes from the study of shock waves from earthquakes. Some of the waves are reflected by layers of the Earth and this reflection shows up the crust, the mantle and the core very clearly.

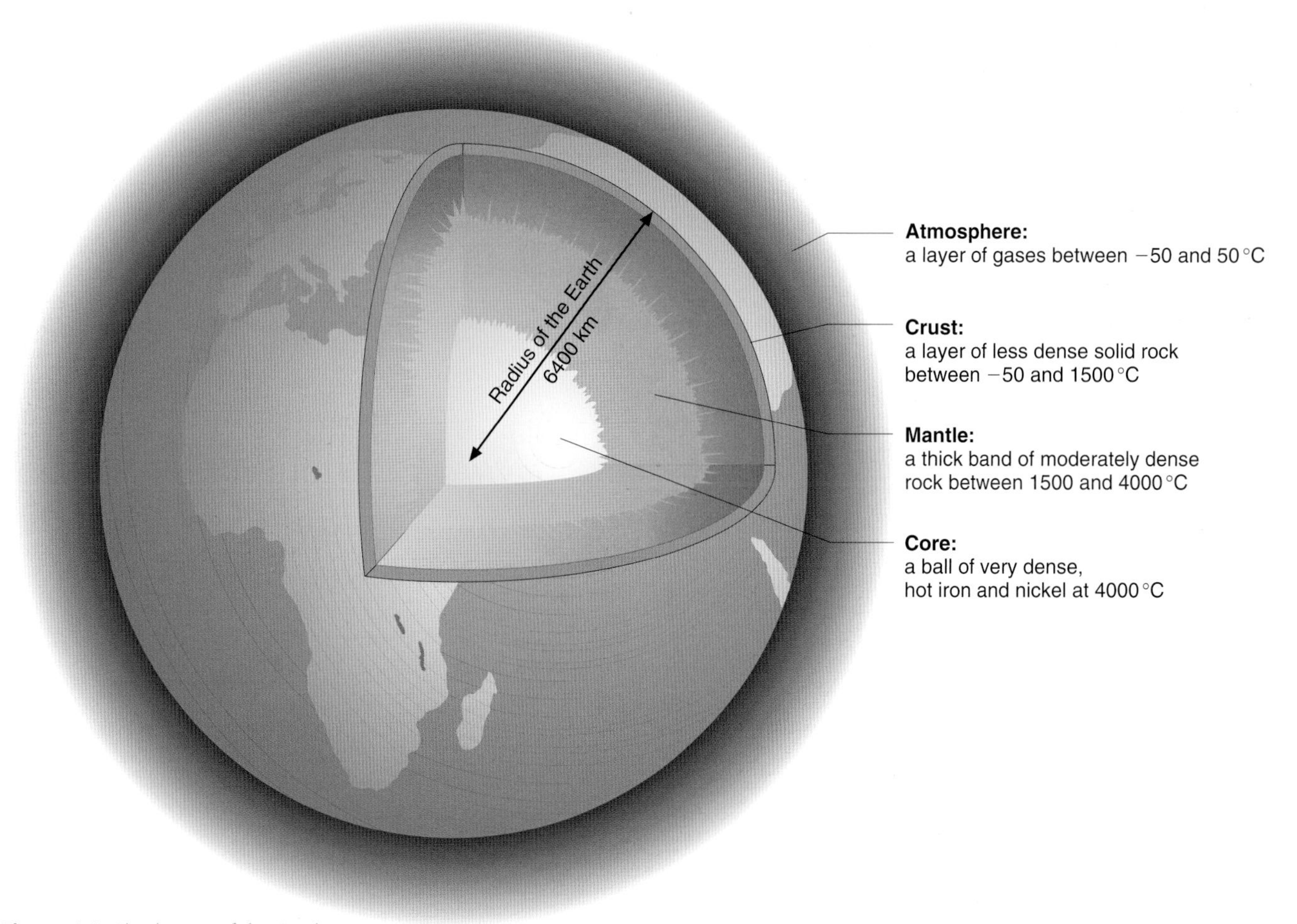

Figure 4.3 The layers of the Earth

❷ a) How do the temperatures of the different layers in the Earth change towards the centre?
b) How do the densities of the layers in the Earth change towards the centre?

More detailed studies have suggested that the core can be split into two – an outer core, which is liquid, and an inner core, which is solid.

Outside and above the Earth there is a fourth layer – the atmosphere, which is a layer of different gases about 100 km thick.

Why is the Earth's core so hot?

One theory about the origin of the Earth is shown in Figure 4.4.

Although the crust and mantle insulate the Earth's core, there is a second, more important, effect that helps to maintain high temperatures inside the Earth. Some rocks in the Earth, particularly granite, contain radioactive atoms of elements such as uranium and potassium. As the nuclei of these elements break up (decay), energy is released as heat and electromagnetic radiation. This energy helps to maintain temperatures inside the Earth.

3 The theory summarised in Figure 4.4 suggests that the Earth began as a giant molten ball. As this cooled down, the crust and other layers formed.

a) Why did iron and nickel move into the core when layers formed in the Earth?

b) Volcanoes are responsible for some features on the surface of the Earth. Suggest two ways in which other features, such as valleys and mountains, formed on the Earth's surface.

c) In the last few years, average temperatures on the surface of the Earth have risen slightly. Why do you think this has happened?

d) Do you think that temperatures on the Earth will continue to rise or will they eventually fall? Explain your answer.

1. The Earth began as a giant ball of molten rock and metal.

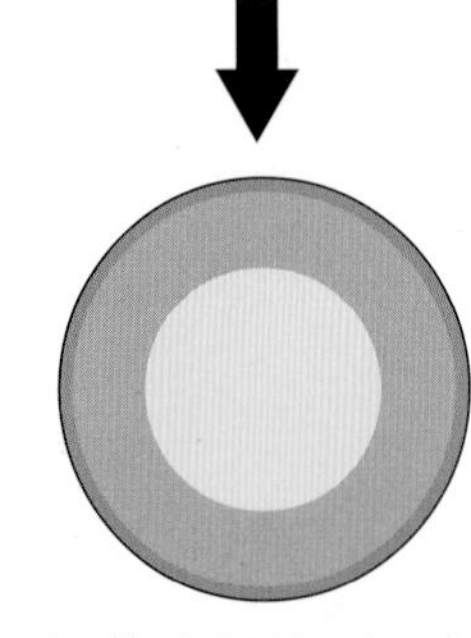

2. As the Earth lost heat and cooled, a crust of solid rock formed and denser molten material sank to the core.

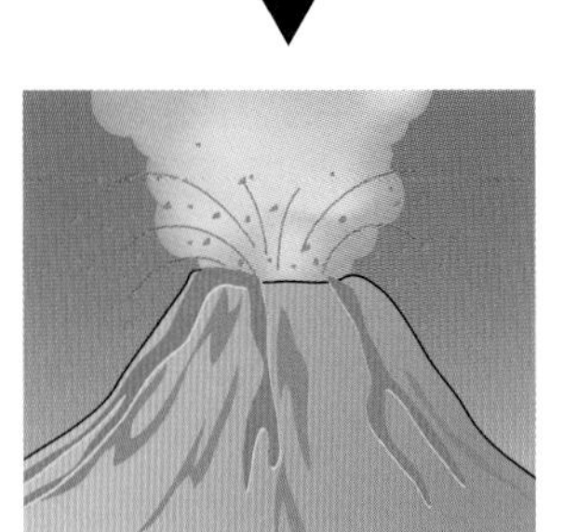

3. Where the crust was thin, cracks appeared and volcanoes erupted. Lava and gases poured out from the volcanoes and the first atmosphere formed.

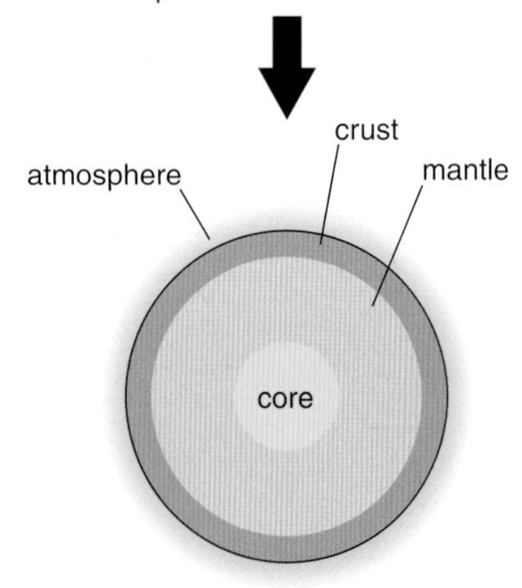

4. As the crust and mantle became thicker, these outer layers insulated the core and its temperature remained high.

Figure 4.4 A theory to explain the high temperatures inside the Earth

4.3 How is the Earth changing?

Before 1915, most scientists thought that:

- the Earth contracted and shrunk as it cooled down;
- the shrinking caused the Earth's crust to wrinkle (Figure 4.5);
- the shrinking and wrinkling of the Earth's crust created mountains, valleys and other features on the Earth's surface;
- the hard, dense solid rocks in the Earth's crust prevented any movement of the continents.

In 1915, the German meteorologist (weatherman) Alfred Wegener suggested that the continents were in fact moving. At first, people laughed at his ideas because they seemed impossible. But, as people listened to Wegener's evidence, his ideas didn't seem quite so impossible after all.

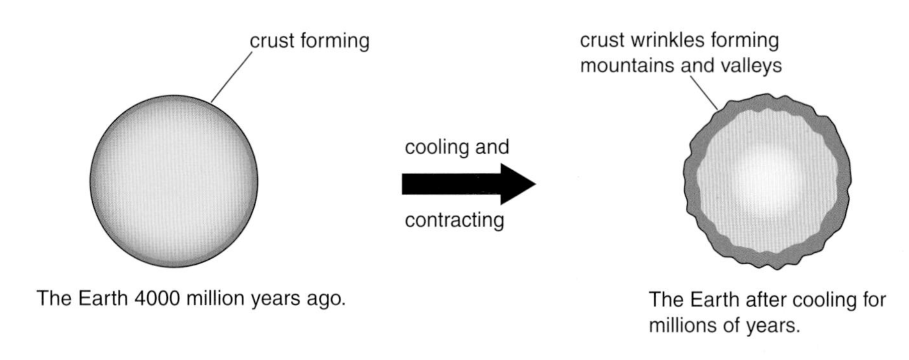

Figure 4.5 Early ideas about the formation of mountains on the Earth's surface

Activity – Wegener's evidence for moving continents

Continental fit

Wegener noticed that the continents looked like pieces of a jigsaw puzzle that could fit together. For example, it was easy to see that South America might fit into the coastline of West Africa (Figure 4.6).

1. Use tracing paper to trace the outline of South America in Figure 4.6. Carefully, cut out your outline. How well does it fit into the coastline of West Africa?

2. Most geographers and geologists thought that this was just a coincidence. How on Earth could continents move sideways by thousands of kilometres! What do you think?

Mountain chains

Wegener also noticed that mountains with similar rock types and structures occurred in different continents, but these mountain chains came close

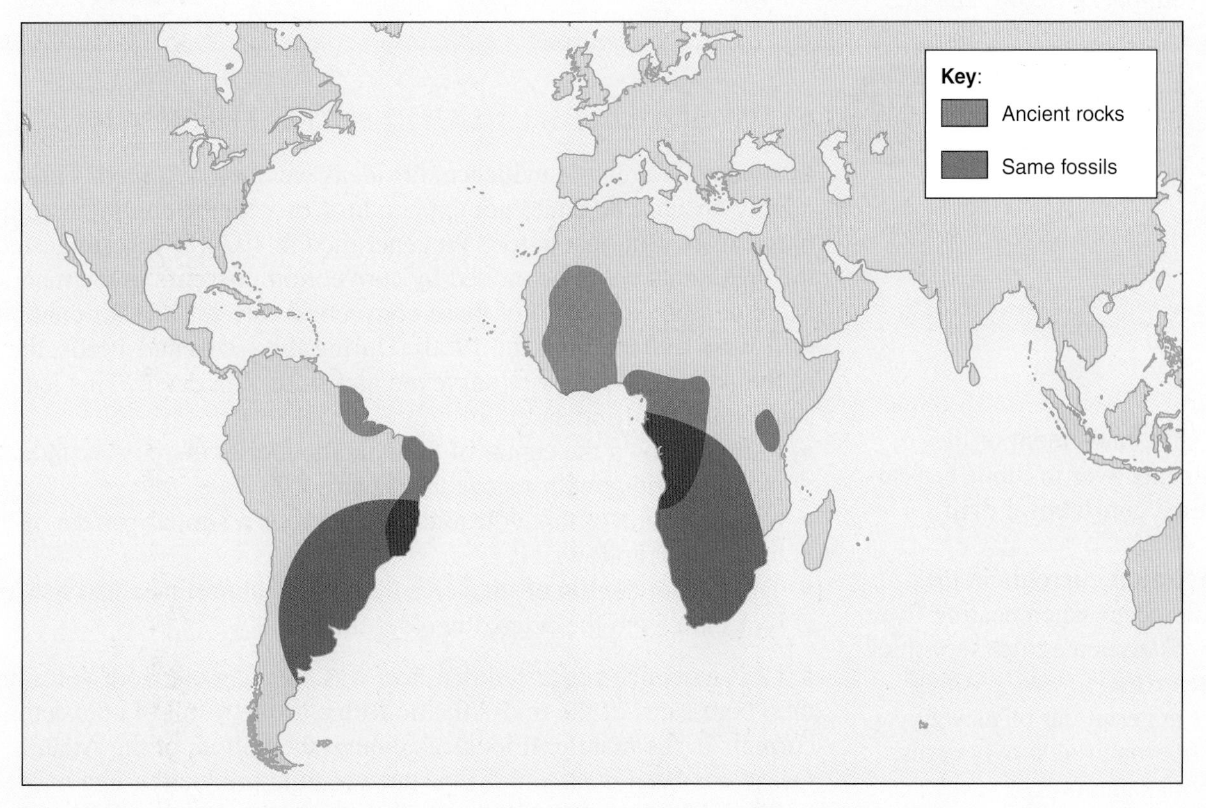

Figure 4.6 Wegener's evidence for the movement of South America and Africa

together when the continents formed one giant jigsaw (Figure 4.6).

3. Shade the areas with ancient rocks on your outline of South America. Fit your outline as snugly as possible into the west coast of Africa. How neatly do the areas with ancient rocks in South America and West Africa align?
4. Experts told Wegener that the alignment of these ancient rocks was not exact. Is that right?

Fossil records

Animals and plants can only survive in areas where the climate and habitat suits them. Yet Wegener noticed that fossils of the same animals and plants had been discovered on different continents, sometimes with different climates and habitats. Just like his observation of mountain chains, these areas with the same fossils were close together when his 'giant jigsaw' was completed.

5. Shade the area with the same fossils on your outline of South America. Now fit your outline again into the west coast of Africa. How well do the areas with the same fossils in South America and West Africa align?
6. Those who disagreed with Wegener's ideas suggested there was probably once a land bridge connecting South America and West Africa. Is this a possibility?

From observations similar to those you have just made, Wegener suggested that:

- about 200 million years ago, all the continents were joined together as one supercontinent which he called Pangaea from the Greek words meaning 'all lands';
- then Pangaea broke up and over millions of years the continents slowly moved apart. Wegener called this movement **continental drift**.

7. How do you think Wegener was able to date the break up of Pangaea to 200 million years ago?
8. Why do you think that Wegener's theory of crustal movement (continental drift) was not accepted for many years?

4.4 Why should the continents move apart?

In spite of Wegener's evidence, his ideas were not accepted. This was mainly because he could not explain how or why the continents should move apart. But, just before Wegener died in 1930, it was suggested that the continents could be moved by **convection currents** in the mantle. More definite evidence for these convection currents and for continental drift wasn't found until the 1950s. During the 1950s and 1960s, the floor of the Atlantic Ocean was surveyed and studied in detail. This led to some startling discoveries.

- Running down the centre of the Atlantic Ocean there is an undersea mountain ridge with volcanoes (Figure 4.7).
- On either side of this volcanic ridge there is a similar pattern of humps and hollows.
- The rock at the top of the volcanic ridge is almost new and the further rocks are from the ridge, the older they get.

All this evidence suggested that lava was spewing out from volcanoes onto both sides of the mid-Atlantic Ridge as the result of convection currents in the mantle. It looks as though the bottom of the Atlantic Ocean has been made on a stop/start volcanic production line over millions of years.

The slow movement of the continents over millions of years is called **continental drift**.

Convection currents in the mantle occur when heating from below causes magma (very thick molten rock) to rise, pushing other material out of its way. The hot material slowly cools, and sinks again.

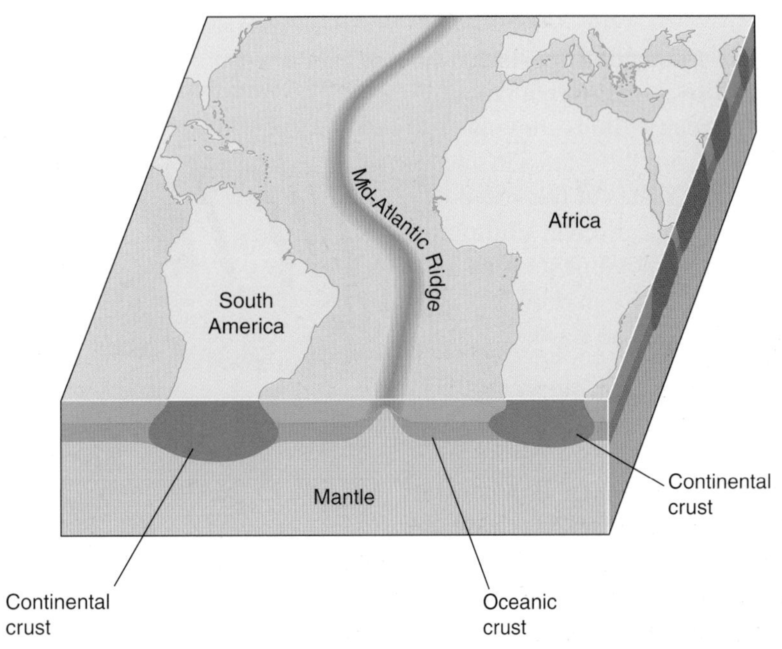

Figure 4.7 The mid-Atlantic Ridge

The final proof of Wegener's ideas about moving continents came in 1963.

During the 1950s, geologists began to study the magnetism in rocks. As molten rocks solidify, any bits of iron in them become magnetised in line with the Earth's magnetic field. To their surprise, the geologists found that every half-million years or so the Earth's magnetic field flips – the North Pole becomes a South Pole and vice versa.

Then, in 1963, magnetic surveys along the mid-Atlantic Ridge showed that the magnetism of rocks on one side of the ridge was an exact reflection of the other side (Figure 4.8). Rocks now hundreds of kilometres apart must have been formed at the same time, as magma spewed out of the volcano. These rocks were pushed further and further apart over millions, if not billions, of years as new rock formed. And as these rocks moved apart, the continents of South America and Africa moved with them.

Wegener's ideas about moving continents resulted in the study of plate tectonics.

How are the continents moving?

The Earth's *shape* is like an orange – spherical but slightly flattened at the poles. Its *structure* is like a badly cracked egg. The 'cracked' shell is like the Earth's thin crust, the 'egg white' is the mantle and the 'yolk' is the core.

The Earth's crust and the upper part of the mantle are cracked into a number of vast pieces called **tectonic plates**.

The crust and the solid upper parts of the mantle are sometimes called the Earth's lithosphere. The lithosphere is not one continuous solid shell. It is cracked and broken into a number of massive pieces called **tectonic plates**.

Measurements from space satellites have shown that the American Plate is moving away from the Eurasian and African Plates at the speed of 5 cm per year. Over millions of years, movements such as this can cause massive changes on the Earth.

4. How many *centimetres* will the American plate have moved away from the Eurasian and African Plates in one million years?
5. How many *kilometres* will it have moved away in one million years?
6. At this rate of separation, how much wider will the Atlantic Ocean be in one million years?

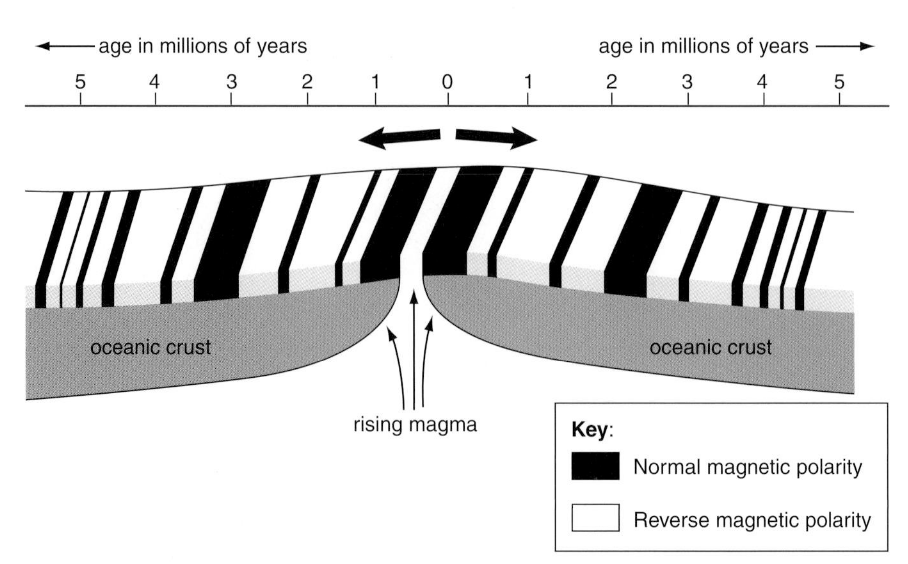

Figure 4.8 Magnetic patterns in the rocks on both sides of the mid-Atlantic Ridge

The Earth's core is incredibly hot, 4000 °C and higher, due to the heat released by natural radioactive processes. This causes slow convection currents in the liquid mantle and slow movements of a few centimetres per year of the tectonic plates. Figure 4.9 shows a map of the world with the main tectonic plates and their direction of movement.

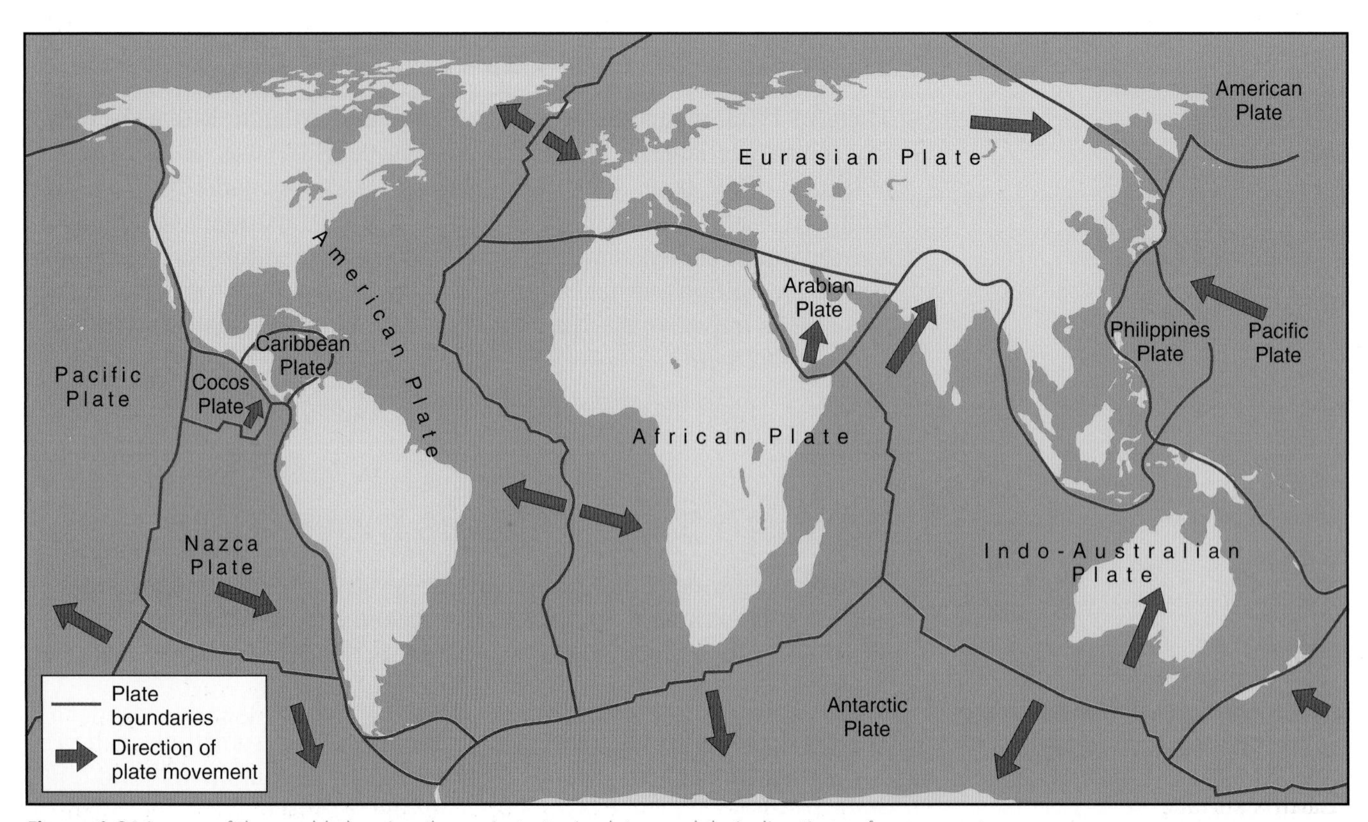

Figure 4.9 A map of the world showing the main tectonic plates and their directions of movement

Plate tectonics explain many of the features on the Earth's surface including volcanoes, earthquakes and even the formation of mountain ranges.

To appreciate the importance of plate tectonics, let's look at what happens when plates slide past each other, when they move apart and when they move towards each other. Although movements of the tectonic plates are normally very slow, they can sometimes be sudden and disastrous.

What happens when plates slide past each other?

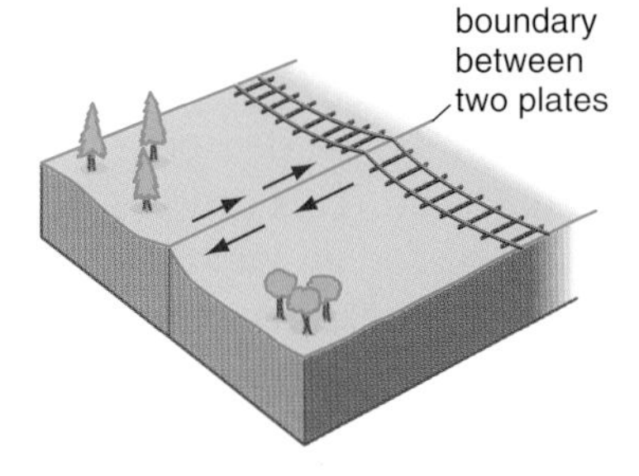

a) Plates in the Earth's crust are bent as they slide past each other.

When two plates slide past each other, stresses and strains build up in the Earth's crust. Massive forces are involved, due to convection currents in the mantle pushing tectonic plates one way or another. The forces are so great that they can cause the plates to bend. In some cases, the stresses and strains build up and are then suddenly released. The Earth moves, the ground shakes violently in an **earthquake** and breaks appear in the ground (Figure 4.10).

These breaks in the ground when plates slide past each other horizontally are called tear faults (Figure 4.10b)). The San Andreas fault in California and the Great Glen fault in Scotland are examples of tear faults. In fact, the map of Scotland would look very different if the Great Glen tear fault did not exist (Figure 4.11).

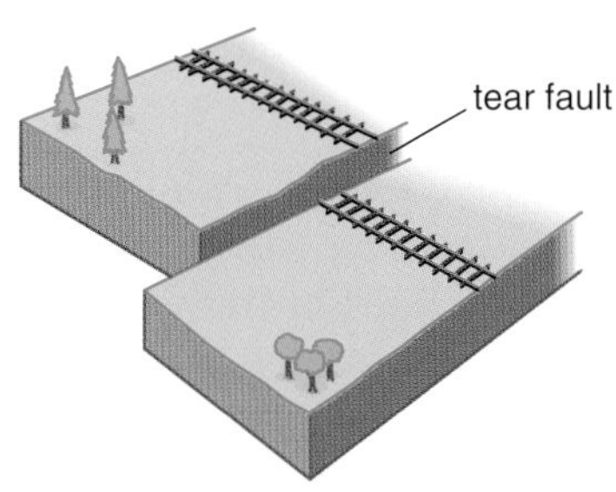

b) Stresses in the bent plates are suddenly released, the earth moves and the ground shakes violently in an earthquake. Breaks appear in the ground and a tear fault has formed.

Figure 4.10 How an earthquake occurs

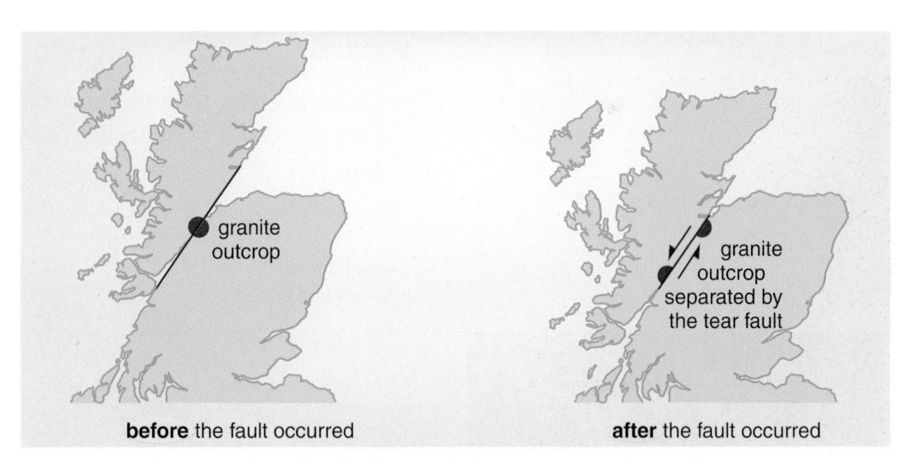

Figure 4.11 The map of Scotland before and after the Great Glen Fault

What happens when plates move apart?

When plates move apart, the crust is stretched and cracks may appear in the Earth's surface. In some cases, hot molten magma rises from deep within the Earth and escapes through the cracks as a **volcanic eruption**. If the plates move further apart, surface rocks sink and may get buried. This results in vertical faults. These vertical faults produced by stretching (tension) forces are called normal faults. When two vertical faults occur alongside each other, a rift valley is formed (Figure 4.12).

In many cases, volcanoes erupt without causing too much damage. But in some cases, a supervolcano erupts with catastrophic results. One of

An **earthquake** occurs when two tectonic plates suddenly move past each other.

In a **volcanic eruption** hot molten magma rises up and escapes through a crack in the Earth's surface.

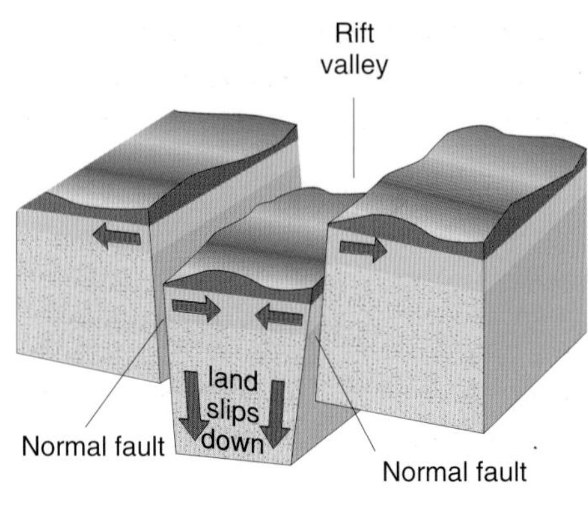

Figure 4.12 As plates move apart, the land may sink into the crack. If there are two normal (vertical) faults near each other, a rift valley may form.

the most famous supervolcanoes is Krakatoa on an island between Sumatra and Java, in Indonesia. Krakatoa had been dormant for centuries but then came 'back to life' in 1883. On 26 August 1883, two thirds of the island was blown away. Ash rose 27 km into the atmosphere and the explosions could be heard 4600 km away in Australia!

What happens when plates move towards each other and collide?

The Earth's crust that lies under the continents is called continental crust. The crust that lies under the oceans is called oceanic crust. The continental crust is usually thicker and contains less dense rocks, such as granite. The oceanic crust contains denser rocks like basalt.

When a continental plate and an oceanic plate collide, the denser oceanic plate sinks below the lighter continental plate. This process is called subduction. Part of the oceanic crust may be forced into the mantle where it forms liquid magma. At the same time, the continental crust gets squashed and pushed into folds.

Over millions of years, this results in the formation of mountain ranges. This is what happened and continues to happen in South America as the Nazca Plate collides with the American Plate to create the Andes Mountains (Figure 4.14).

7 In California, most of the orange groves have trees growing in straight lines. In some groves, the lines of trees are kinked, but they were not planted this way. Why are the lines of trees now kinked?

Figure 4.13 When an earthquake occurs below the seabed, it can set off a giant wave called a tsunami. This is what happened in the Indian Ocean on 26 December 2004, causing the loss of more than 200 000 lives and enormous damage in the areas affected.

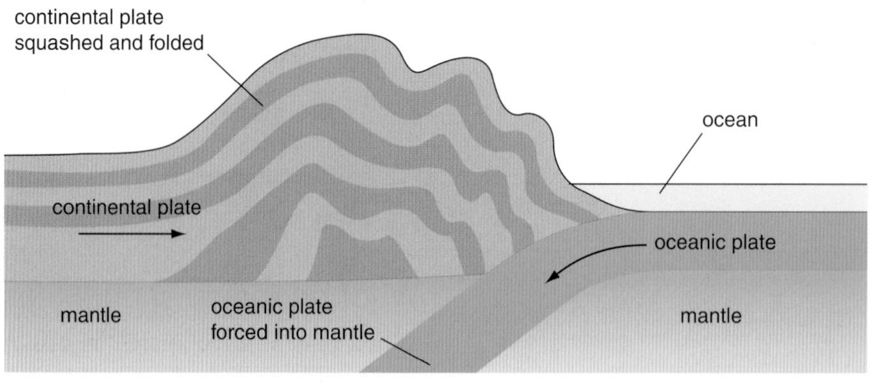

Figure 4.14 A continental plate and an oceanic plate moving towards each other and colliding

If two continental plates or two oceanic plates move towards each other, they tend to collide head on. Layers in the plates become tilted, folded and even turned upside down. This process has been happening for millions of years during the formation of the Alps, the Pyrenees and the Himalayas.

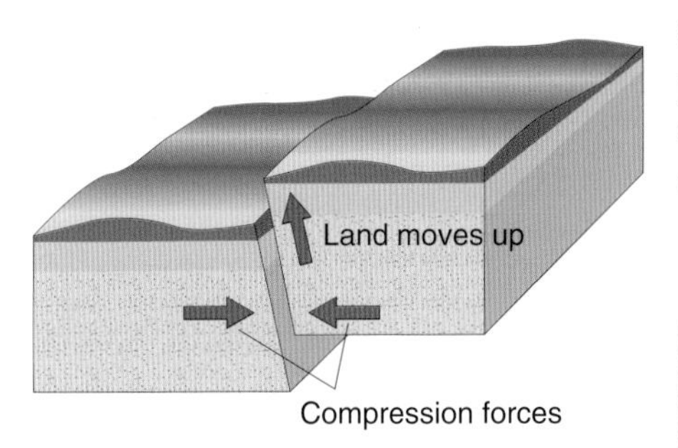

Figure 4.15 When plates collide and push against each other, the Earth may crack forming a reverse fault.

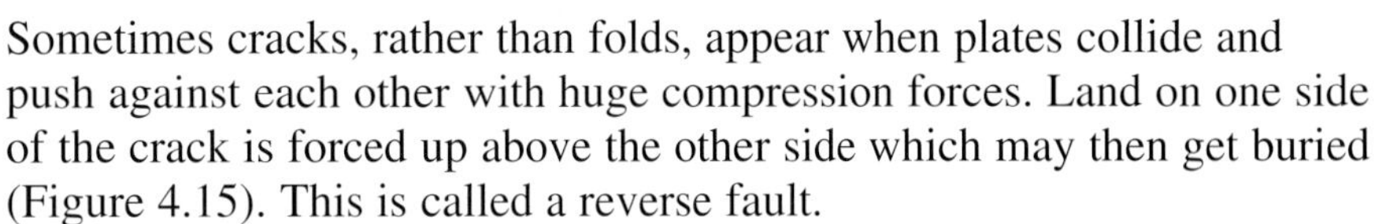

Sometimes cracks, rather than folds, appear when plates collide and push against each other with huge compression forces. Land on one side of the crack is forced up above the other side which may then get buried (Figure 4.15). This is called a reverse fault.

8 The photo in Figure 4.16 is taken looking east.

a) Which directions did the forces come from to create the fold?
b) Was the force involved a compression or a tension force?
c) Draw a sketch of the fold and explain how it formed.

Figure 4.16 Folds in the Earth's crust at Stair Hole, Lulworth Cove, in Dorset. Folds like this have been created by the movement of the Eurasian Plate.

Breaking news – Massive Earthquake

We report the latest situation
BBC News 24

For further information go to:
www.geo.ed.ac.uk and search on 'quakes' or www.eqnet.org

Activity – Earthquake

Use the internet websites given above to find out about a recent large earthquake.

Now imagine you are the senior reporter for a news programme on TV.

You are about to report live about the earthquake. Compose what you will say.

Your report must not last more than 2 minutes.

Important things that you should report on are:

- the time and place of the earthquake;
- injuries to people, the number of people missing and any loss of life;
- damage to homes and other buildings in the area;
- the size of the quake on the Richter Scale.

Other things that you may wish to include are:

- any indications that the earthquake may recur with aftershocks;
- whether the earthquake might have been predicted;
- the dates of any previous earthquakes in the area;
- one or two personal experiences of people affected by the earthquake.

4.6 Can we predict when earthquakes and volcanic eruptions will occur?

Most earthquakes occur in predictable areas of the world along or near the boundaries between tectonic plates. One of the areas most at risk from earthquakes is along the San Andreas fault in California, where the Pacific plate slides past the North American plate.

Although we can predict *where* earthquakes are likely to occur, we cannot predict *when* they will occur with any accuracy. If we could predict when an earthquake will occur, even to the nearest day or so, people could be evacuated from their homes to safety and many lives could be saved.

Figure 4.17 A seismograph used to monitor for the risk of earthquakes in the Philippines

Predicting earthquakes

Earthquakes occur suddenly and without warning. Predicting them is difficult but there are signs to look for and monitoring instruments that scientists can use.

- **Seismometers** are used continuously in risky earthquake areas to monitor for any possible smaller shocks before the main earthquake occurs.
- **The GPS (global positioning system)**, which locates positions on the Earth to within a few centimetres, can be used to follow the movement of plates.
- **Strainmeters** are used to measure the forces in rocks.
- **Tiltmeters** are used to record tiny movements and bulges in rocks.
- **Animals often behave strangely** before an earthquake. Snakes come out of hibernation, rats leave their holes and cattle become restless.

In some cases, scientists have been remarkably successful in predicting earthquakes. For example, in 1975 Chinese scientists in Haicheng City recorded increased seismic activity and had reports that snakes were coming out of hibernation. People living in the city were warned and told to leave their homes. At 7.30p.m. an earthquake occurred, destroying most of the buildings in Haicheng, but thousands of people's lives were saved.

Figure 4.18 After being quiet for nearly 400 years, the volcano on the Caribbean island of Montserrat erupted in the summer of 1997. The capital, Plymouth, had to be abandoned because of the flows of hot rocks, ash and gases.

Predicting volcanic eruptions

It is much easier to predict when a volcano will erupt than when an earthquake will occur. Unlike earthquakes, volcanoes usually give a variety of warning signs months, sometimes years, before they erupt. Here are some of the warning signs that scientists pick up prior to a volcanic eruption.

- **Seismometers** will detect any small earth tremors caused by the movement of magma deep inside a volcano.
- **The GPS** will detect any changes in ground level around a volcano as pressure builds up inside it.

❾ Write three sentences to explain why scientists cannot predict accurately when an earthquake will occur.

❿ List three signs which would suggest that a volcano may soon erupt.

⓫ Why are volcanic eruptions easier to predict than earthquakes?

⓬ Do you think that, in time, scientists will be able to predict accurately when a volcano will erupt? Explain your answer.

- **Air monitoring equipment** can be used to monitor the levels of sulfur dioxide near the mouth of a volcano. An increase in the concentration of sulfur dioxide indicates that the magma is rising.
- **Infra-red cameras** mounted on satellites will detect any increase in ground temperature near a volcano as magma rises towards the Earth's surface.

Although these instruments can give us clues that a volcano might soon erupt, just as with earthquakes, predicting an eruption is only an intelligent guess.

In 1985, the ground rose by 2 metres in just a few days at Potsuoli near Naples in Italy, close to the volcano on Mount Vesuvius. Buildings were in danger and 40 000 people were evacuated from the area. Scientists are still waiting for the eruption!

4.7 How has the Earth's atmosphere changed?

Figure 4.19 The early atmosphere on Earth was mainly carbon dioxide and water vapour.

Using radioactive dating techniques scientists estimate that the oldest rocks on the Earth were formed about 4500 million (4.5 billion) years ago. This is usually taken to be the age of the Earth.

- During the first billion (1000 million) years of the Earth's existence, there was intense volcanic activity. Rocks decomposed, elements reacted and gases were released to form the first atmosphere.

This early atmosphere was mainly carbon dioxide and water vapour with smaller proportions of methane (CH_4) and ammonia (NH_3) (Figure 4.19).

- As the molten rocks on the Earth's surface cooled and the temperature dropped further, most of the water vapour condensed to form rivers, lakes and oceans (Figure 4.20).
- Plants evolved and first appeared on the Earth 3500 million years ago. As plants slowly colonised most of the Earth's surface, further changes occurred in the atmosphere. Plants took in water and carbon dioxide for photosynthesis and released oxygen (Figure 4.21).

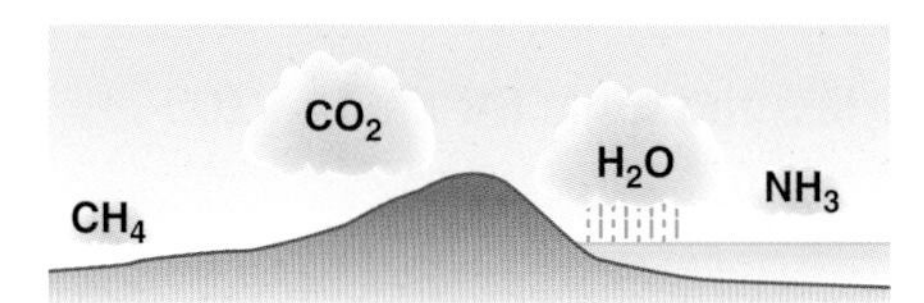

Figure 4.20 As the Earth cooled down, most of the water vapour in the early atmosphere condensed to form rivers, lakes and oceans.

Figure 4.21 When plants appeared on the Earth, carbon dioxide and water were taken up during photosynthesis and oxygen was produced.

⓭ The early atmosphere on Earth contained little or no oxygen. What would have happened to any oxygen which did form? Remember that the Earth would be very hot.

⓮ Where in the Solar System today might you find an atmosphere like the Earth's early atmosphere? Explain your answer.

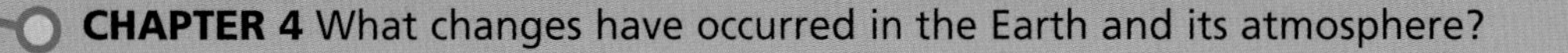

Figure 4.22 Methane and ammonia burnt in the oxygen from photosynthesis producing more water, more carbon dioxide and nitrogen.

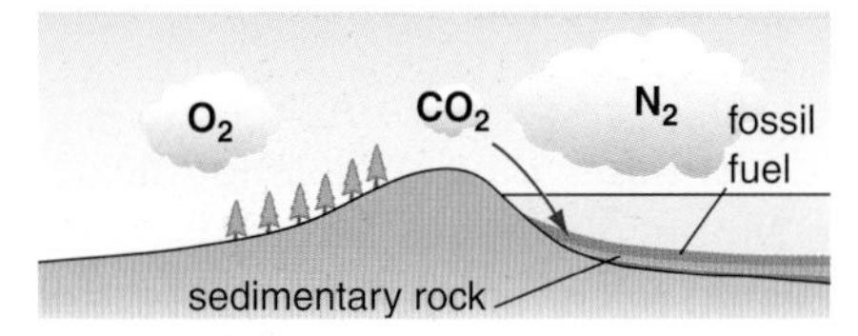

Figure 4.23 Over billions of years, carbon dioxide became locked up as fossil fuels and in sedimentary rocks as carbonates.

- As oxygen collected in the atmosphere, flammable gases, like methane and ammonia, burnt in this oxygen producing more water, more carbon dioxide and nitrogen (Figure 4.22).
- At the same time, carbon dioxide in the changing atmosphere was being removed by two other processes:
  1. the formation of fossil fuels from carbon compounds in plants and sea creatures;
  2. the deposition of carbonates as sedimentary rocks, following erosion by rivers and from the shells and bones of sea creatures.

So, over billions of years, most of the carbon dioxide in the air gradually became locked up as fossil fuels and in sedimentary rocks as carbonates (Figure 4.23).

15. Write a word equation for the reaction which occurred when methane reacted with oxygen in the Earth's early atmosphere.
16. In the early atmosphere, ammonia reacted with oxygen to form nitrogen and water vapour. Copy and balance the following equation for this reaction.

$$\_\_ NH_3 + \_\_ O_2 \rightarrow \_\_ N_2 + \_\_ H_2O$$

17. What two substances do the Earth's surface and atmosphere contain which are essential for all living things?
18. How did these two substances get into the atmosphere and onto the Earth's surface?

4.8 Our atmosphere today

Our atmosphere has remained more or less the same for the last 200 million years. It is composed of:

- about four fifths (80%) nitrogen;
- about one fifth (20%) oxygen;
- small proportions of other gases including carbon dioxide, water vapour and noble gases.

Gas	Percentage
Nitrogen	78.1
Oxygen	20.9
Argon	0.9
Carbon dioxide	less than 0.1
Neon	
Krypton	
Xenon	

Table 4.1 The percentages of gases in dry air

Accurate percentages of these gases in dry air are shown in Table 4.1.

The Earth is the only planet in our Solar System with oxygen in its atmosphere and abundant surface water in rivers, lakes and oceans. Other planets, such as Mars, do however have some water vapour and polar ice caps.

The noble gases

The noble gases are in Group 0 on the extreme right of the Periodic Table (Figure 4.24).

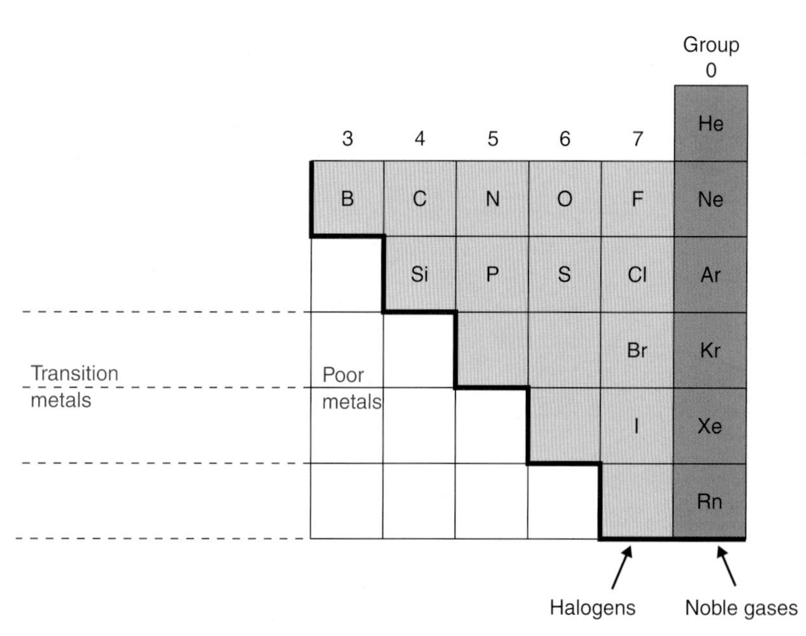

Figure 4.24 The position of the noble gases in the Periodic Table

The **noble gases** are in Group 0 of the Periodic Table. They are all chemically unreactive.

The noble gases are all colourless and odourless with very low melting points and boiling points. They all exist as separate single atoms. Other gaseous elements (hydrogen H_2, oxygen O_2, nitrogen N_2 and the halogens) all exist as diatomic molecules.

Until 1962, there were no known compounds of the noble gases. Chemists thought they were completely unreactive. Because of this, they were called the inert gases. Today, several of their compounds are known and have been produced. They are not inert, so we now call them the **noble gases**. The word 'noble' was chosen because unreactive metals like gold and silver are called noble metals.

Uses of the noble gases

Helium is used in balloons and airships because it has a low density and is non-flammable.

The noble gases produce a coloured glow when their atoms are bombarded by a stream of electrons. The stream of electrons can be produced either from a high-voltage discharge across the terminals of a discharge tube or from a laser. Neon and argon are used in discharge tubes to create fluorescent advertising signs. Neon tubes give a red colour and argon tubes give a blue colour.

Figure 4.25 A technician releases a weather balloon filled with helium.

Argon and krypton are used in electric filament lamps (light bulbs). If there is a vacuum inside the lamps, metal atoms evaporate from the superhot tungsten filament. To reduce this evaporation and prolong the life of the filament, the bulb is filled with an unreactive gas which cannot react with the hot tungsten filament.

19 Hydrogen was once used for inflating balloons.
a) What was the special advantage of using hydrogen?
b) Why was the use of hydrogen dangerous?

20 a) Give two reasons why argon and krypton are used in light bulbs (filament lamps).
b) Why is air unsuitable to use inside filament lamps?

Figure 4.26 The bright lights of Piccadilly

Global warming is the name given to the increase in average temperatures of the Earth.

Activity – What changes are occurring in the Earth's atmosphere?

Table 4.2 shows the concentration of carbon dioxide in the atmosphere, the average global temperature and the global population at intervals of 50 years since 1750.

	Year					
	1750	**1800**	**1850**	**1900**	**1950**	**2000**
Concentration of carbon dioxide in the atmosphere as % by volume	0.0278	0.0282	0.0288	0.0297	0.0310	0.0368
Average global temperature in °C	13.3	13.4	13.4	13.6	13.8	14.4
Global population in millions	350	500	1000	1500	3000	5500

Table 4.2

1 It is widely believed that the burning of fossil fuels has increased the level of carbon dioxide in the atmosphere.
a) Natural gas is probably the simplest fossil fuel containing mainly methane with traces of ethane. Copy and complete the following equation for the burning of methane.

$$CH_4 + \_\_\_\_ O_2 \rightarrow CO_2 + \_\_\_\_$$

b) State two major processes which involve burning fossil fuels.
c) Name one important process that removes carbon dioxide from the atmosphere.

❷ Use the data in Table 4.2 to plot a graph which will show conclusively that the level of carbon dioxide in the atmosphere is increasing.
   a) In your graph, what was i) the independent variable; ii) the dependent variable?
   b) From your graph, describe the way in which carbon dioxide is increasing.
   c) Which rows of data in Table 4.2 show that there might well be a link between carbon dioxide and global warming?
   d) Name one other gas that contributes to global warming.

❸ Many people believe that the increasing levels of carbon dioxide in the atmosphere are responsible for **global warming**.
   a) Why should we worry about global warming?
   b) How does carbon dioxide contribute to global warming? (You may wish to refer to Section 2.4 to help you with this question.)

❹ Some people believe that global warming is largely the result of an increasing global population using ever-increasing amounts of fossil fuels. What are your views about this suggestion?

❺ What should we in the UK be doing about global warming:
   a) at a personal / family level;
   b) at a national / government level?

Summary

✓ The Earth and its atmosphere provide the raw materials for everything we need.

✓ The Earth is nearly spherical with a layered structure comprising:
- a thin **crust**;
- a **mantle** extending almost halfway to the Earth's centre, which has all the properties of a solid except that it can flow very slowly;
- a central **core** of about half the Earth's diameter, made of iron and nickel, the outer part of which is liquid and the inner part of which is solid.

✓ The Earth's crust and the upper part of the mantle are cracked into a number of pieces called **tectonic plates**.

✓ The decay of natural radioactive materials inside the Earth releases heat which produces **convection currents** in the mantle. These convection currents cause the tectonic plates to move at relative speeds of a few centimetres per year.

✓ This movement of tectonic plates (called **continental drift**) is normally very slow, but on occasions it can be sudden and disastrous. At the boundaries of the plates, it can result in **earthquakes** and **volcanic eruptions**.

✓ For the last 200 million years, proportions of the different gases in the Earth's atmosphere have been more or less the same:
- about four fifths (80%) nitrogen;
- about one fifth (20%) oxygen;
- small proportions of other gases including carbon dioxide, water vapour and the noble gases.

✓ The **noble gases** occupy Group 0 of the Periodic Table. They are chemically unreactive. Helium is much less dense than air and is used in balloons. Neon and argon are used in fluorescent advertising signs and argon and krypton are used in electric filament lamps (light bulbs).

✓ During the first billion years of the Earth's existence, there was intense volcanic activity. This activity released gases that formed the early atmosphere – mainly carbon dioxide and water vapour with smaller proportions of methane and ammonia.

✓ As the Earth cooled down, the water vapour condensed to form rivers, lakes and oceans.

✓ When plants evolved, carbon dioxide and water were used up in photosynthesis and oxygen was produced. Although the reverse happened when plants respired, the proportion of oxygen in the atmosphere slowly increased.

✓ During the next 3 billion years, most of the carbon from the carbon dioxide in the air gradually became locked up in fossil fuels and in sedimentary rocks as carbonates.

✓ During the last 150 years or so, the level of carbon dioxide in the atmosphere has slowly increased. This is due to the burning of fossil fuels in industry (particularly power stations), in our homes and in our vehicles.

✓ The increase in the level of carbon dioxide in the atmosphere has resulted in **global warming**.

EXAMQUESTIONS

❶ a) Figure 4.27 shows the layered structure of the Earth. Copy and complete the figure by adding the three missing labels.

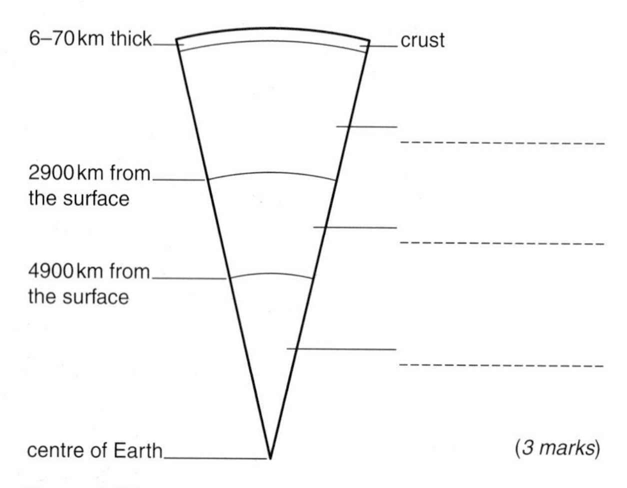

(3 marks)

Figure 4.27

b) The crust and upper mantle are cracked into a number of pieces called tectonic plates.
   i) Why do these tectonic plates move? *(2 marks)*
   ii) Explain how the movement of the tectonic plates can lead to earthquakes. *(4 marks)*

❷ a) Figure 4.28 shows a cross-section through two tectonic plates A and B. Copy the diagram and add the following labels: ocean; oceanic crust; continental crust; folds; mountains; sediments; mantle. *(7 marks)*

b) Put an arrow on your diagram to show the movement of plate B. *(1 mark)*

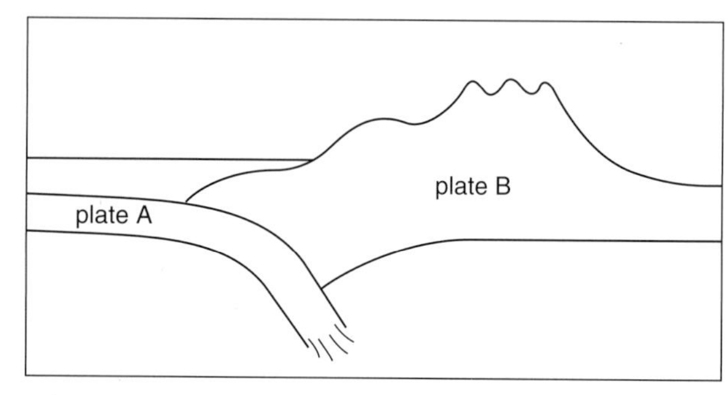

Figure 4.28

c) In one year, will plate A move a few millimetres, a few centimetres or a few metres? *(1 mark)*

d) Why does plate A sink below plate B? *(1 mark)*

e) What is the scientific name for this process? *(1 mark)*

f) Mark a point, X, on your diagram where solid rock may be forming a viscous liquid. *(1 mark)*

❸ Match the terms A, B, C and D with the spaces 1–4 in the sentences.

A continental drift
B land mass
C radioactive processes
D tectonic plates *(4 marks)*

In 1915, Alfred Wegener put forward the idea that millions of years ago there was a single large ..... (1) ..... This broke up and the smaller parts, which we now call ..... (2) ....., moved apart. This process is called ..... (3) ..... and the heat required for the movement comes from ..... (4) ..... in the Earth's core and mantle.

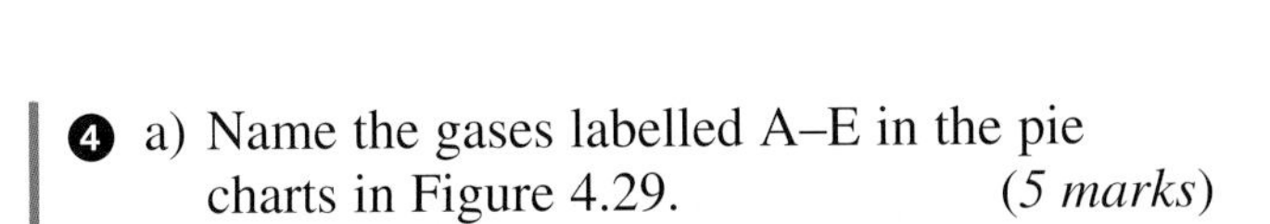

4 a) Name the gases labelled A–E in the pie charts in Figure 4.29. *(5 marks)*

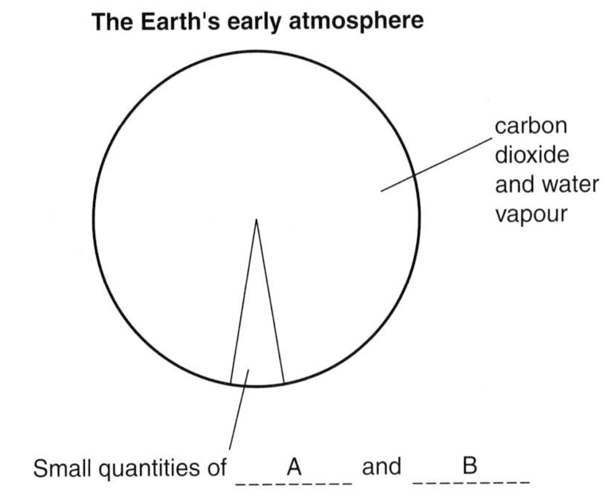

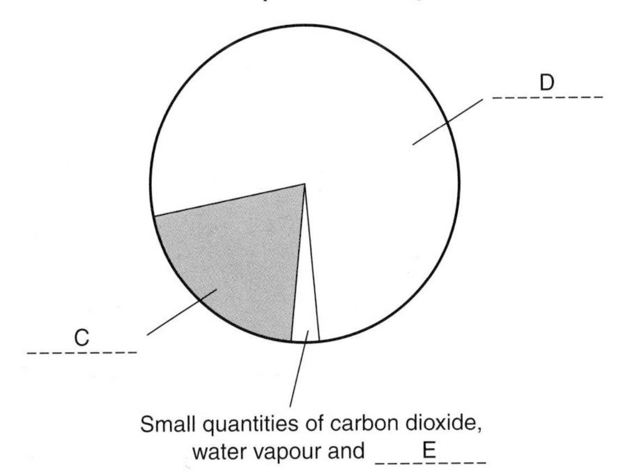

Figure 4.29

b) Over the last 150 years, the amount of carbon dioxide in the atmosphere has increased from 0.03% to 0.04%.

i) Why is the amount of carbon dioxide gradually increasing? *(2 marks)*

ii) Suggest two effects that increasing levels of carbon dioxide may have on the environment. *(2 marks)*

c) The percentage of carbon dioxide in country air is often lower than that in city air. Why is this? *(1 mark)*

5 The gases in our present atmosphere evolved over billions of years from the gases produced by erupting volcanoes.

a) The gas from erupting volcanoes contained a high proportion of water vapour, but our present atmosphere contains only a very small proportion of water vapour. Why is this? *(4 marks)*

b) Volcanic gases also contained a high proportion of carbon dioxide. Suggest three processes which led to a reduction of carbon dioxide in the atmosphere. *(3 marks)*

c) Oxygen did not appear in the Earth's atmosphere for about 1 billion (1 000 000 000) years. Why was this? *(2 marks)*

d) Scientists believe that nitrogen was formed in our atmosphere when ammonia (NH_3) from the volcanic gases reacted with oxygen. Copy and complete the following equation for this reaction.

$$\_\_ NH_3 + \_\_ O_2 \rightarrow \_\_ N_2 + \_\_ H_2O$$

(2 marks)

Chapter 5
How do sub-atomic particles explain the structure and reactions of substances?

At the end of this chapter you should:

- ✓ know the relative masses and relative charges of protons, neutrons and electrons;
- ✓ understand the terms atomic number and mass number and how they relate to isotopes and relative atomic masses;
- ✓ appreciate that elements are arranged in the modern Periodic Table in order of their atomic number;
- ✓ be able to calculate relative formula masses and the percentages of elements in compounds;
- ✓ be able to calculate empirical formulae;
- ✓ be able to calculate the amount of product obtained from a given amount of reactant;
- ✓ understand how chemical bonding involves either transferring or sharing electrons in the outer shells (highest occupied energy levels) in atoms;
- ✓ understand how electron transfer results in the formation of positive and negative ions held together by ionic bonds;
- ✓ understand how electron sharing results in the formation of uncharged molecules in which atoms are held together by covalent bonds;
- ✓ be able to represent the electronic structures of the atoms and ions of the first 20 elements in the Periodic Table;
- ✓ be able to represent the covalent bonds in simple molecules.

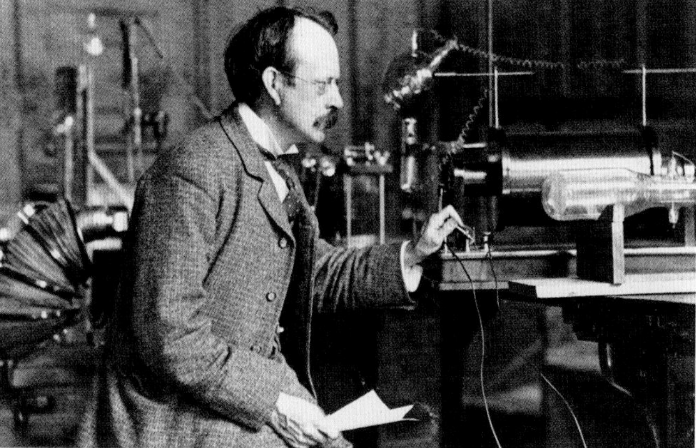

Figure 5.1 The work of scientists with sub-atomic particles has changed dramatically since J. J. Thomson first discovered electrons in 1897 (top right) and Rutherford and Geiger (bottom right) found evidence for protons. Today, scientists working with sub-atomic particles use much more sophisticated and complex equipment like the detector in the particle accelerator (above) at the European Centre for Nuclear Research (CERN) in Geneva. Particle accelerators can accelerate sub-atomic particles almost to the speed of light.

What's inside atoms?

Just over a century ago, scientists thought that atoms were solid, indestructible particles like tiny snooker balls. Then, between 1897 and 1932, experiments showed that atoms contained three smaller particles – electrons, protons and neutrons. You should already know something about these three sub-atomic particles. But how were they discovered and how did this change our ideas about atoms?

Activity – Thomson's big surprise

In 1897, J. J. Thomson was investigating the way that gases conduct electricity. When he connected 15 000 volts across the terminals of a tube containing air, he observed rays travelling in straight lines from the negative terminal (Figure 5.2). Further study of a narrow beam of the rays surprised Thomson. He found that the rays consisted of tiny negative particles about 2000 times lighter than hydrogen atoms. Thomson called these tiny negative particles electrons.

1. Why was Thomson surprised by his discovery of electrons? (Hint: In 1897, scientists thought that atoms were indivisible.)
2. How did the narrow beam move after the parallel plates were charged?
3. Why did Thomson conclude that particles in the beam were negatively charged? (Hint: Why was the beam deflected when the plates were charged?)
4. Before his experiment, Thomson removed most of the air from the tube using a vacuum pump. Why do you think he did this?
5. Thomson obtained electrons even with different gases in the tube and when the terminals were made of different substances. What could he conclude from this?
6. Our TV screens today work like the one used by Thomson. Why do you think the material on the screen fluoresces when the beam of electrons hits it?

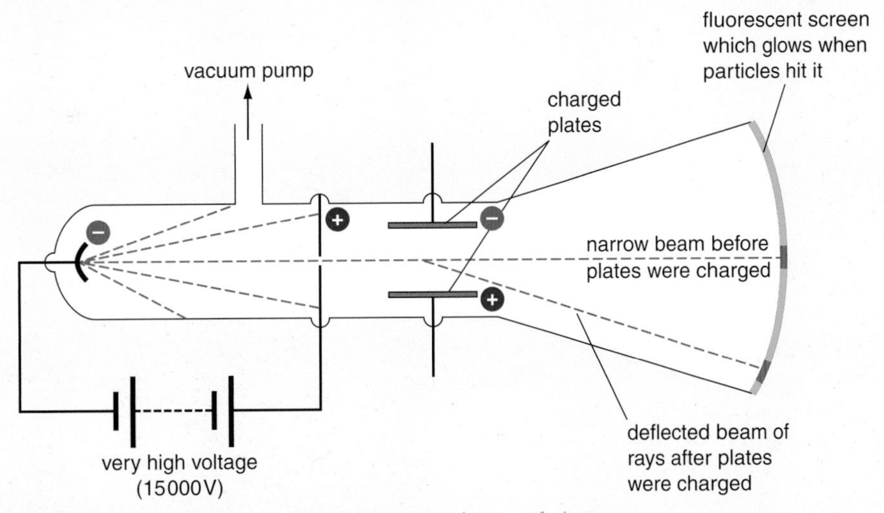

Figure 5.2 The effect of charged plates on a beam of electrons

Electrons, protons and neutrons

Thomson's experiments led scientists to conclude that all atoms contained electrons. So, as atoms are neutral, they must contain sufficient positive charge to cancel the negative charge on their electrons.

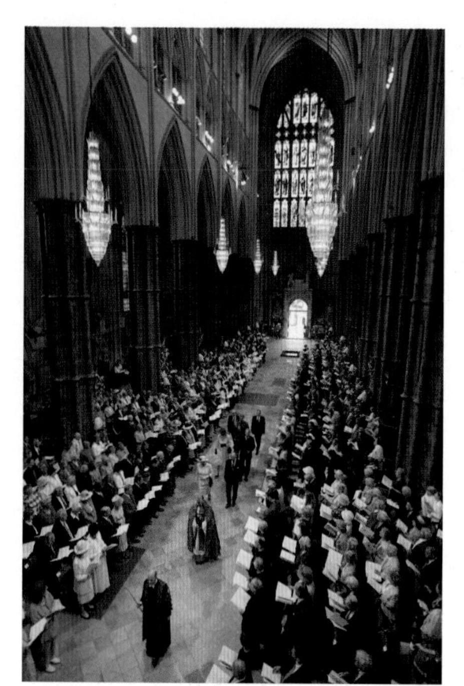

Figure 5.3 If the nucleus of a hydrogen atom is enlarged to the size of a marble and put in the centre of Westminster Abbey, the atom's one electron would be whizzing around well outside the walls of the Abbey.

In the early twentieth century, Ernest Rutherford and his colleagues found a way of probing inside atoms, using a form of radioactivity that had just been discovered – alpha particles from radioactive substances. These experiments led to the discovery of the atomic nucleus.

Through Rutherford's work we know that:

- atoms have a small positive nucleus surrounded by a much larger region of empty space in which there are tiny negative electrons;
- the positive charge of the nucleus is due to positive particles which Rutherford called protons;
- protons are about 2000 times heavier than electrons;
- atoms have equal numbers of protons and electrons, so the positive charges on the protons cancel the negative charges on the electrons;
- the smallest atoms are those of hydrogen with one proton and one electron. The next smallest atoms are those of helium with two protons and two electrons, then lithium atoms with three protons and three electrons, and so on.

Although Rutherford was successful in explaining many aspects of atomic structure, one big problem remained. If hydrogen atoms contain one proton and helium atoms contain two protons, then the relative masses of hydrogen and helium atoms should be one and two, respectively. But the mass of helium atoms relative to hydrogen atoms is four and not two.

In 1932, James Chadwick, one of Rutherford's colleagues, discovered where the extra mass in helium came from. Chadwick showed that the

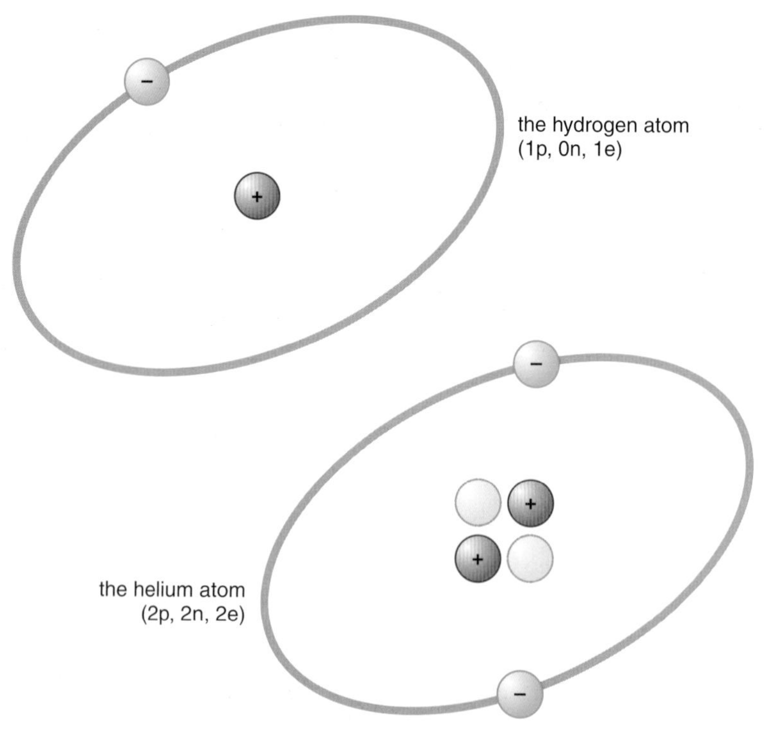

Figure 5.4 Protons, neutrons and electrons in a hydrogen atom and a helium atom

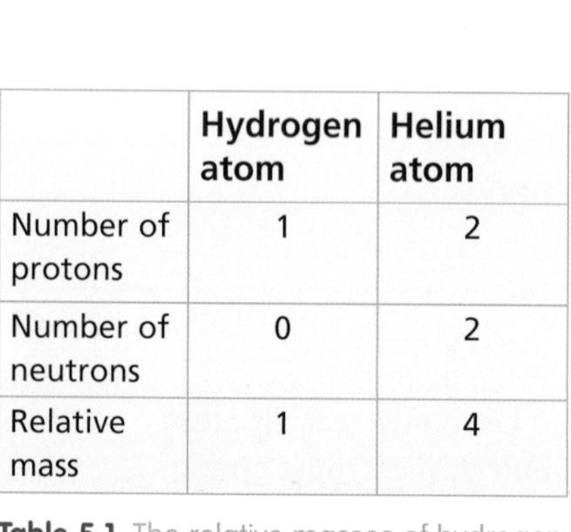

	Hydrogen atom	Helium atom
Number of protons	1	2
Number of neutrons	0	2
Relative mass	1	4

Table 5.1 The relative masses of hydrogen and helium atoms

nuclei of atoms contain uncharged particles as well as positively charged protons. Chadwick called these uncharged particles neutrons. Experiments showed that neutrons have the same mass as protons. This helped Chadwick to explain the problem concerning the relative masses of hydrogen and helium atoms (Table 5.1).

Hydrogen atoms have one proton and no neutrons, so a hydrogen atom has a relative mass of one unit. Helium atoms have two protons and two neutrons, so a helium atom has a relative mass of four units. This makes a helium atom four times as heavy as a hydrogen atom (Figure 5.4).

We now know that all atoms are made up from protons, neutrons and electrons. The relative masses, relative charges and positions within atoms of these sub-atomic particles are summarised in Table 5.2.

Particle	Relative mass (atomic mass units)	Relative charge	Position within atoms
Proton	1	+1	Nucleus
Neutron	1	0	Nucleus
Electron	$\frac{1}{2000}$	−1	In space outside the nucleus

Table 5.2 The relative masses, charges and positions within atoms of protons, neutrons and electrons

1. Lithium atoms have three protons and a relative mass of 7 compared to hydrogen atoms.
   a) How many electrons are there in one lithium atom?
   b) How many neutrons are there in one lithium atom?
   c) How many protons, neutrons and electrons are there in one lithium ion, Li^+?
2. Oxygen atoms have eight protons. How many protons and how many electrons are there in:
   a) one O atom;
   b) one O^{2-} ion;
   c) one O_2 molecule;
   d) one H_2O molecule?

Atoms and ions

From the structure of atoms, we can understand how ions form from atoms by the loss or gain of electrons. A helium atom (He) has two protons each with one positive charge, two neutrons, and two electrons each with one negative charge. If an electron is removed from a helium atom, it leaves an ion (charged particle) with two positive charges and only one negative charge. This gives an overall charge on the ion of one positive charge. So, we can write the symbol He^+ for this particle.

If two electrons are removed from a helium atom, the remaining particle has two positive charges and no negative charges. This can be represented by the symbol He^{2+} (Figure 5.5).

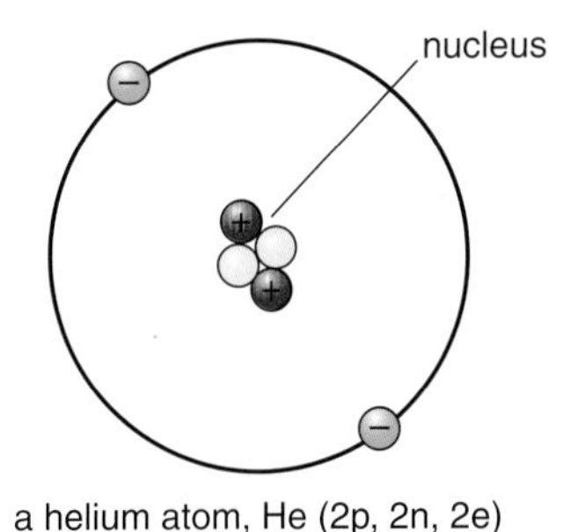

a helium atom, He (2p, 2n, 2e)

a helium ion, He^+ (2p, 2n, 1e)

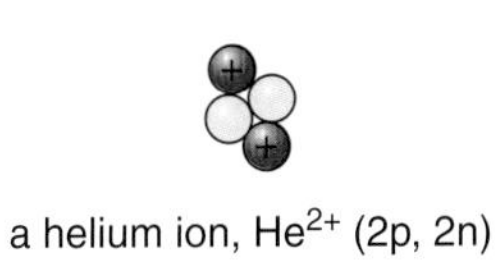

a helium ion, He^{2+} (2p, 2n)

Figure 5.5 Protons, neutrons and electrons in a helium atom and helium ions

The **atomic number** is the number of protons in an atom.

The **mass number** is the total number of protons plus neutrons in an atom.

Atomic number and mass number

Atoms of different elements have different numbers of protons. Hydrogen atoms are the only atoms with one proton. Helium atoms are the only atoms with two protons. Lithium atoms are the only atoms with three protons, and so on. This means that the number of protons in an atom determines which element it is. Scientists call this the **atomic number** or proton number. So, hydrogen has an atomic number of 1, helium has an atomic number of 2, lithium has an atomic number of 3, and so on.

Notice that hydrogen, the first element in the Periodic Table, has atoms with one proton. Helium, the second element in the Periodic Table, has atoms with two protons. Lithium, the third element in the Periodic Table, has atoms with three protons, and so on. So, the position of an element in the Periodic Table tells us how many protons it has.

Aluminium is the thirteenth element in the Periodic Table. This means it has13 protons and its atomic number (proton number) is therefore 13. The number of protons in an atom can tell you which element it is, but it cannot tell you its mass. The mass of an atom depends on the number of protons plus the number of neutrons. This number is called the **mass number** of the atom.

mass number = 23

Na

atomic number = 11

Figure 5.6 The mass number and atomic number shown with the symbol for sodium

Hydrogen atoms (with one proton and no neutrons) have a mass number of 1. Helium atoms (two protons and two neutrons) have a mass number of 4, and sodium atoms (11 protons and 12 neutrons) have a mass number of 23.

We can write the symbol $^{23}_{11}\text{Na}$ (Figure 5.6) to show the mass number and the atomic number of a sodium atom. The mass number is written at the top left and the atomic number at the bottom left of the symbol.

3 Look at Figure 5.6.

a) How many protons, neutrons and electrons are there in:
   i) one Na atom;
   ii) one Na^+ ion?

b) What do 27, 13, 3+ and Al mean with reference to the symbol $^{27}_{13}\text{Al}^{3+}$?

5.2 Comparing the masses of atoms

Relative atomic masses tell us the relative masses of the atoms of different elements.

A single atom is so small that it cannot be weighed on a balance. However, the mass of one atom can be compared with that of another atom using an instrument called a mass spectrometer (Figure 5.7).

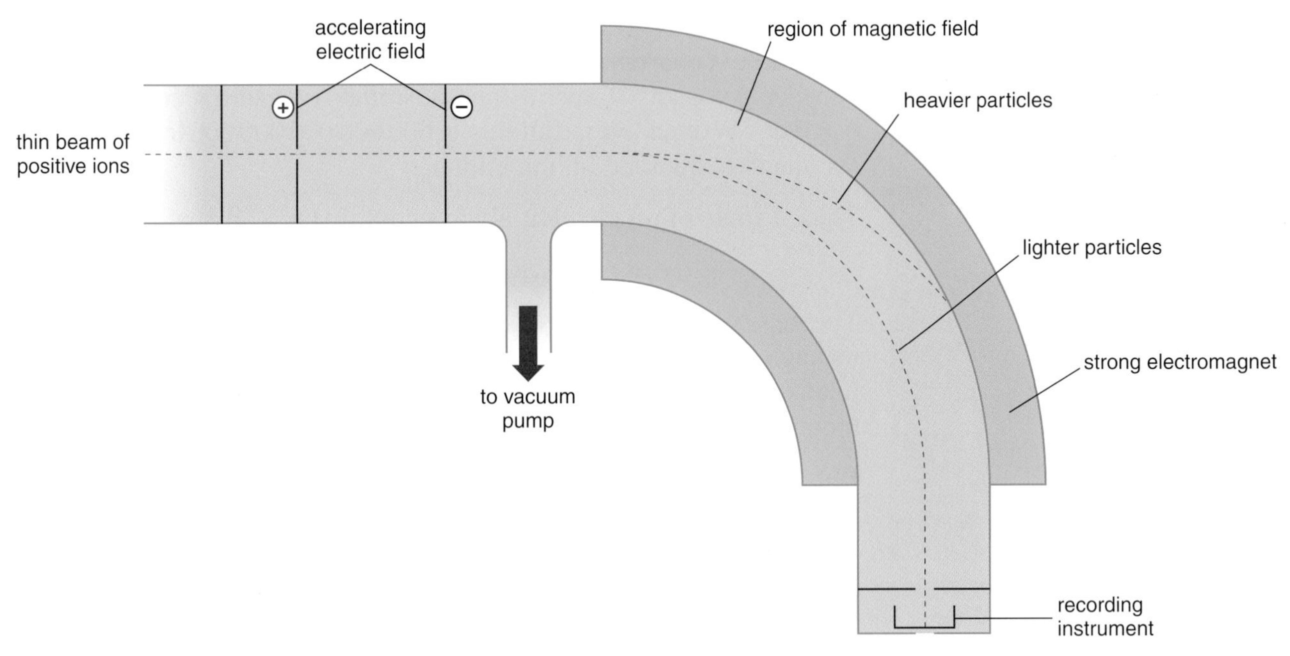

Figure 5.7 A mass spectrometer produces positive ions of the elements being analysed. A thin beam of these ions passes through an electric field, which accelerates the ions. The ions then pass through a magnetic field, which deflects them. How much they are deflected depends on the mass of the particles – lighter particles are deflected more than heavier particles. From the amount of deflection, we can compare the masses of different atoms and make a list of their relative masses.

One **mole** of an element has a mass equal to its relative atomic mass in grams.

❹ Why is the relative atomic mass of helium four and that of hydrogen one?

A mass spectrometer must be calibrated before it is first used. Atoms of known mass are analysed in the instrument to find how much they are deflected. Scientists can then find the mass of other atoms from their extent of deflection. This is an example of **calibration**, which involves marking a scale on a measuring instrument.

Relative atomic masses

The relative masses of different atoms are called **relative atomic masses**. As you might expect, the relative atomic mass of hydrogen is 1.0, the relative atomic mass of helium is 4.0 and the relative atomic mass of carbon is 12.

The symbol for relative atomic mass is A_r. So, we can write A_r (C) = 12.0, A_r (H) = 1.0 and so on, or simply C = 12.0 and H = 1.0 for short.

The relative atomic mass of an element in grams is sometimes called one **mole**.

So, 12 g of carbon = 1 mole
1 g of hydrogen = 1 mole
and 24 g of carbon = 2 moles

Notice that: $\text{number of moles} = \dfrac{\text{mass}}{\text{relative atomic mass}}$

Relative atomic masses show that one atom of carbon is 12 times as heavy as one atom of hydrogen. This means that 12 g of carbon will contain the same number of atoms as 1 g of hydrogen. An atom of oxygen is 16 times as heavy as an atom of hydrogen, so 16 g of oxygen will also contain the same number of atoms as 1 g of hydrogen.

The relative atomic mass in grams (i.e. one mole) of every element (1 g of hydrogen, 12 g carbon, 16 g oxygen, etc.) will contain the same

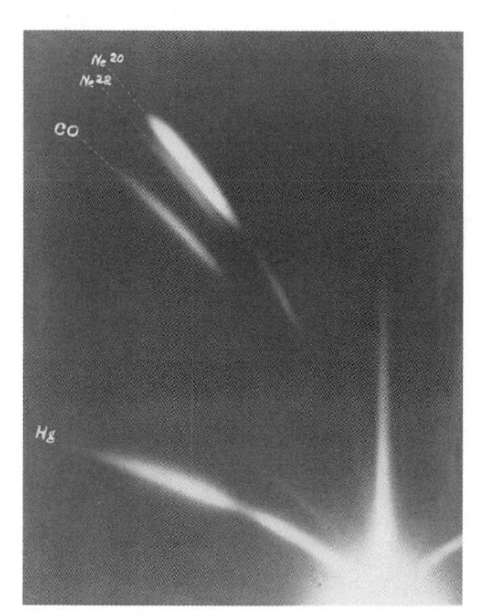

Figure 5.8 This photo shows evidence for the two isotopes in neon, neon-20 and neon-22.

number of atoms. This number is called Avogadro's constant. The term 'Avogadro's constant' was chosen in honour of the Italian scientist, Amedeo Avogadro. Experiments show that Avogadro's constant is 6×10^{23}. Written out in full this is 600 000 000 000 000 000 000 000 or six hundred thousand billion billion.

Thus, **1 mole of an element always contains 6×10^{23} atoms.**

Isotopes and atomic mass

The mass of an atom depends on the number of protons and neutrons in its nucleus. As both these particles have a relative mass of 1.00, the relative atomic masses of many elements are very close to whole numbers.

But, some elements have relative atomic masses that are nowhere near whole numbers. For example, the relative atomic mass of chlorine is 35.5 and that of copper is 63.5. At one time, scientists could not understand why the relative atomic masses of these elements were not close to whole numbers. The answer soon became clear when the first mass spectrometer was built in 1919. When atoms of elements such as chlorine or copper were ionised and passed through a mass spectrometer, the beam of ions separated into two or more paths (Figure 5.8). This suggested that one element could have atoms with different masses. These atoms of the same element with different masses are called **isotopes**.

Isotopes are atoms of an element with the same atomic number, but different mass numbers.

All the isotopes of one element have the same number of protons. So, they have the same atomic number and the same number of electrons. This gives them the same chemical properties because chemical properties depend upon the number of electrons in an atom.

Isotopes have the same:
• number of protons; • number of electrons; • atomic number; • chemical properties.
Isotopes have different: • numbers of neutrons; • mass numbers; • physical properties.

Table 5.3

Isotopes do, however, have different numbers of neutrons, and this gives them different masses. This means that isotopes have the same atomic number but different mass numbers (Table 5.3).

For example, neon has two isotopes (Figure 5.9). Each isotope has 10 protons and 10 electrons and therefore an atomic number of 10. But one of these isotopes has 10 neutrons and the other has 12 neutrons. Their mass numbers are therefore 20 and 22. They are sometimes called neon-20 and neon-22.

These two isotopes of neon have the same chemical properties because they have the same number of electrons. But they have different physical properties because they have different masses. So, ${}^{20}_{10}Ne$ and ${}^{22}_{10}Ne$ have different densities, different melting points and different boiling points.

At one time, the standard for relative atomic masses was hydrogen. Because of the existence of isotopes, it became necessary to choose one particular isotope as the standard. Today, the isotope carbon-12 (${}^{12}_{6}C$) is chosen as the standard and given a relative mass of exactly 12.

5 Look closely at Figure 5.8.
a) The trace from neon-20 is much stronger than that from neon-22. What does this tell you about the two isotopes?
b) A beam of CO has been deflected less than the neon isotopes. What substance is CO? Why is CO deflected less than the neon isotopes?

Element	Symbol	Relative atomic mass
Carbon	C	12.0
Hydrogen	H	1.0
Helium	He	40
Oxygen	O	16.0
Magnesium	Mg	24.0
Sulfur	S	32.1
Iron	Fe	55.8
Copper	Cu	63.59
Gold	Au	197.0

Table 5.4 The relative atomic masses of a few elements

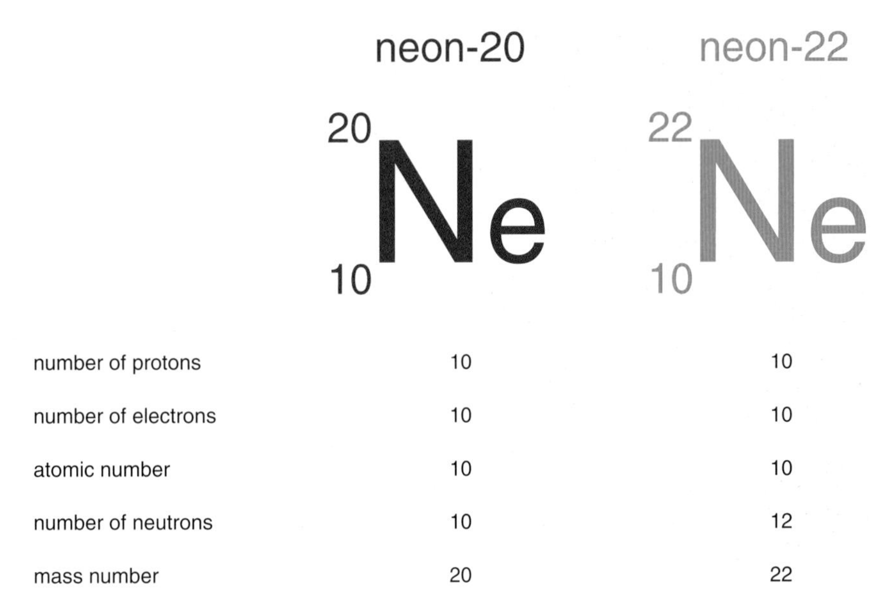

Figure 5.9 The two isotopes of neon

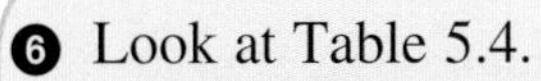

6 Look at Table 5.4.
a) How many times heavier are carbon atoms than hydrogen atoms?
b) How many times heavier are magnesium atoms than carbon atoms?
c) Which element has atoms almost four times as heavy as oxygen atoms?
d) Write the following elements in order of increasing deflection in a mass spectrometer – helium, iron, gold.

The relative masses of other atoms are then obtained by comparison with carbon-12. On this scale, the relative atomic mass of hydrogen is still 1.0 and that of helium is 4.0. A few relative atomic masses are listed in Table 5.4.

Measuring relative atomic masses

Most elements contain a mixture of isotopes. This explains why their relative atomic masses are not whole numbers. Using modern mass spectrometers, it is possible to obtain print-outs showing the relative amounts of the different isotopes in a sample of an element and the relative atomic masses of each isotope. One of these print-outs is shown in Figure 5.10. The print-out shows that chlorine consists of two isotopes, $^{35}_{17}Cl$ and $^{37}_{17}Cl$, with relative atomic masses of 35 and 37.

The relative atomic mass of an element is the average mass of one atom of the element. This average must take into account the relative amounts of the different isotopes. For example, if chlorine contained 100% $^{35}_{17}Cl$, its relative atomic mass would be 35.0. If it contained 100% $^{37}_{17}Cl$, then its relative atomic mass would be 37.0. A 50 : 50 mixture of the isotopes $^{35}_{17}Cl$ and $^{37}_{17}Cl$ would have a relative atomic mass of 36.0

Figure 5.10 shows that naturally occurring chlorine contains three times as much $^{35}_{17}Cl$ as $^{37}_{17}Cl$, i.e. ¾ or 75% of chlorine is $^{35}_{17}Cl$ and ¼ or 25% is $^{37}_{17}Cl$.

So, the relative atomic mass can be calculated as:

75% chlorine-35 + 25% chlorine-37

$$= \frac{75}{100} \times 35.0 + \frac{25}{100} \times 37.0 = 26.25 + 9.25 = 35.5$$

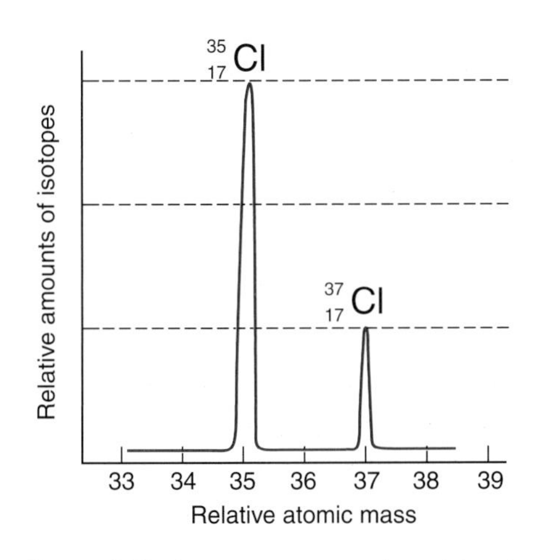

Figure 5.10 A mass spectrometer print-out for chlorine

5.3 Using relative atomic masses to find out about compounds

Relative formula masses can be used to compare the masses of different compounds.

Comparing the masses of molecules

In the last section, we learned that relative atomic masses can be used to compare the masses of different atoms. Relative atomic masses can also be used to compare the masses of molecules in different compounds. The relative masses of compounds are called **relative formula masses** (symbol M_r). The relative formula mass of a compound is the sum of the relative atomic masses of all the atoms in its formula.

For example, the relative formula mass of water, $M_r(H_2O)$

= 2 × relative atomic mass of hydrogen + relative atomic mass of oxygen

$= (2 \times 1.0) + 16.0$

$= 18.0$

The relative formula mass of red iron oxide, $M_r(Fe_2O_3)$

= 2 × relative atomic mass of iron + 3 × relative atomic mass of oxygen

$= 2 \times 55.8 + 3 \times 16.0 = 159.6$

The relative formula mass of a compound in grams is known as one mole of that substance.

So, 1 mole of water is 18.0 g and

0.1 mole of water = $0.1 \times 18.0 = 1.8$ g

One mole of red iron oxide is 159.6 g, and so

5 moles of red iron oxide = $5 \times 159.6 = 798.0$ g

Figure 5.11 The chemical formula for table sugar is $C_{12}H_{22}O_{11}$. The relative formula mass of sugar, $M_r(C_{12}H_{22}O_{11})$:
$= 12 \times A_r(C) + 22 \times A_r(H) + 11 \times A_r(O)$
$= 12 \times 12.0 + 22 \times 1.0 + 11 \times 16$
$= 144 + 22 + 176$
$= 342$

Finding the percentage composition of compounds

Relative atomic masses can be used to calculate the percentage of different elements in a compound. First, you work out the relative amounts of each element in the relative formula mass of the compound. For example, carbon dioxide (CO_2) contains 12.0 g of carbon and 2 × 16.0 g (32.0 g) of oxygen in 44.0 g of carbon dioxide.

So, it contains 12/44 parts carbon and 32/44 parts oxygen. Changing to percentages, this is:

$$\frac{12}{44} \times 100 = 27\% \text{ carbon and } \frac{32}{44} \times 100 = 73\% \text{ oxygen}$$

7 What is the relative formula mass of:
a) carbon dioxide, CO_2;
b) sulfuric acid, H_2SO_4?

8 Calculate the mass of:
a) 3 moles of carbon dioxide;
b) 0.6 moles of carbon dioxide;
c) 5 moles of sulfuric acid;
d) 0.2 moles of sulfuric acid.

Figure 5.12 The yellow substance in 'no parking' double lines is lead chromate, $PbCrO_4$.

Figure 5.13 This sulfur is being stored after mining. 32.1 g of sulfur can be reacted with oxygen and water to manufacture 98.1 g of sulfuric acid.

Finding the masses of reactants and products in a reaction

In industry, it is vitally important to know the amounts of reactants that are needed for a chemical process and the amount of product that can be obtained. Industrial chemists need to know how much product they can obtain from a given amount of starting material.

To calculate masses of reactants and products, chemists use relative formula masses and equations. As an example, let's calculate how much iron we could obtain by reducing 1 kg (1000 g) of pure iron ore (red iron oxide) with carbon monoxide in a blast furnace.

The equation for the reaction is:

$$Fe_2O_3 + 3CO \rightarrow 2Fe + 3CO_2$$

From this equation:

$$1 \text{ mole } Fe_2O_3 \rightarrow 2 \text{ moles Fe}$$

Using relative formula masses for Fe_2O_3 and 2Fe:

$$(2 \times 55.8 + 3 \times 16) \text{ g } Fe_2O_3 \rightarrow 2 \times 55.8 \text{ g Fe}$$

$$159.6 \text{ g } Fe_2O_3 \rightarrow 111.6 \text{ g Fe}$$

$$1.0 \text{ g } Fe_2O_3 \rightarrow \frac{111.6}{159.6} \text{ g Fe} = 0.7 \text{ g Fe}$$

$$\Rightarrow 1.0 \text{ kg } Fe_2O_3 \rightarrow 0.7 \text{ kg Fe}$$

So, 1 kg of pure red iron oxide produces 0.7 kg of iron.

The amount of product that we actually make in a chemical process doesn't always match the theoretical (calculated) amount. You will learn more about this in Section 7.1.

9 a) Look at Figure 5.12. What elements are present in lead chromate besides lead?
b) What is the relative formula mass of lead chromate?
c) What is the percentage of lead in lead chromate? (Pb = 207, Cr = 52, O = 16)

10 Copy and complete the following calculation to find how much lime (calcium oxide) can be obtained by heating 1 kg of pure limestone (calcium carbonate). (Ca = 40, C = 12, O = 16.)

The equation for the reaction is:

calcium carbonate → ____ + ____

$CaCO_3 \rightarrow$ ____ $+ CO_2$

$\therefore$ 1 mole $CaCO_3 \rightarrow$ ____ mole CaO

$\therefore$ ____ g $CaCO_3 \rightarrow$ ____ g CaO

$\therefore$ 1000 g $CaCO_3 \rightarrow$ ____ g CaO

Activity – Extracting tin from tinstone

The production manager at a tin smelter has to meet certain production targets. The smelter must produce 595 tonnes of tin every month. In order to achieve this, the manager must calculate how much purified tinstone (tin oxide) he must buy to produce 595 tonnes of tin. To do this he uses the equation:

tin oxide + carbon (coke) → tin + carbon monoxide

SnO_2 + 2C → Sn + 2CO

The equation shows that 1 mole of SnO_2 produces 1 mole of Sn. (The relative atomic masses of tin, carbon and oxygen are Sn = 119, C = 12, O = 16.)

1. Now complete the following statement. __ g SnO_2 produces __ g Sn.
2. What mass of tinstone is needed to meet the production target of 595 tonnes of tin per month?
3. What mass of carbon (coke) is needed each month?
4. What concerns should the production manager have about the emissions from the smelter?
5. The smelter normally operates all through the day and night. It can produce a maximum of 1 tonne of tin per hour. Is the production target possible? Explain your answer.
6. In the first six months of 2006, the amount of tin produced each month, to the nearest five tonnes, was: 650, 630, 575, 585, 560 and 600.
   a) What is the range in production over the six months?
   b) What is the mean (average) production per month?
   c) Has the monthly target been achieved on average over the first six months of 2006?
   d) What should the production manager do to get a more reliable value for the average production per month?
   e) What should the production manager do to get a more accurate value for the average production per month?
7. In 2006, tin from the smelter was sold at £3500 per tonne. What is the value of the tin produced by the smelter in the first six months of 2006?
8. Why did the smelters have tall chimneys?
9. Why is tin no longer mined and produced in Cornwall?

Figure 5.14 A disused tin smelting building in Cornwall

An **empirical formula** shows the simplest whole number ratio for the atoms of different elements in a compound.

Finding formulae using relative atomic masses

All formulae of compounds are obtained by doing experiments to find the relative number of moles of the different elements that react to form a compound. These formulae are called **empirical formulae**. When water is decomposed into hydrogen and oxygen, results show that:

18.0 g of water give 2.0 g of hydrogen + 16.0 g of oxygen

= 2 moles of hydrogen + 1 mole of oxygen

= $2 \times 6 \times 10^{23}$ atoms of hydrogen + 6×10^{23} atoms of oxygen

This shows that 12×10^{23} hydrogen atoms combine with 6×10^{23} oxygen atoms, so two hydrogen atoms combine with one oxygen atom. Therefore, the formula of water is H_2O. These results are set out in Table 5.5.

	H	O
Masses reacting	2.0 g	16.0 g
Mass of 1 mole	1 g	16 g
∴ moles reacting	2	1
Ratio of atoms	2	1
⇒ formula = H_2O		

Table 5.5 Finding the formula of water

Activity – Finding the formula of red copper oxide

After finding that the formula of black copper oxide was CuO, George and Meera decided to investigate the formula of red copper oxide. They took a weighed amount of red copper oxide and reduced it to copper (Figure 5.15). They carried out the experiment five times, starting with different amounts of red copper oxide. Their results are shown in Table 5.6.

Experiment number	Mass of red copper oxide in g	Mass of copper in the oxide in g
1	1.43	1.27
2	2.10	1.87
3	2.72	2.54
4	3.55	3.15
5	4.29	3.81

Table 5.6

1. Look at Figure 5.15. What safety precautions should George and Meera take during the experiment?
2. Write a word equation for the reduction of copper oxide to copper using methane (CH_4) in natural gas. (Hint: The only solid product is copper.)
3. What steps could George and Meera take to ensure that all the copper oxide is reduced to copper?
4. Make a table similar to Table 5.6 but with three extra columns.
   a) In the first of these extra columns, write the mass of oxygen in each of the samples of red copper oxide.
   b) In the second extra column, calculate the number of moles of copper in the oxide (Cu = 63.5).
   c) In the third extra column, calculate the number of moles of oxygen in the oxide (O = 16.0).
5. Use the results in the last two columns of your table to plot a graph of moles of copper (*y*-axis) against moles of oxygen (*x*-axis). Draw the line of best fit through the points on your graph.
6. Which of the points is anomalous and should be disregarded in drawing the line of best fit?
7. Look at your graph.
   a) Find the average value for

   $$\frac{\text{moles of copper}}{\text{moles of oxygen}}$$

   b) How many moles of copper combine with one mole of oxygen in red copper oxide?
   c) What is the formula for red copper oxide?
8. How did George and Meera improve the reliability of their results?
9. Is their result for the formula of red copper oxide valid? State 'Yes' or 'No' and explain your opinion.
10. Use the results in Table 5.6 to explain why a balance reading to only one decimal place would have been useless for this investigation.
11. a) What was i) the independent variable and ii) the dependent variable in this investigation?
    b) Are these categoric, continuous or ordered variables?

Results are **reliable** if they can be reproduced.

Results are **valid** if they are reliable and the measurements taken are affected by a single independent variable.

How **precise** a measurement is depends on the smallest scale division on the measuring instrument. For greater precision you need to use a scale with smaller divisions.

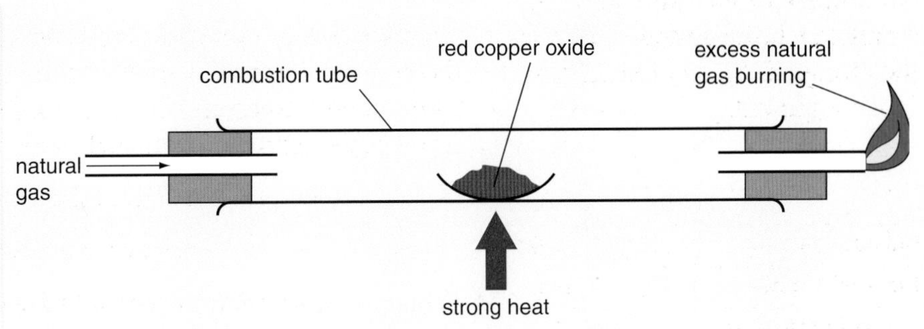

Figure 5.15 Reducing red copper oxide by heating in natural gas

5.4 How are the electrons arranged in atoms?

Evidence is data that has been checked and found to be valid.

In science, ideas change all the time. New **evidence** is used to improve theories so that they give a better explanation of what we observe.

In 1911, Rutherford put forward some key ideas about atomic structure (see Section 5.1). He said that atoms had a small positive nucleus containing protons surrounded by lots of empty space. Rutherford compared atoms to the Solar System and said that electrons orbited the nucleus like planets orbiting the Sun.

In 1913, the Danish scientist Niels Bohr suggested a more detailed model for the arrangement of electrons in atoms. Bohr said that electrons were grouped together in layers or shells at particular energy levels. Bohr's ideas about electrons in layers or shells helped us to understand what happens when elements react.

Electron shells are the regions occupied by electrons around the nucleus of an atom.

Electron structures show the number of electrons in each shell (energy level). The structures can be drawn as shell diagrams, or written as numbers of electrons in each shell, for example 2, 8.

Filling the shells

Different atoms have different **electron structures**. Atoms with larger atomic numbers have more electrons as the number of electrons is the same as the number of protons. The electrons in an atom always occupy the lowest energy levels first, starting with the **electron shell** nearest the nucleus.

The electrons in the first shell are strongly attracted to the nucleus. The first shell can hold only two electrons. When it contains two electrons, it is full and the electrons in it are stable (Figure 5.16).

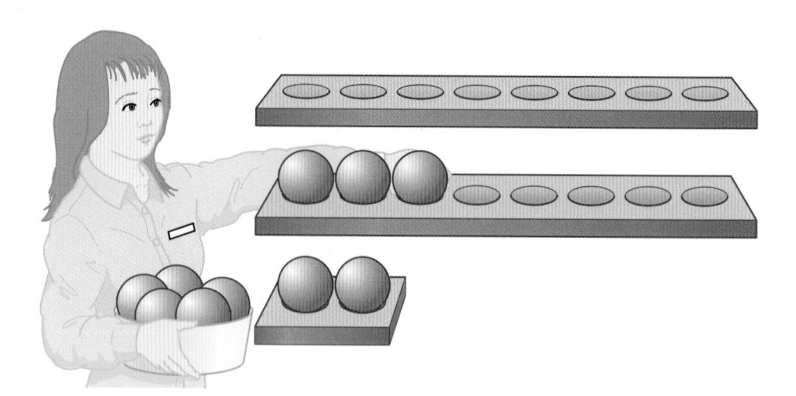

Figure 5.16 Filling shells with electrons is like filling shelves in a shop. The lowest shells (shelves) are filled first. Each shell (shelf) only holds a limited number of electrons (items).

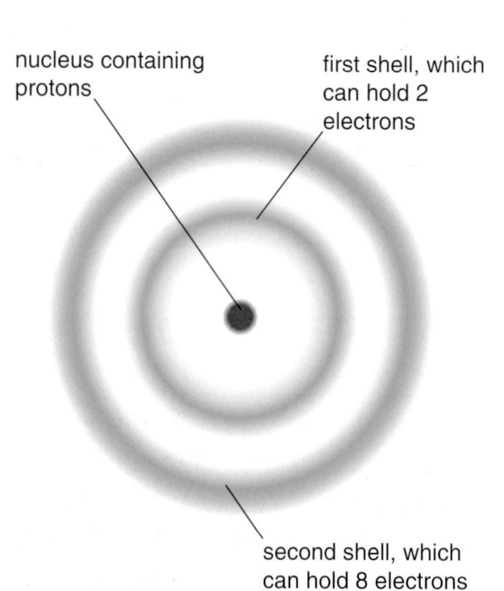

Figure 5.17 A model for the arrangement of electrons in the first and second shells

When the first electron shell is full, the second shell starts to fill. This shell is further from the nucleus and the electrons in it have more energy. The second shell can hold a maximum of eight electrons. When the second shell contains eight electrons, it is full (Figure 5.17). This electron structure is very stable.

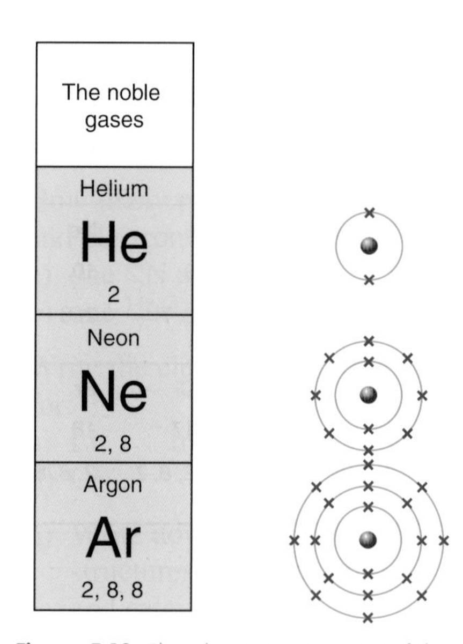

Figure 5.18 The electron structures of the noble gases. The diagrams show another way of representing these structures.

Once the second shell is full, the third shell starts filling. This shell is further from the nucleus than the second shell and at a higher energy level. The third shell is also very stable when it contains eight electrons.

Electron structures and the noble gases

Bohr's ideas about electrons in shells or energy levels have helped us to understand why the noble gases are so unreactive. The first shell is full and the electrons in it are stable when the shell contains two electrons. This corresponds exactly with the structure of helium. Helium has two electrons. They both go into the first shell, which is then full and stable.

The second shell is full and the electrons in it are stable when the shell contains eight electrons. So, the next element to be unreactive like helium will have two electrons filling the first shell and then eight electrons filling the second shell. This corresponds to neon, the tenth element in the Periodic Table. Because neon has two electrons in the first shell and eight electrons in the second shell, we say its electron structure is 2, 8 (Figure 5.18).

The third shell is also stable when it contains eight electrons. So, the next very unreactive element will have two electrons filling the first shell, eight filling the second shell and eight in the third shell. This corresponds exactly with the electron structure of argon, which we can write as 2, 8, 8 (Figure 5.18).

Figure 5.19 Electric light bulbs are filled with argon or krypton. These noble gases are so unreactive and their electrons are so stable that the metal filament can be white hot at 1000 °C without reacting with them.

Electron structures and chemical reactivity

Figure 5.20 shows the first 20 elements in their order in the Periodic Table. The atomic number and electron structure of each element are written below its symbol. When the first shell is full at helium, further electrons go into the second shell. So, the electron structure of lithium is 2, 1, beryllium is 2, 2, boron is 2, 3 and so on. When the second shell is full at neon, electrons start to fill the third shell.

Summary

- ✓ Atoms have a small nucleus containing positive protons and uncharged neutrons around which there are tiny negative electrons.
- ✓ In all atoms, the positive charges on the protons are exactly cancelled by the negative charges on an equal number of electrons.
- ✓ **Ions** are formed from atoms by the loss or gain of one or more electrons.
- ✓ The **atomic number** is the number of protons in an atom.
- ✓ The **mass number** is the total number of protons plus neutrons in an atom.
- ✓ **Isotopes** are atoms of the same element with the same atomic number, but different mass numbers.
- ✓ **Relative atomic masses** are used to compare the relative masses of atoms.
- ✓ Atoms of the isotope carbon-12 are given a relative atomic mass of exactly 12. The relative atomic masses of other atoms are obtained by comparison with carbon-12.
- ✓ The relative atomic mass of an element is the average mass of one atom, taking into account the relative amounts of its different isotopes.
- ✓ Using relative atomic masses and equations for reactions, it is possible to calculate the masses of reactants and products.
- ✓ An **empirical formula** shows the simplest whole number ratio for the atoms of different elements in a compound.
- ✓ The **relative formula mass** of a compound is the sum of the relative atomic masses of all the atoms in its formula.
- ✓ The relative atomic mass of an element in grams and the relative formula mass of a compound in grams are sometimes called one **mole**.
- ✓ Using relative atomic masses it is possible to calculate the percentage of different elements in a compound.
- ✓ The electrons in an atom determine its chemical properties. Electrons are grouped together in **electron shells** at different distances from the nucleus and at particular **energy levels**. When the shells are full, the atoms or ions are stable. **Electron structures** show the number of electrons in each shell.
- ✓ Elements in the same group of the Periodic Table have the same number of electrons in their outermost shell (highest energy level). This gives them similar chemical properties.
- ✓ When atoms react, they lose, gain or share electrons in order to get a more stable electron structure. This idea is called the **electronic theory of chemical bonding**.
- ✓ **Ionic bonds** are formed when metals react with non-metals. Electron transfer occurs. Metals lose electrons to form positive ions. Non-metals gain electrons to form negative ions. The ionic bond results from the electrical attraction between ions with opposite charges.
- ✓ **Covalent bonds** are formed when non-metals react with each other. The non-metal atoms share electrons. The positive nucleus of each atom attracts the shared negative electrons and this forms the covalent bond.

EXAMQUESTIONS

1 Many processed foods contain chemicals as additives.

a) Sodium carbonate is used as an acidity regulator in some tinned foods. Sodium carbonate contains sodium ions (Na^+) and carbonate ions (CO_3^{2-}). Which one of the following is the formula of sodium carbonate?

A $NaCO_3$ B Na_2CO_3
C $Na(CO_3)_2$ D $Na_3(CO_3)_2$ *(1 mark)*

b) The electron structure of a sodium atom can be written as 2, 8, 1. Which one of the following is the electron structure of a sodium ion?

A 2, 2 B 2, 8, 2 C 2, 8 D 2, 8, 8 *(1 mark)*

c) Calcium chloride is used as a firming agent in some tinned foods. Calcium chloride contains calcium ions (Ca^{2+}) and chloride ions (Cl^-). Which one of the following is the formula of calcium chloride?

A CaCl B Ca_2Cl C Ca_2Cl_2 D $CaCl_2$ *(1 mark)*

d) Copy and complete the following sentence.

A calcium ion (Ca^{2+}) is formed when a calcium atom ______ two ______. *(2 marks)*

❷ If a person is anaemic and has a low red blood cell count, he or she may be prescribed iron tablets (Figure 5.28). These contain iron sulfate.

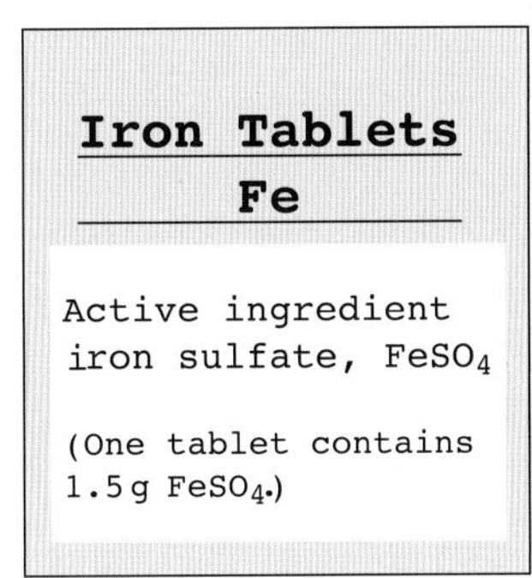

Figure 5.28

a) What elements are present in iron sulfate? *(3 marks)*

b) Iron sulfate contains iron ions (Fe^{2+}) and sulfate ions. Write the symbol and charge for a sulfate ion. *(2 marks)*

c) Calculate the relative formula mass (M_r) of iron sulfate. *(2 marks)*
(Fe = 55.8, S = 32.1, O = 16.0)

d) Calculate the percentage of iron in iron sulfate. *(2 marks)*

e) Calculate the mass of iron in one tablet. *(1 mark)*

f) Why do some people need to take iron tablets? *(1 mark)*

❸ Toothpaste sometimes contains calcium fluoride, CaF_2, which helps to produce strong teeth and bones.

a) A fluoride ion can be written as ${}^{19}_{9}F^-$. How many protons, neutrons and electrons are there in one ${}^{19}_{9}F^-$ ion? *(3 marks)*

b) Why does one calcium ion combine with two fluoride ions in calcium fluoride? *(2 marks)*

c) Figure 5.29 shows the outer shell electrons and the full electron structures in brackets for a calcium atom and two fluorine atoms before reacting to form calcium fluoride. Copy and complete the diagram showing the outer shell electrons, charges and electron structures of the calcium and fluoride ions which form. *(8 marks)*

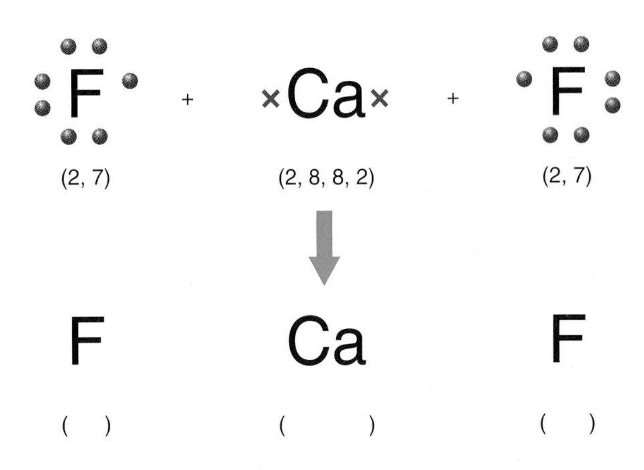

Figure 5.29

d) Why does children's toothpaste often contain calcium fluoride? *(2 marks)*

❹ Read the newspaper cutting below and then answer the following questions.

Toxic gas scare at local factory

Workers were forced to flee from the 'Cupromets' factory in Woodside Avenue last Tuesday. The factory was evacuated when fumes of highly toxic nitrogen dioxide filled the building. The gas was produced when concentrated nitric acid spilled onto thin copper pipes.

a) What does 'toxic' mean? *(1 mark)*

b) Explain why the reaction in the factory was more dangerous because the acid spilled onto thin copper pipes rather than thick copper bars. *(2 marks)*

c) Water was sprayed onto the copper and nitric acid to slow down the reaction. Why would this slow down the reaction? *(2 marks)*

d) Copy and balance the following equation for the reaction: *(2 marks)*

$$Cu + \_\ HNO_3 \rightarrow Cu(NO_3)_2 + \_\ H_2O + 2NO_2$$

❺ Using the equation for the reaction in question **4**, calculate the mass of nitrogen dioxide produced when 254 g of copper reacts completely with concentrated nitric acid.
(Cu = 63.5, N = 14.0, O = 16.0) *(2 marks)*

EXAMQUESTIONS

tions

er 6
do the structures of substances influence their properties and uses?

At the end of this chapter you should:

- ✓ know that the arrangement of particles in a substance determines its structure and the forces between particles determine its bonding;
- ✓ understand how the structure and bonding of a substance determine its properties and that these properties determine its uses;
- ✓ know that the particles in substances may be atoms, ions or molecules;
- ✓ understand how these three different particles give rise to four different structures (giant metallic, giant covalent, giant ionic and simple molecular) with very different properties;
- ✓ appreciate that the bonds between atoms in metals, between atoms in molecules and between ions in ionic compounds are strong, but the forces between molecules are much weaker;
- ✓ be able to relate the properties of substances to their uses;
- ✓ be able to identify the structure of a substance from its properties;
- ✓ understand that nanomaterials have different properties from normal-sized chunks of the same material because of their very small size;
- ✓ be able to evaluate the benefits and risks of using nanomaterials.

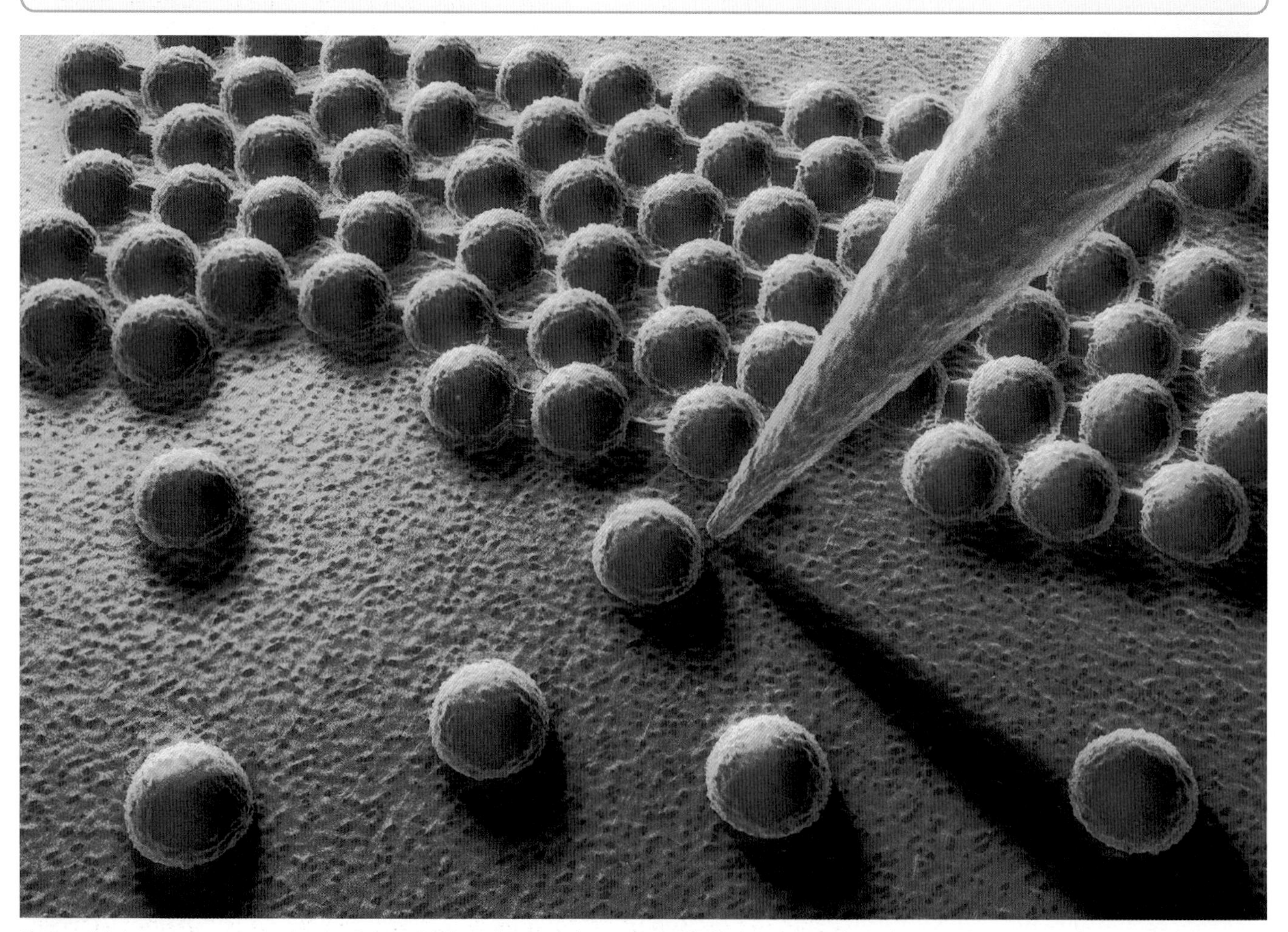

Figure 6.1 In recent years, scientists have developed new substances called nanomaterials with special properties because of the ultra-small size of their particles. This photo shows individual nanoparticles being moved around.

6.1 Studying the structures of materials

All substances are made up of particles – atoms, ions or molecules. If we know how these particles are arranged in materials, we can begin to understand their structure and properties.

Figure 6.2 Crystals of rock salt (sodium chloride, NaCl)

The study of crystals and their shapes has helped scientists to understand the structures and properties of materials. Sometimes crystals grow unevenly and their shapes are not perfect. Even so, it is usually easy to see their general shape.

Careful study shows that all the crystals of a substance have similar shapes. For example, if you look carefully at Figure 6.2, you will see that individual crystals of rock salt are cubic even though they interlock with each other.

The similar shapes suggest that particles in the crystals of a substance always pack in the same way to give the same overall shape. A similar thing happens with piles of apples and oranges, which tend to form pyramid shapes.

Figure 6.3 Crystals of quartz. Quartz is silicon dioxide (SiO_2).

Figure 6.4 shows how cubic crystals and hexagonal crystals can form. If the particles are always in parallel lines or at 90° to each other, the crystals will be cubic. If the particles are arranged in hexagons, the final crystal will be hexagonal.

The shape of a crystal only gives a clue to the way in which the particles are arranged. Using X-rays, it is possible to get much better evidence for the arrangement of particles.

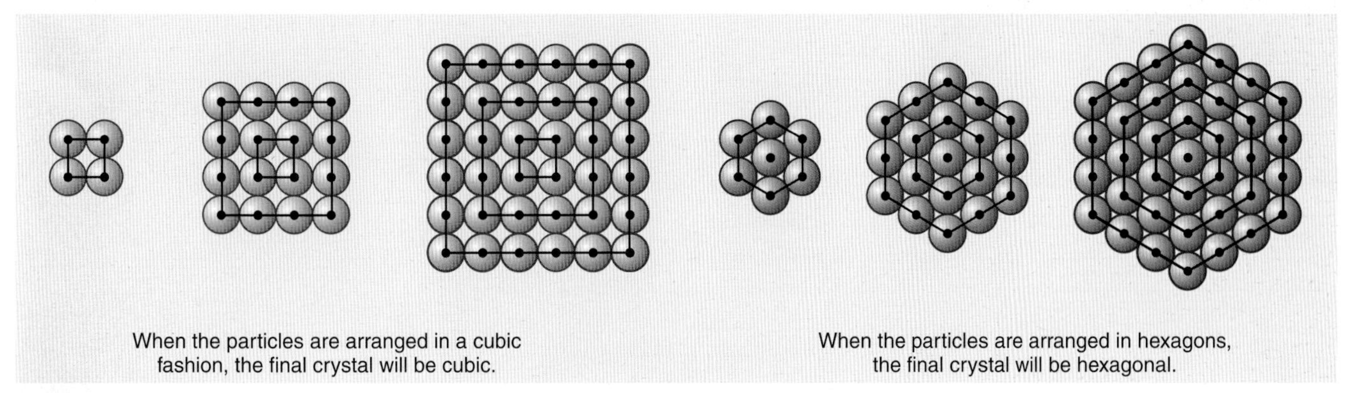

Figure 6.4

Using X-rays to study structures

Look through a piece of thin stretched cloth at a small bright light. You will see a pattern of bright dots. The pattern forms because the light gets deflected as it passes through the tightly spaced threads of the fabric.

From the pattern of bright dots which you can see, it is possible to work out the pattern of threads in the fabric which you cannot see.

Look at the crystals of rock salt and quartz in Figures 6.2 and 6.3.

1. Do most of the rock salt crystals have roughly the same shape?
2. Do most of the quartz crystals have roughly the same shape?
3. Describe the shape of:
   a) the rock salt crystals;
   b) the quartz crystals.

The same idea is used to find out how particles are arranged in substances. A thin beam of X-rays is directed at a crystal of the substance being studied (Figure 6.5).

The X-rays are deflected by particles in the crystal onto X-ray film. When the film is developed, a regular pattern of spots appears. From the pattern of spots which we can see, it is possible to work out the pattern of particles in the crystal which we cannot see. A regular arrangement of spots on the film indicates a regular arrangement of particles in the crystal.

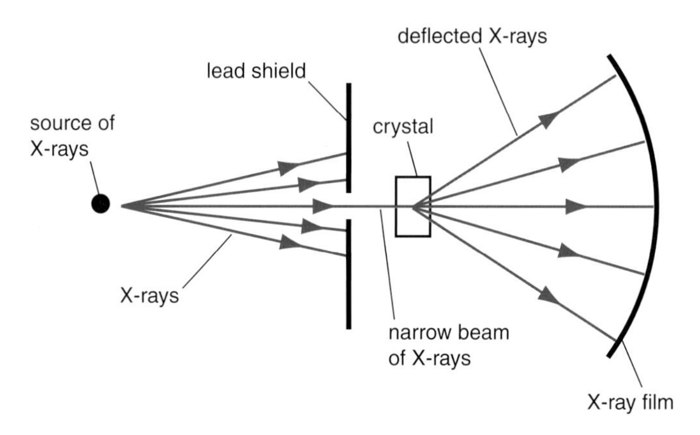

Figure 6.5 Using X-rays to study the particles in a crystal

A **lattice** is a regular arrangement of particles in a solid, such as a metal or an ionic compound.

The regular arrangement of particles in a crystal is called a **lattice**. X-rays have been used in this way to study the structures of thousands of different solids. An X-ray photo is shown in Figure 6.6. Photos like this give accurate information about the structure of different substances.

Figure 6.6 An X-ray photo of crystals of a protein found in egg white. Notice the general pattern in the dots.

4. Look at the photo of a snowflake in Figure 6.7.
   a) What substance makes up snowflakes?
   b) What are the particles in snowflakes?
   c) What is (are) the angle(s) between branches on the snowflake?
   d) How do you think the particles are arranged in snowflake crystals?

Figure 6.7 Crystals of a snowflake

5. Explain the following words:
   a) crystal;
   b) lattice.
6. Why do all crystals of one substance have roughly the same shape?

6.2 From structure to properties and uses

If we know how the particles in a substance are arranged (the structure) and how they are held together (the bonding), we can explain the properties of the substance.

For example, copper is composed of close-packed atoms with freely moving outer electrons. These electrons move through the structure when copper is connected to a battery, so it is a good conductor of electricity. Atoms in the close-packed structure of copper can slide over each other and because of this copper can be drawn into wires. These properties of copper lead to its use in electrical wires and cables.

Notice how:
- the structure and bonding of a substance determine its properties;
- the properties of a substance determine its uses.

Figure 6.8 Wet clay is soft and easily moulded by the potter. After she has moulded the clay, it is heated (fired) in a furnace. This makes it hard and rigid and it can be used for pots and crockery.

The links from structure and bonding to properties help us to explain the uses of substances and materials. They explain why metals are used as conductors, why graphite is used in pencils and why clay is used to make pots.

From Chapter 5 we know that all substances are made up from only three kinds of particle – atoms, ions and molecules.

These three particles give rise to four different solid structures:
- giant metallic;
- simple molecular;
- giant covalent (macromolecular);
- giant ionic.

7 Use the words given below to complete the sentences.

bond hard molecules moulded three two water

Wet clay is soft and easily _____ because _____ molecules can get between its flat _____-dimensional particles. When the clay is fired, all the water _____ are driven out. Atoms in one layer _____ to those in the layers above and below. This gives the clay a _____-dimensional structure making it _____ and rigid.

Table 6.1 shows the particles in these four structures, the types of substances they form and examples of these substances. In the following sections, we will look at each of these structures in more detail.

Type of structure	Particles in the structure	Types of substance	Examples
Giant metallic	Atoms	Metals and alloys	Sodium, iron, copper, steel, brass
Simple molecular	Small molecules containing a few atoms	Non-metals or non-metal compounds	Iodine (I_2), oxygen (O_2), water (H_2O), carbon dioxide (CO_2)
Giant covalent (macromolecular)	Very large molecules containing thousands or millions of atoms	Non-metals or non-metal compounds	Diamond, graphite, polythene, sand (silicon dioxide, SiO_2)
Giant ionic	Ions	Compounds of metals with non-metals	Sodium chloride (salt, Na^+Cl^-), calcium oxide (quicklime, $Ca^{2+}O^{2-}$), magnesium chloride ($Mg^{2+}(Cl^-)_2$).

Table 6.1 The four types of solid structures and the particles they contain

8. Match each of the following substances to one of the solid structures in Table 6.1:
   a) chlorine (Cl_2);
   b) limestone (calcium carbonate, $CaCO_3$);
   c) gold (Au);
   d) PVC (polyvinylchloride);
   e) copper sulfate ($CuSO_4$);
   f) methane in natural gas (CH_4).
9. Look at the photo of a lead sulfide crystal in Figure 6.9.

Figure 6.9 A natural crystal of lead sulfide

   a) What elements does lead sulfide contain?
   b) Is each of these elements a metal or a non-metal?
   c) What particles does lead sulfide contain – atoms, ions or molecules?
   d) How do you think the particles are arranged in lead sulfide? Explain your answer.

6.3 Giant metallic structures

Metals are very important and useful materials. Just look around you and notice the uses of different metals – vehicles, girders, bridges, pipes, taps, radiators, cutlery, pans, jewellery and ornaments.

X-ray studies show that the atoms in most metals are packed together as close as possible. This arrangement is called close packing, which you met in your study of alloys in Section 1.7.

Figure 6.10 shows a model of a few atoms in one layer of a metal crystal.

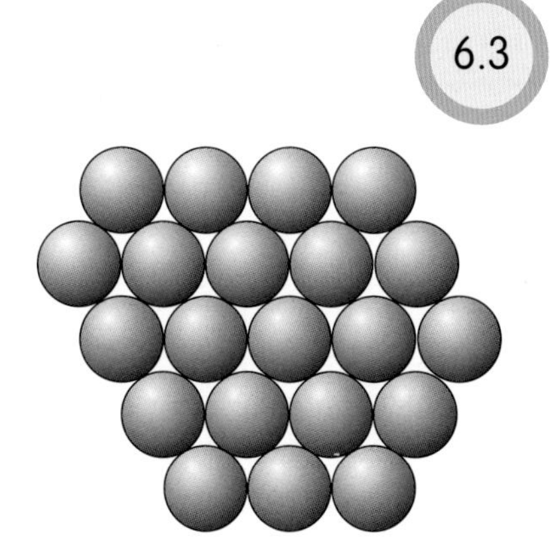

Figure 6.10 The close packing of atoms in one layer of a metal

Notice that each atom in the middle of the layer 'touches' six other atoms in the same layer. When a second layer is placed on top of the first, atoms in the second layer sink into the dips between atoms in the first layer (Figure 6.11).

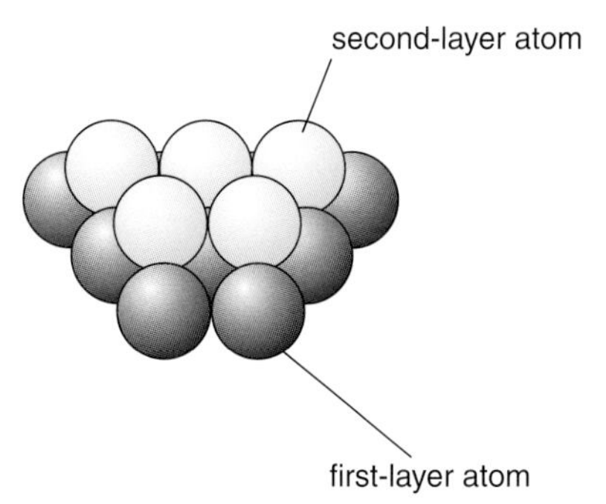

Figure 6.11 Atoms in two layers of a metal crystal

This close packing allows atoms in one layer to get as close as possible to those in the next layer. So, metals consist of **giant metallic structures** of closely packed atoms in a regular pattern.

In this giant lattice structure, electrons in the outer shell of each metal atom are free to move through the whole structure. These electrons are described as 'delocalised'. The metal consists of a structure of positive ions with negative electrons moving around and between them (Figure 6.12). The electrostatic attractions between the 'sea' of delocalised electrons and the positive ions result in strong forces between the metal atoms.

In **giant metallic structures**, metal atoms are close-packed in a giant lattice through which the outer shell electrons can move freely.

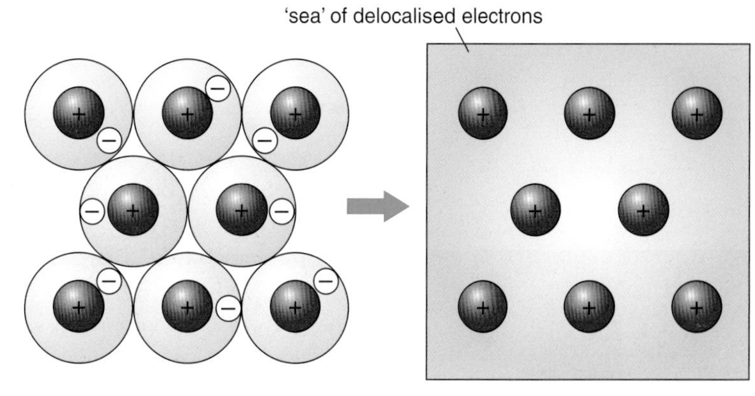

Figure 6.12 The attractions between negative delocalised electrons and positive ions result in strong forces between metal atoms.

In general, metals:

- have high densities,
- have high melting points and high boiling points,
- are good conductors of heat and electricity,
- are malleable (can be bent or hammered into different shapes).

The properties of metals

All the properties of metals can be explained by their close-packed structure.

- High density – The close packing means that all the atoms get as close as possible to atoms in their own layer and to atoms in the layers above and below them, resulting in a high density.
- High melting points and high boiling points – The atoms are closely packed with strong forces between them. So, it takes a lot of energy to separate the particles and change state.
- Good conductors of heat – When a metal is heated, energy is transferred to the electrons. The delocalised electrons move around faster and conduct the heat (energy) rapidly to other parts of the metal.
- Good conductors of electricity – When a metal is connected in a circuit, the delocalised electrons move towards the positive terminal (Figure 6.13). This flow of electrons through the metal forms an electric current.

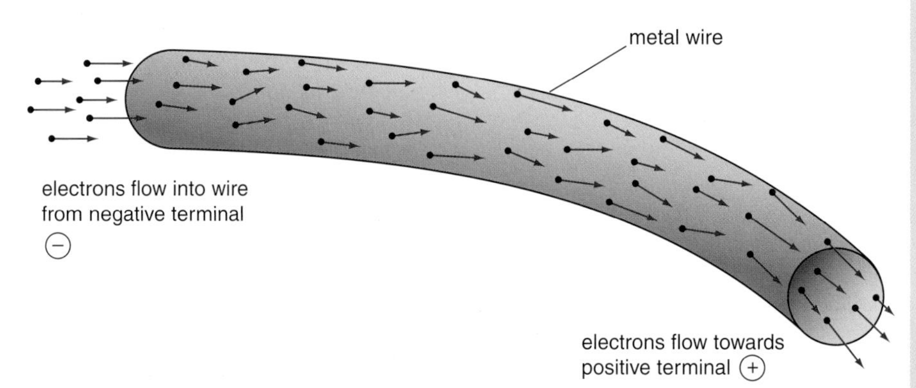

Figure 6.13 Electrons flowing along a metal wire form an electric current.

- Malleable – The bonds between metal atoms are strong but they are not fixed and rigid. When a force is applied to a metal, the layers of atoms can 'slide' over each other. This is known as 'slip'. After slipping, the atoms settle into close-packed positions again. Figure 6.15 shows the positions of atoms before and after slip. This is what happens when a metal is bent or hammered into different shapes.

Figure 6.14 Blacksmiths rely on the malleability of metals to hammer and bend them into useful shapes.

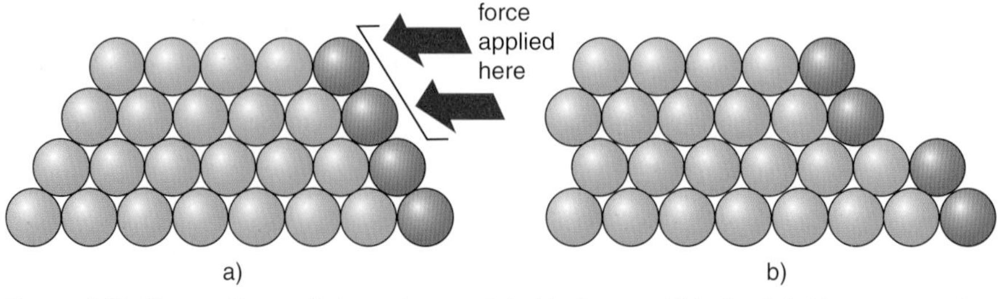

Figure 6.15 The positions of atoms in a metal a) before and b) after 'slip' has occurred

Figure 6.16 Steel was used to make suits of armour in the Middle Ages.

10 Explain in terms of atoms why blacksmiths usually get metals red hot before hammering them into the shapes they want.

11 Which of the following statements provide evidence for classifying copper as a metal? Copper:
A burns to form an oxide.
B reacts with non-metals.
C reacts only with non-metals.
D is non-magnetic.
E conducts electricity.
F has two common isotopes.

12 Look at the suit of armour in Figure 6.16.
a) What particles make up the structure of steel in the armour?
b) How are these particles arranged in steel?
c) What properties of steel are important for its use in the armour?
d) Explain, in terms of particles, why steel has one of the properties you gave in answer to part c).

13 Why are the electrons in the outermost shell of metal atoms described as 'delocalised'?

Activity – Choosing metals for different uses

Various properties of five metals are shown in Table 6.2.

Metal	Density in g/cm^3	Relative strength compared with iron	Melting point in °C	Relative electrical conductivity compared with iron	Relative thermal conductivity compared with iron	Rate of corrosion	Cost per tonne in £
Aluminium	2.7	0.33	660	3.7	3.0	Very slow	950
Copper	8.9	0.62	1083	5.8	4.8	Very slow	1100
Iron	7.9	1.00	1535	1.0	1.0	Quick	130
Silver	10.5	0.39	962	6.1	5.2	Very slight	250 000
Zinc	7.1	0.51	420	1.6	1.4	Very slow	730

Table 6.2 The properties of five commonly used metals

1 Use the information in Table 6.2 to explain the following statements.
a) The frames and poles of tents are made from aluminium.
b) Electrical cables in the National Grid are made of aluminium although copper is a better conductor.
c) Bridges are built from steel which is mainly iron, even though it corrodes (rusts) faster than the other metals.
d) Metal gates and dustbins are made from iron coated (galvanised) with zinc.

e) Silver is no longer used to make our coins.
f) High-quality saucepans have copper bottoms rather than steel (iron).

2 a) What is the range in values for i) densities, ii) melting points in Table 6.2?
b) Do you think any of the values for density or melting point are anomalous (unusual or irregular)? Explain your answer.

3 Luke and Kathy were working together on a metals project. Luke said, 'If the atoms in a metal pack closer, the density should be higher, the bonds between atoms should be stronger and so the melting point should be higher.'
Kathy agreed with Luke and predicted that there could be a pattern (relationship) between the densities and melting points of metals.
a) Use the data in Table 6.2 to plot a graph of density against melting point for the five metals.
b) Is there a pattern or relationship between the densities and melting points of the metals in Table 6.2? Say 'Yes' or 'No' and explain your answer.

4 The explanation of both electrical and thermal conductivity in metals relies on the idea of delocalised electrons. This suggests that there should be a pattern (relationship) between electrical and thermal conductivities for metals. Use the relative electrical and thermal conductivities in Table 6.2 to check whether there is a pattern.
a) Describe briefly what you did to investigate whether there is a pattern.
b) Is there a pattern? Say 'Yes' or 'No' and explain your answer.

Simple molecular substances

Simple molecular substances are small molecules with just a few atoms, compared to polymers or giant molecules.

Intermolecular forces are weak forces between one molecule and another.

Most non-metals and compounds of non-metals are **simple molecular substances**. Oxygen, hydrogen, chlorine, water, hydrogen chloride and methane are good examples of simple molecular substances. They have simple molecules containing a few atoms. Their formulae and structures are shown in Figure 6.17. Sugar ($C_{12}H_{22}O_{11}$) has much larger molecules than the substances in Figure 6.17, but it still counts as a simple molecule.

In these simple molecular substances, the atoms are held together in each molecule by strong covalent bonds (Figure 6.18). But there are only weak forces between the separate molecules (Figure 6.19).

The weak forces between the separate molecules in simple molecular substances are called **intermolecular forces** ('inter' means between).

The properties of simple molecular substances

The properties of simple molecular substances can be explained in terms of their simple structures and the weak forces between their molecules. In simple molecular substances, there are no ions (like ionic compounds) or delocalised electrons (as in metals). These substances consist of simple molecules with no overall charge. So, there are no obvious electrical forces holding the molecules together. However, some simple molecular substances, such as sugar, water and iodine, do exist as solids and liquids, so there must be some forces holding their molecules together.

Name and formula	Structural formula	Model of structure
Hydrogen, H_2	H—H	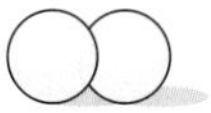
Oxygen, O_2	O=O	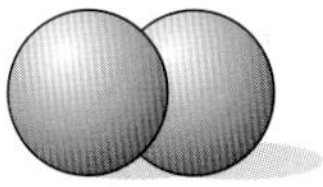
Water, H_2O	H–O–H (bent)	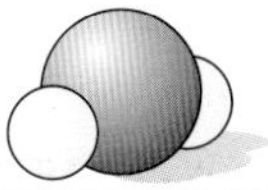
Methane, CH_4	H–C–H with H above and H below C	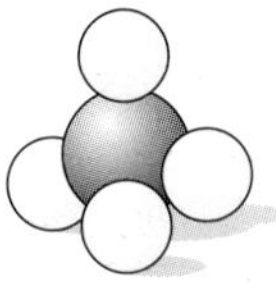
Hydrogen chloride, HCl	H—Cl	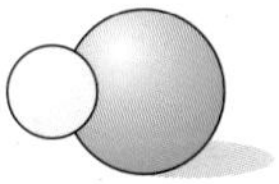
Chlorine, Cl_2	Cl—Cl	
Carbon dioxide, CO_2	O=C=O	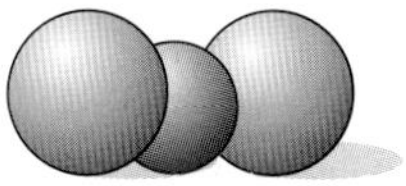
Iodine, I_2	I—I	

Figure 6.17 The formulae and structures of some simple molecular substances

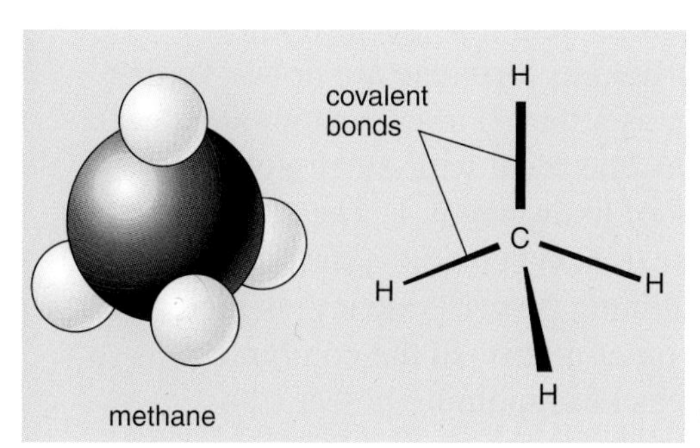

Figure 6.18 Methane (CH_4) is a simple molecular substance. In methane, the carbon atom and four hydrogen atoms are held together by strong covalent bonds.

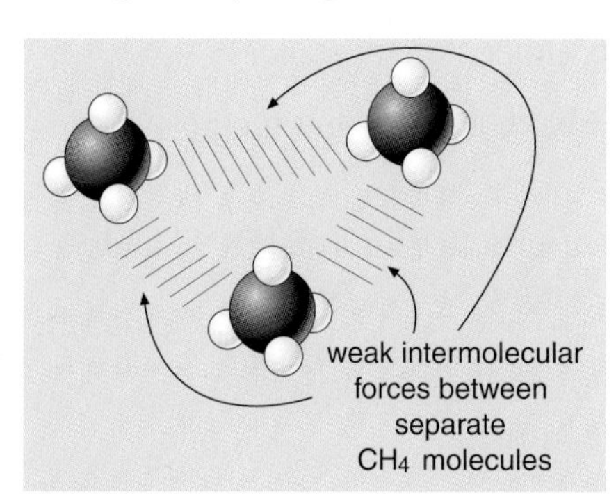

Figure 6.19 Intermolecular forces between molecules of methane

- Low melting points and low boiling points – There are only weak forces between the molecules in simple molecular substances. These weak intermolecular forces are overcome when simple molecular substances melt or boil on heating. It takes much less energy to separate the simple molecules than to separate the oppositely charged ions in ionic compounds or the atoms in metals.
- Soft – The separate molecules in simple molecular substances are usually further apart than atoms in metals and ions in ionic structures. The forces between the molecules are weak and the molecules are easy to separate. Because of this, simple molecular solids, such as iodine and sugar, are softer than other materials.
- Non-conductors of electricity – Simple molecules have no overall electric charge. They have no mobile (delocalised) electrons like metals and no ions like ionic compounds. This means they cannot conduct electricity.

Figure 6.20 This butcher is using 'dry ice' (solid carbon dioxide) to keep meat cool during mincing. 'Dry ice' is a simple molecular substance. After mincing, the 'dry ice' sublimes (changing directly from solid to gas) without spoiling the meat.

The formulae of molecular compounds

The **combining power** or **valency** of an atom is the number of covalent bonds that it forms with other atoms.

Figure 6.17 shows the formulae and structures of some well-known simple molecular compounds. The structural formulae are drawn so that the number of covalent bonds (drawn as a line ?) to each atom are clear. Notice that each hydrogen atom forms one bond with other atoms (H–). So, the **combining power** or **valency** of hydrogen is 1. The combining powers of chlorine and iodine are also 1. Oxygen atoms form two bonds to other atoms (–O– or O=). Its combining power (valency) is therefore 2. Carbon atoms form four bonds to other atoms, so the combining power of carbon is 4. Using these ideas of combining power (the number of bonds which the atoms form), we can predict the formulae of most (but not all) molecular compounds.

Figure 6.21 Butter contains a mixture of simple molecular substances.

14. Look at the photo in Figure 6.21. What properties does butter have which show that it contains simple molecular substances?
15. Simple molecular substances often have a smell, but metals never have a smell. Why is this?
16. Solid X is a poor conductor of electricity. It melts at 150 °C and boils at 350 °C. Which of the following could X be?
    A a metal
    B a non-metal
    C a compound of a metal with a non-metal
    D a compound of non-metals
    E an ionic compound
    F a simple molecular compound

⑰ Assuming the usual combining power (valency) of the elements, draw structural formulae for the following compounds (show each bond as a line –):

a) dichlorine oxide (Cl_2O);
b) tetrachloromethane (CCl_4);
c) hydrogen peroxide (H_2O_2);
d) ethane (C_2H_6).

6.5 Giant covalent structures

Giant covalent structures contain very large numbers of atoms joined to each other by strong covalent bonds.

Atoms that share electrons in covalent bonds can also form **giant covalent structures** or macromolecules. One important group of compounds with giant covalent structures are polymers, such as polythene and polyvinylchloride (PVC). Polythene and PVC consist of long, thin molecules made by addition reactions between smaller monomer molecules. The polymers consist of hundreds and often thousands of atoms joined by strong covalent bonds.

Diamond, graphite and silicon dioxide are other examples of giant covalent structures, but these form three-dimensional lattices with billions of atoms.

Diamond and graphite are both made of pure carbon, but the two solids have very different properties and uses. Diamond is hard and clear, whereas graphite is soft and black. Diamonds are used to cut stone and engrave glass whereas graphite is used by artists to achieve a soft, shaded effect for their sketches.

Figure 6.22 This artist is using a soft graphite pencil.

Diamond and graphite have different properties and different uses because they have different structures. They both contain carbon, but the carbon atoms are packed in different ways. The arrangements of carbon atoms in diamond and graphite have been studied by X-ray analysis.

Diamond

In diamond, carbon atoms are joined to each other by strong covalent bonds. Inside the diamond structure (Figure 6.23), each carbon atom forms covalent bonds with four other carbon atoms. Check this for yourself in Figure 6.23. The strong covalent bonds extend through the whole diamond, forming a giant three-dimensional structure. A perfect diamond, without flaws or cracks, is a single giant molecule (macromolecule) with covalent bonds linking one carbon atom to the next.

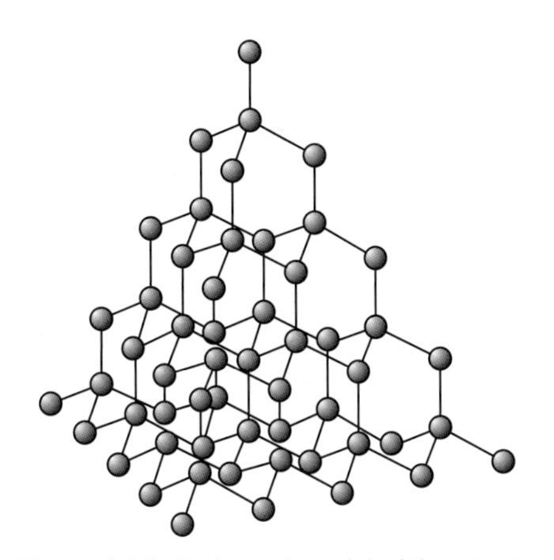

Figure 6.23 An 'open' model of the giant structure in diamond. Each black ball represents a carbon atom and each stick is a covalent bond.

Only a small number of atoms are shown in the model in Figure 6.23. In a real diamond, this arrangement of carbon atoms is extended billions and billions of times.

The properties and uses of diamond

- Hard – It is difficult to break the strong covalent bonds between carbon atoms in diamond. Another reason for diamond's hardness is that the atoms are not arranged in layers. So they cannot slide over one another like the atoms in metals. Most industrial uses of diamond depend on its hardness. Diamonds which are not good enough for jewellery are used in glass cutters, in diamond-studded saws and as abrasives for smoothing very hard materials.
- Very high melting point – Carbon atoms in diamond are linked together in the giant structure by very strong covalent bonds. This means that the atoms cannot vibrate fast enough to break away from their neighbours until the temperature is about 3800 °C.
- Non-conductor of electricity – Unlike metals, diamond has no free electrons. All the electrons in the outer shell of each carbon atom are held firmly in covalent bonds. There are no free electrons in diamond to form an electric current.

Graphite

Figure 6.24 shows a model of part of the structure of graphite. Notice that the carbon atoms are arranged in layers. Each layer contains billions and billions of carbon atoms arranged in hexagons. Each carbon atom is held in its layer by strong covalent bonds to three other carbon atoms. So, every layer is a giant covalent structure. The distance between carbon atoms in the same layer is only 0.14 nm, but the distance between the layers is 0.34 nm – more than twice as great.

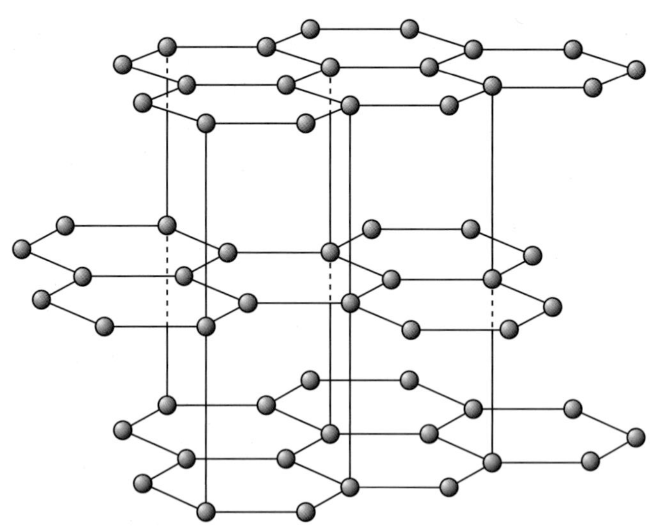

Figure 6.24 A model of the structure of graphite. Notice how graphite has layers of carbon atoms. Within the layers, the carbon atoms are arranged in hexagons

The properties and uses of graphite

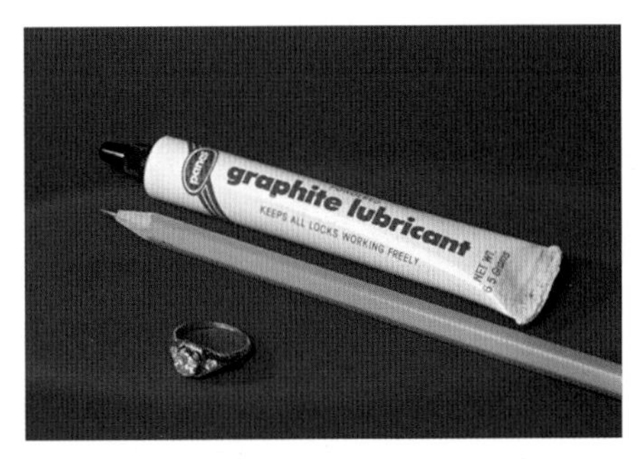

Figure 6.25 A tube of lubricating oil containing graphite. The layers in graphite slide over one another very easily. This makes graphite an excellent material to improve the lubricating action of oils.

- Soft and slippery – In graphite, each carbon atom is linke[illegible] covalent bonds to three other atoms in its own layer. But, the [illegible] are more than twice as far apart as carbon atoms in the same layer[illegible] This means that the forces between the layers are weak. If you rub graphite, it feels soft and slippery as the layers slide over each other and onto your fingers. This property has led to the use of graphite as the 'lead' in pencils and as a lubricant in engine oils.
- High melting point – Although the layers of graphite move over each other easily, it is difficult to break the strong covalent bonds between carbon atoms within one layer. Because of this, graphite does not melt until 3730 °C and it is used to make crucibles for hot molten metals.
- Conducts electricity and heat – Carbon atoms have four outer-shell electrons. In graphite, three of these electrons form covalent bonds with electrons from three other carbon atoms in the same layer. This leaves one unshared electron on each carbon atom. These unshared electrons are delocalised over the atoms in their layer. So, they allow graphite to conduct electricity and heat in a similar way to metals. Because of this property, graphite is used for electrodes in industry and as the positive terminal in dry cells.

Figure 6.26 Graphite fibres are used to reinforce the shafts of badminton rackets, golf clubs and even broken bones. The strong bonds between carbon atoms in layers of graphite mean that graphite fibres can be very strong.

18 a) Why is diamond called a 'giant molecule'?
b) Diamond is sometimes described as a three-dimensional giant structure and graphite as a two-dimensional giant structure. Why is this?

19 The largest natural diamond is the Cullinan diamond. This weighs about 600 g.
a) How many moles of carbon does it contain? (C = 12)
b) How many atoms of carbon does it contain? (1 mole of carbon is 6×10^{23} atoms.)

20 a) Make a table to show at least three similarities and three differences between diamond and graphite.
b) Why do diamond and graphite have some differences?

21 a) Why will a zip-fastener move more freely after being rubbed with a soft pencil?
b) Why is it better to use a pencil rather than oil to make a zip-fastener move more freely?

22 Why is graphite better than oil for lubricating the moving parts of hot machinery?

6.5 Giant covalent structures

d by strong
layers

·anomaterials

Nanomaterials and nanotechnology belong to the twenty-first century. Very few people had heard of these terms before the year 2000. Since then, a wide range of materials have been developed and used in the form of **nanoparticles** between 1 and 100 nanometres (nm) in size. That's between one millionth and one ten thousandth of a millimetre in size.

Nanoparticles are groups of just a few hundred atoms. They include metals, carbon, plastics and polymers. Nanoparticles show different properties from the same material in bulk (large pieces) because of the precise way in which the atoms in them are arranged and because of their high surface area-to-volume ratio.

Figure 6.27 These polymer nanoparticles have been designed to carry drug molecules into the body. Just 100 nm to 500 nm across, they can pass across vein walls, and so the drug can be released into tissues in exactly the part of the body that needs treating.

Nanotechnology is already being used to manufacture products that take advantage of the ability to make layers of material just a few atoms thick or particles with a very large surface area per unit mass. Here are some of the current applications:

- One of the biggest developments for nanotechnology is in stain-resistant clothing. Fabrics can be coated with nano-sized polymer particles. Any liquid spilt on the fabric turns into 'beads' and rolls off without staining the fabric.
- Ultra-thin, yet extremely strong coatings of scratch-resistant polymers are now being used on lenses for spectacles, cameras and contact lenses. The coatings are so thin that the polymer material is transparent.
- New solar cells have been developed with nano-sized metal crystals on their surface. These have a vast surface area on which to collect the Sun's energy more efficiently.
- Ordinary tennis rackets are weakened by tiny gaps between the carbon fibres in their structure. These tiny gaps can be plugged and reinforced with nano-sized crystals of silicon dioxide.

- Nanoparticles have also led to important developments in catalysts because of their large surface area compared with bulk materials. These developments include the use of gold particles to oxidise poisonous carbon monoxide to carbon dioxide. The gold particles are totally inactive until they are less than about 8 nm in diameter. Their high surface area means you only need a tiny amount of gold – which is cheaper!

Future nanotechnology applications could include:

- smaller and more sensitive sensors in electronic devices and potentially smaller and faster computers and disc drives with layers of magnetic nanoparticles to increase data storage capacity;
- lighter yet stronger construction materials made from composite materials containing nanotubes. Carbon nanotubes are 10 times lighter than steel but 250 times stronger;
- 'smart' clothing with nano-sized sensors to measure pulse rate and temperature.

Are nanomaterials safe?

Although nanomaterials open up exciting possibilities for new materials, there are concerns about the effects of nanomaterials when they get into the human body. Nanoparticles could be more reactive and so more toxic than larger masses of the same material. Although most areas of nanotechnology pose no risks, scientists agree that more research of the health risks of nanomaterials is urgent. They also suggest:

- Nanoparticles should be treated as 'new chemicals'.
- An independent scientific safety committee should approve new substances at the nano-scale before they are used in consumer products.
- Nanoparticles that have direct contact with our bodies such as those in some cosmetic and sunscreen creams are the most likely to cause long-term damage.
- There needs to be workplace regulations to protect workers in industry who are exposed to nanoparticles.

Activity – Nanomaterials and the future

1. Explain what is meant by nanoscience in no more than 25 words.
2. Scientific and technological developments offer different opportunities to different groups of people. Which groups of people stand to benefit most from the development of nanomaterials?
3. Describe three benefits to society from the development of nanomaterials.
4. Describe two potential risks from their development.
5. Should scientists or society make decisions about the development of nanomaterials?

If you would like more information on these issues, go to www.news.bbc.co.uk and search on 'nanomaterials' and 'nanotechnology'.

Giant ionic structures

Ionic compounds are formed when metals react with non-metals. During these reactions, electrons are transferred from the metal to the non-metal forming positive metal ions and negative non-metal ions (see Section 5.6). For example, when calcium reacts with oxygen to form calcium oxide, two electrons are transferred from each calcium atom to each oxygen atom.

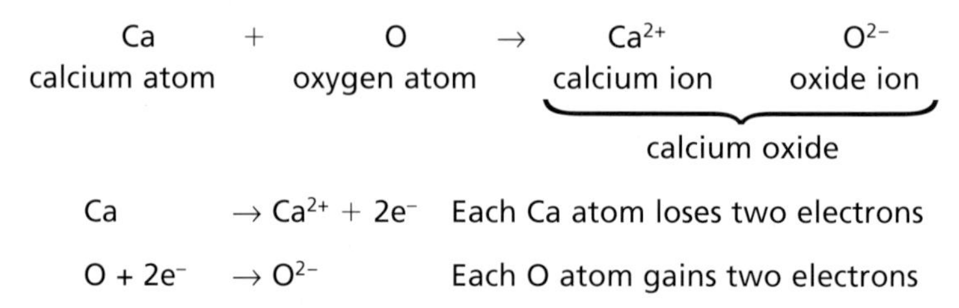

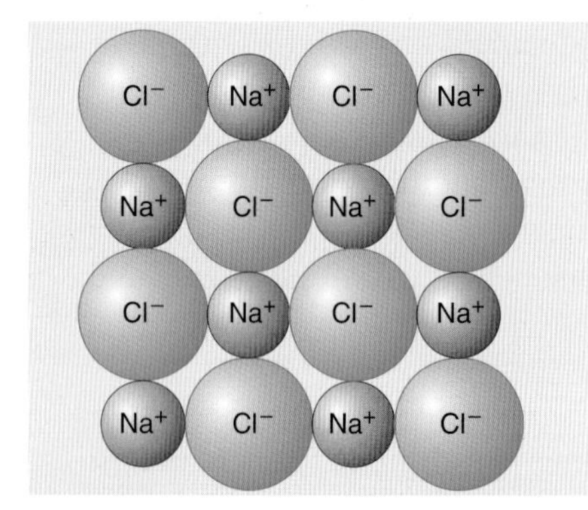

Figure 6.28 The arrangement of ions in one layer of a sodium chloride (salt) crystal

The structure and properties of ionic compounds

In ionic compounds, large numbers of positive and negative ions are packed together in a regular pattern. These **giant ionic lattices** containing billions of ions are further examples of giant structures.

Figure 6.28 shows how the ions are arranged in one layer of sodium chloride (NaCl) and Figure 6.29 is a three-dimensional model of its structure. Notice that Na^+ ions are surrounded by Cl^- ions and that Cl^- ions are surrounded by Na^+ ions. This means that there are strong electrostatic forces in all directions in the lattice between oppositely charged ions. These forces of attraction between the oppositely charged ions are called ionic bonds (see Section 5.6).

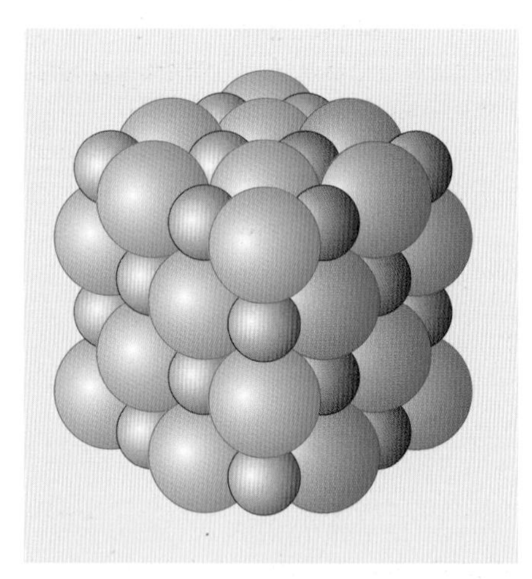

Figure 6.29 A three-dimensional model of the structure of sodium chloride. The larger green balls represent Cl^- ions (Cl = 35.5). The smaller red balls represent Na^+ ions (Na = 23.0).

Giant ionic lattices form a regular structure of oppositely charged ions.

Strong ionic bonds hold the ions firmly together in ionic compounds. This explains why ionic compounds:

- are hard substances;
- have high melting points and high boiling points;
- do not conduct electricity when solid, because their ions cannot move away from fixed positions in the giant structure;
- conduct electricity when they are melted or dissolved in water because the charged ions are then free to move. Positive ions move towards the negative terminal and negative ions move towards the positive terminal and they carry the current through the liquid. The conduction of electricity by ionic compounds is discussed in more detail in Chapter 8.

The formulae of ionic compounds

The formulae of ionic compounds can be obtained by balancing the charges on positive and negative ions. For example, the formula of calcium chloride is $Ca^{2+}(Cl^-)_2$, or simply $CaCl_2$. Here, the two positive charges on one calcium ion (Ca^{2+}) are balanced by the single negative charges on two chloride ions (Cl^-).

Name of compound	Formula
Calcium nitrate	$Ca^{2+}(NO_3^-)_2$ or $Ca(NO_3)_2$
Zinc sulfate	$Zn^{2+}SO_4^{2-}$ or $ZnSO_4$
Magnesium carbonate	$Mg^{2+}CO_3^{2-}$ or $MgCO_3$
Potassium iodide	K^+I^- or KI
Iron(III) chloride	$Fe^{3+}(Cl^-)_3$ or $FeCl_3$
Copper(II) bromide	$Cu^{2+}(Br^-)_2$ or $CuBr_2$

Table 6.3 The names and formulae of some ionic compounds

The number of charges on an ion is a measure of its combining power or valency (Section 6.4). Na^+ has a combining power of 1, whereas Ca^{2+} has a combining power of 2. Na^+ can combine with only one Cl^- to form Na^+Cl^-, whereas Ca^{2+} can combine with two Cl^- ions to form $Ca^{2+}(Cl^-)_2$.

Elements such as iron, which have two different ions (Fe^{2+} and Fe^{3+}), have two valencies. So, iron can form two different compounds with chlorine – iron(II) chloride, $FeCl_2$, and iron(III) chloride, $FeCl_3$.

Table 6.3 shows the names and formulae of some other ionic compounds. Notice that the formula of calcium nitrate is $Ca(NO_3)_2$. The brackets around NO_3^- show that it is a single unit containing one nitrogen and three oxygen atoms with one negative charge. Thus, two NO_3^- ions balance one Ca^{2+} ion. Other ions, such as SO_4^{2-}, CO_3^{2-} and OH^-, should also be put in brackets when there are two or more of them in a formula.

23 Which of the following substances conduct electricity:
a) when liquid;
b) when solid?
A diamond B potassium chloride
C copper D sulfur

24 Look carefully at Figures 6.28 and 6.29.
a) How many Cl^- ions surround one Na^+ ion in one layer of the NaCl crystal?
b) How many Cl^- ions surround one Na^+ ion in the three-dimensional crystal?
c) How many Na^+ ions surround one Cl^- ion in the three-dimensional crystal?

25 Use the symbols and the charges on ions given in Section 6.7 and Table 6.3 to write formulae for the following compounds:
a) iron(II) nitrate;
b) sodium carbonate;
c) iron(III) oxide;
d) zinc chloride;
e) magnesium sulfate;
f) copper(I) bromide.

26 Sodium fluoride (NaF) and magnesium oxide (MgO) have the same crystal structure and similar distances between ions. The melting point of NaF is 992 °C but that of MgO is 2640 °C.
a) Write the formula for sodium fluoride showing charges on the ions.
b) Write the formula for magnesium oxide showing charges on the ions.
c) Why is there such a big difference in the melting points of NaF and MgO?

Activity – Using models in science

Models are very important in science. They help us to think about the way substances behave and how things work. They also help us to test new ideas (hypotheses) and make predictions. The kinetic theory is a model that helps us to understand how the particles in solids, liquids and gases move and interact. Sometimes the models that scientists use are complex computer programs like those that help weather forecasters. In this chapter, we have been using models of the structures of different substances to help us understand their properties and uses.

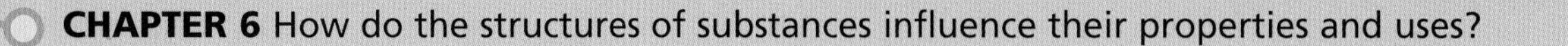

Atomic models for giant metallic structures

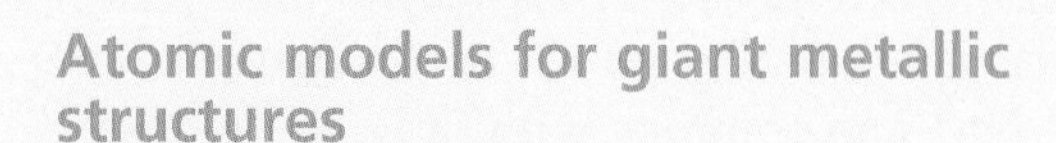

You will need about 25 marbles or 25 polystyrene spheres for this activity. The marbles or polystyrene spheres represent atoms.

Put three books flat on a bench so that they form a triangle around 10 atoms. Move the books carefully so that they enclose a layer of close-packed atoms like those in Figure 6.11.

Now add another layer of atoms on top of the first one. Then put further layers on top of this.

Look carefully at your structure, which represents a very good model for most metals.

1. How many atoms does an atom in the middle of the structure touch?
2. Why is this called a close-packed structure?
3. Why is this a giant structure?
4. How does this model help us to understand metals?

The structure that you have built for metals is sometimes described as a space-filling model because it shows all the space occupied by metal atoms.

Earlier in this chapter, we also used space-filling models to represent the structures of simple molecular substances in Figure 6.17.

These simple molecular substances can also be studied by using ball and stick models like that in Figure 6.23. Ball and stick models are useful because the 'sticks' show the bonds which hold the atoms together.

Building a ball and stick model for silicon dioxide

Silicon dioxide (silica, SiO_2) has a giant covalent structure (Figure 6.30). It can occur as large, beautiful crystals of quartz (Figure 6.3) or more often as impure gritty bits of sand.

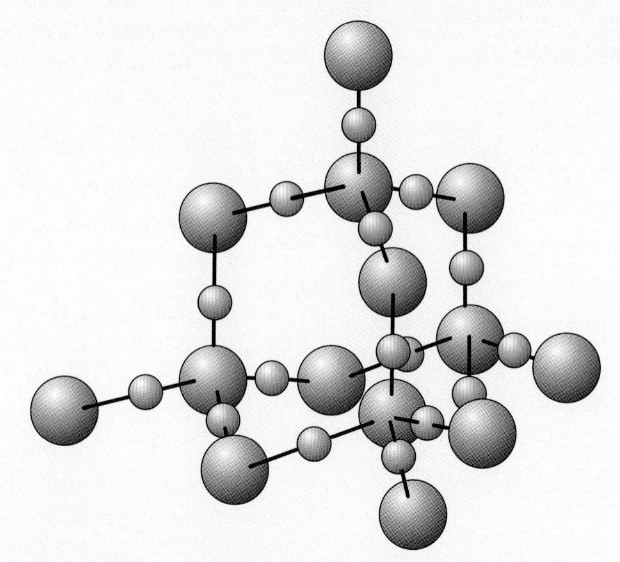

Figure 6.30 A ball and stick model of silicon dioxide, SiO_2

5. Use a ball and stick model kit to build a structure of silicon dioxide like that in Figure 6.30. Use black balls (normally carbon atoms) from the kit to represent silicon atoms and red balls to represent oxygen atoms.
   a) How many covalent bonds are formed by silicon atoms in the middle of your structure?
   b) How many covalent bonds are formed by oxygen atoms in your structure?
6. Explain why silicon dioxide:
   a) is hard;
   b) has a very high melting point;
   c) does not conduct electricity.

Silicon is the second most common element in the Earth's crust. Most rocks including sandstones and clays contain silicon, and the structures of these silicate rocks are based on that of silicon dioxide.

7. Why are models important in science?

Although models are usually helpful, it is important to remember that they are not real representations. For example, metal atoms are not really like marbles or polystyrene spheres and silicon dioxide doesn't have tiny balls and sticks.

Summary

- ✓ The structures of materials can be studied using X-rays.
- ✓ The regular arrangement of particles in a crystal is called a **lattice**.
- ✓ The structure and bonding of a substance determine its properties and these properties determine its uses.
- ✓ All substances are made up from only three kinds of particle – atoms, ions and molecules.
- ✓ These three particles result in four different types of solid structures – **giant metallic**, **giant ionic**, **giant covalent** and **simple molecular**.
- ✓ **Simple molecular substances** are small molecules with just a few atoms, compared with polymers or giant molecules.
- ✓ **Intermolecular forces** are weak forces between one molecule and another.
- ✓ The **combining power** or **valency** of an atom is the number of covalent bonds that it forms with other atoms.
- ✓ **Nanoparticles** are manufactured particles with sizes measured in nanometres (nm).
- ✓ The atoms in most metals are packed together as close as possible. This arrangement is called close packing.
- ✓ A **giant structure** is an arrangement of very large numbers (billions and billions) of atoms or ions held together by strong forces in a regular pattern.
- ✓ The particles, bonding, structure and properties of the four different solid structures are shown in Figure 6.31.

Type of structure	Particles in the structure	Type of substance	Structure	Bonding	Properties
Giant metallic	Atoms close-packed	Metals, e.g. Na, Fe, Cu and alloys such as steel		Atoms are held in a close-packed giant structure by the attraction of positive ions for delocalised electrons.	• High melting points and boiling points • Conduct electricity • High density • Hard but malleable
Giant covalent (macro-molecular)	Very large molecules containing thousands or billions of atoms (giant molecules)	A few non-metals (e.g. diamond, graphite) and some non-metal compounds (e.g. polythene, PVC, silicon dioxide (sand))		Large numbers of atoms are joined together by strong covalent bonds to give a giant 3D lattice or a very long, thin molecule.	• High melting points and boiling points • Do not conduct electricity (except graphite) • Hard but brittle (3D structures) or flexible polymers
Simple molecular	Small molecules containing a few atoms	Most non-metals and non-metal compounds, e.g. O_2, I_2, H_2O, CO_2, sugar		Atoms are held together in small molecules by strong covalent bonds. The bonds between molecules are weak.	• Low melting points and boiling points • Do not conduct electricity • Soft when solid
Giant ionic	Ions	Metal / non-metal compounds, e.g. Na^+Cl^-, $Ca^{2+}O^{2-}$, $Mg^{2+}(Cl^-)_2$		Positive and negative ions are held together by the attraction between their opposite charges – ionic bonds.	• High melting points and boiling points • Conduct electricity when melted or dissolved in water • Hard but brittle

Figure 6.31 The particles, structure, bonding and properties of the four different solid structures

EXAMQUESTIONS

1 Table 6.4 shows some properties of diamond and graphite.

Diamond	Graphite
Colourless crystals	Shiny black solid
Hardest natural substance	Soft and slippery

Table 6.4

a) Why would you expect diamond and graphite to have the same properties? *(1 mark)*
b) Why do diamond and graphite have different properties? *(1 mark)*
c) Explain why graphite is soft and slippery. *(2 marks)*
d) Explain why diamond is hard. *(2 marks)*

2 This question is about aluminium oxide and how it is formed from atoms of aluminium and oxygen.

a) Copy and complete the following sentences by choosing the most suitable words from the box below to fill in the blank spaces.

atoms covalent giant hard high
ionic ions low simple soft

Aluminium oxide is a substance with _____ bonding. It has a _____ structure composed of _____. Strong forces between particles in the structure of aluminium oxide make it very _____ with a _____ melting point. *(5 marks)*

b) The electron structure in an aluminium atom is 2, 8, 3 and the electron structure in an oxygen atom is 2, 6. Describe the changes in electron structure of the aluminium atoms and the oxygen atoms when aluminium oxide forms. *(4 marks)*
c) Explain why the formula of aluminium oxide is Al_2O_3. *(2 marks)*

3 Copy and complete Table 6.5 for methane and sodium chloride. *(5 marks)*

	Methane	Sodium chloride
Formula		NaCl
Particles in the substance	Molecules	
Appearance at room temperature		White solid
Type of bonds	Covalent	
One use		In cooking

Table 6.5

4 Substance Y melts at a high temperature and the liquid which forms conducts electricity.

a) Which of the following could Y be?
calcium chloride starch copper
polythene brass carbon disulfide *(3 marks)*
b) Explain your answers to part a). *(6 marks)*

5 a) Liquid and gaseous carbon dioxide at high pressure are used in fire extinguishers. When the extinguisher is used, carbon dioxide pours out and smothers the fire. Explain why carbon dioxide smothers and puts out fires. *(4 marks)*
b) Solid carbon dioxide is used for refrigerating ice-cream, soft fruit and meat. The solid carbon dioxide is called 'dry ice' or 'Dricold'. Why is 'dry ice' better than ordinary ice for refrigeration? *(2 marks)*

6 Diamond does not conduct electricity but graphite does.

a) Explain why diamond does not conduct electricity. *(2 marks)*
b) Explain why graphite conducts electricity. *(3 marks)*

Chapter 7
How do we control the rate of chemical reactions and measure energy transfer?

At the end of this chapter you should:

- ✓ be able to use the state symbols (s), (l), (g) and (aq) in equations;
- ✓ understand that in some chemical reactions, not all of the reactants will be converted into products;
- ✓ understand chemical calculations used with atomic mass numbers, including yield and atom economy, concentrations in solution, and reactions involving gas volumes;
- ✓ know that atom economy (atom utilisation) is important for sustainable development;
- ✓ be able to plot graphs of amount of products formed or reactants used over time in a reaction and interpret the graphs in terms of the rate of the reaction;
- ✓ know that the rate of a chemical reaction increases:
  - if the temperature increases;
  - if the concentration of dissolved reactants or the pressure of gases increases;
  - if solid reactants are in smaller pieces;
  - if a catalyst is used;
- ✓ be able to use collision theory to model activation energy and explain how increased energy or increased collision frequency change the rates of reactions;
- ✓ understand the advantages and disadvantages of using catalysts in industrial processes;
- ✓ understand that energy can be transferred both to reactants and from reactants in a chemical change;
- ✓ know about exothermic and endothermic reactions;
- ✓ know that in a reversible reaction the products can react to reform the reactants, so the reaction goes both ways;
- ✓ understand about equilibrium and energy changes in a reversible reaction;
- ✓ understand the effects of changing the conditions of temperature and pressure on a given reaction or process;
- ✓ understand how the conditions used in the Haber process to manufacture ammonia are a compromise between yield, rate of reaction and cost;
- ✓ be able to evaluate the conditions used in industrial processes in terms of energy requirements.

Figure 7.1 Cooking involves many chemical changes. When you turn the oven on, some of the energy supplied starts chemical reactions. These chemical reactions change the flavour and texture of food – making potatoes soft and turning meat brown and tender. But if the temperature is too high, the reactions happen too fast and the meat is brown on the outside, and cold and uncooked inside. Being able to control the speed of reactions is important when making new materials in industry too.

Chemical amounts are measured in moles. The symbol for the unit is **mol**.

1. One mole of sodium hydroxide (NaOH) is reacted with hydrochloric acid according to the equation below.

   $NaOH(aq) + HCl(aq) \rightarrow NaCl(aq) + H_2O(l)$

   What is the theoretical amount of sodium chloride that can be produced? (Na = 23, O = 16, H = 1, Cl = 35.5)
2. One mole of methane was burned in air. Some sooty deposits were noticed. The carbon dioxide produced was absorbed by a chemical which gained 32 g in mass.
   a) What was the percentage yield of carbon dioxide?
   b) Why was the percentage yield so low?

Atom economy (also called atom utilisation) refers to the proportion of atoms in the reactants that ends up in the useful product. Atom economy is calculated as the mass of atoms in the desired product divided by the mass of atoms in all the reactants, expressed as a percentage.

The equation for the reaction is:

$Ca(OH)_2(s)$	+	$2HCl(aq)$	→	$CaCl_2(aq)$	+	$2H_2O(l)$
calcium hydroxide		hydrochloric acid		calcium chloride		water

From the equation, 1 mole of calcium hydroxide neutralises 2 moles of hydrochloric acid.

As 3.30 mol of acid were neutralised, the solid added to the hydrochloric acid contained (3.30/2) = 1.65 moles of calcium hydroxide. What mass of reactant is this? Using the relative formula mass of calcium hydroxide:

$$1.65 \times 74\,g = 122\,g$$

From the answer to part a), the theoretical yield is 148 g. The percentage yield is therefore:

$$\text{percentage yield} = \frac{122\,g}{148\,g} \times 100\% = 82.4\%$$

Possible explanations for the yield being less than 100%:

1. The starting material (marble chips) contained some impurity.
2. Some of the calcium hydroxide was lost in the dissolving, filtering and drying processes.
3. The thermal decomposition was incomplete (not all the calcium carbonate decomposed).

Atom economy

The yield of a chemical reaction is the amount of desired product obtained. This calculation ignores how the product was obtained, for example what starting materials were used and the amount of unwanted products formed. Not all atoms of the reactants always end up in the required product. Sometimes by-products are formed. If you want information about the overall efficiency of a chemical reaction, you would calculate the **atom economy**.

Example

Glucose can be fermented to produce ethanol (this reaction is also used to make ethanol fuel from renewable plant sources):

glucose	$\xrightarrow{\text{yeast}}$	ethanol	+	carbon dioxide
$C_6H_{12}O_6(aq)$		$2C_2H_5OH(aq)$		$2CO_2(g)$

If all the sugar really did turn to alcohol in this reaction, the atom economy would be 100%. But some sugar turns to carbon dioxide.

We can calculate the atom economy for this way of making ethanol.

relative formula mass of glucose = (6 × 12) + (12 × 1) + (6 × 16) = 180 g

relative formula mass of ethanol = (2 × 12) + (6 × 1) + (16) = 46 g

From the equation, 1 mole of glucose produces 2 moles of ethanol.

$$\text{atom economy} = \frac{\text{mass of atoms in required product}}{\text{total mass of atoms in reactants}} \times 100\%$$

$$= \frac{2 \times 46\,\text{g}}{180\,\text{g}} \times 100\% = 51.1\%$$

Nearly half the sugar is 'wasted' in this process. Not all of the reactant atoms end up in molecules of ethanol – some carbon and oxygen atoms are released as carbon dioxide.

In the 1960s, Boots made the anti-inflammatory drug ibuprofen in a six-step reaction that has an atom economy of just 40%. When the patent expired, another company developed a new method. This has three stages and produces far less waste, as more of the reactant atoms end up in the required product.

Atom economy and sustainability – green chemistry

Some reactions where there are several stages and several by-products score low on atom utilisation. Other direct combination or decomposition reactions score highly because there are fewer by-products.

Chemists in the pharmaceutical industry now look for alternative ways to make the same product with a good yield and a higher atom economy. This is more efficient and minimises waste.

High atom economy reactions may be more **sustainable**, since they use fewer resources and reduce the amount of waste material that has to be disposed of. It's better to produce less waste in the first place, than to treat waste afterwards to remove even harmless by-products. This reduces the energy used in waste processing.

A system is **sustainable** if it meets present needs without compromising those of future generations. For example, a reaction will be more sustainable if we can reduce the amount of resources used, including energy.

There are other environmental issues to consider in industrial chemistry. For example, how much energy is used in the process, including that needed to separate the products? Can the starting materials be made from renewable resources? Does the reaction use hazardous substances (such as organic solvents) and does it produce by-products that are toxic to human health and the environment?

3. Aniline ($C_6H_5NH_2$) is an important chemical used to make synthetic dyes. Aniline can be prepared in two ways. In the first method, nitrobenzene ($C_6H_5NO_2$) reacts with iron to produce aniline and iron oxide. In the second method, a catalyst speeds up the conversion of nitrobenzene to aniline, and water is the only by-product.
   a) Explain why the atom economy is higher for the second process.
   b) Explain why most aniline is produced using the catalysed method.

4 The following reactions can be used to reduce titanium oxide to titanium metal. (Na = 23, Ti = 48, O = 16, Mg = 24)

TiO_2(s)	+	4Na(s)	→	$2Na_2O$(s)	+	Ti(s)
TiO_2(s)	+	2Mg(s)	→	2MgO(s)	+	Ti(s)
titanium dioxide		sodium or magnesium metal		sodium or magnesium oxide		titanium metal

a) Calculate the atom economy in reducing titanium oxide to titanium using:
   i) sodium metal;
   ii) magnesium metal.

b) Which is the better process?

7.2 How do we measure the rate of a chemical change?

Chemists need to control how fast a chemical change happens when they make new materials.

The **rate of reaction** is the speed at which a reaction takes place. It can be measured as the rate of formation of a product or the rate of removal of a reactant.

How can you measure a rate of reaction?

A 'rate' is a speed. You calculate a rate as a measured quantity divided by the time taken. For example, your pulse rate is the number of heart beats in 1 minute.

Measuring the time taken for a reaction is straightforward. Use a clock, or a calendar for very slow reactions.

Figure 7.5 The reactants are used up very quickly – an explosion is a fast reaction.

Figure 7.6 Sometimes you have to wait a long time to see the products form in a chemical reaction – rusting is a slow reaction.

Figure 7.7 When cooking an egg you can control the outcome carefully by controlling the temperature and how long you let the reaction continue.

The quantity you measure in a chemical change can be:

- mass lost (if the reaction produces a gas);
- time for all the solid reactants to be used up;
- amount of unreacted solid left at any time;
- temperature change produced (if energy is given out);
- change in colour;
- change in transparency of a solution; or
- change in viscosity (thickness of flow).

In the school laboratory, we usually measure rates of reaction by the amount of reactant used up or the amount of product formed, i.e.

$$\text{rate of reaction} = \frac{\text{amount of reactant used up}}{\text{time taken}}$$

or

$$\text{rate of reaction} = \frac{\text{amount of product formed}}{\text{time taken}}$$

For example, when magnesium reacts with acid, hydrogen gas is formed. How could the rate of reaction be measured?

$$\underset{\text{magnesium}}{Mg(s)} + \underset{\text{sulfuric acid}}{H_2SO_4(aq)} \rightarrow \underset{\text{magnesium sulfate}}{MgSO_4(aq)} + \underset{\text{hydrogen}}{H_2(g)}$$

You could measure the volume of gas collected, or if the gas was allowed to escape, you could measure the mass of the flask at regular intervals (see Figures 7.8 and 7.9).

Graphs of rates of reaction follow the same rules as the distance–time graphs you have studied. A steep graph line shows a fast reaction. A flat graph line shows the reaction has stopped.

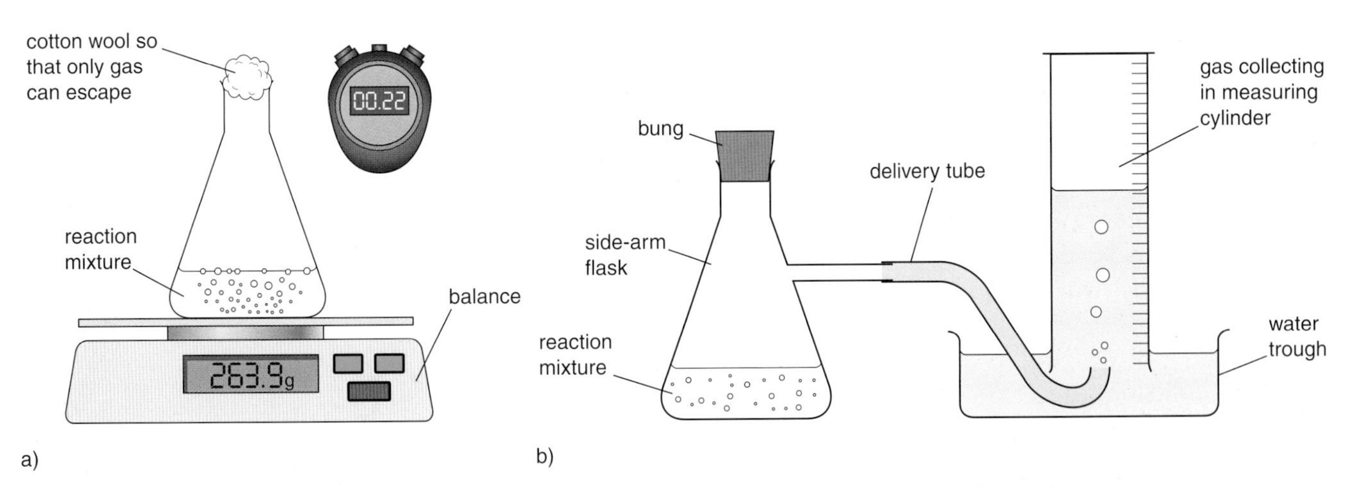

Figure 7.8 The rate of a reaction that produces a gas can be measured by a) recording the mass lost as the gas escapes, or b) collecting the gas produced.

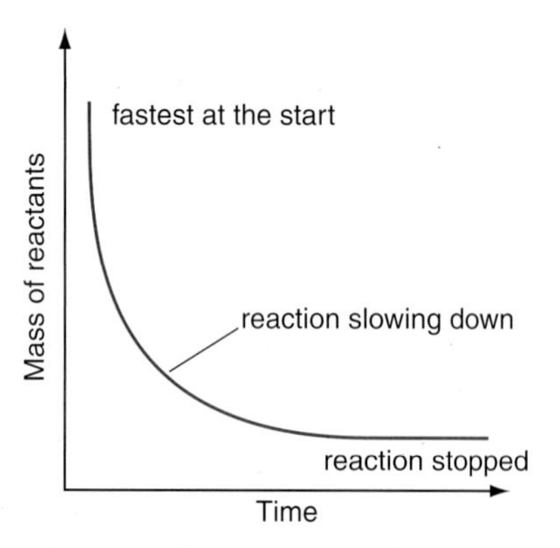

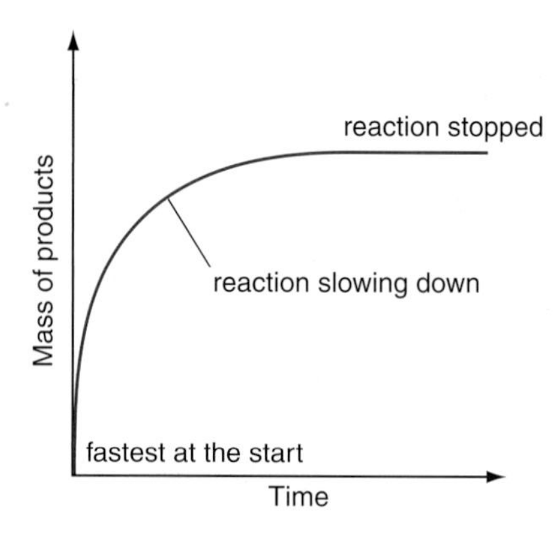

Figure 7.9 a) The rate of a reaction can be measured by recording the change in mass over time. When the graph levels off, one of the reactants has been used up and the reaction stops.

b) This graph shows the volume of gas formed over time. The rate of reaction is highest when the gradient is steepest.

What makes a reaction happen, and how can we control it?

A collision model for chemical reactions

For a chemical change to happen, the groups of atoms reacting have to collide. A model is a simplified picture that scientists construct to fit observed data. Observations of reactions show that the reaction is usually not instantaneous. The collision model says that molecules only react if the collision is hard enough for energy to be transferred to break bonds between atoms. The minimum energy required for a reaction to happen is called the **activation energy**.

The collision model can be summarised as follows:

- Particles can only react if they collide with each other.
- For particles to collide, some of the particles must be in solution, as a liquid or as a gas so they are moving about.
- In solids, only the particles on the surface can react.
- Not all collisions result in a reaction. There has to be enough energy transferred when the particles collide for a reaction to happen. These are the **effective collisions**.

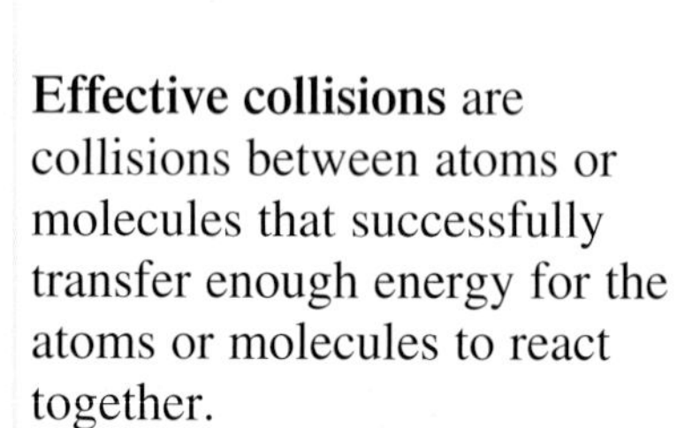
The **activation energy** is the minimum energy needed for molecules to react when they collide.

Effective collisions are collisions between atoms or molecules that successfully transfer enough energy for the atoms or molecules to react together.

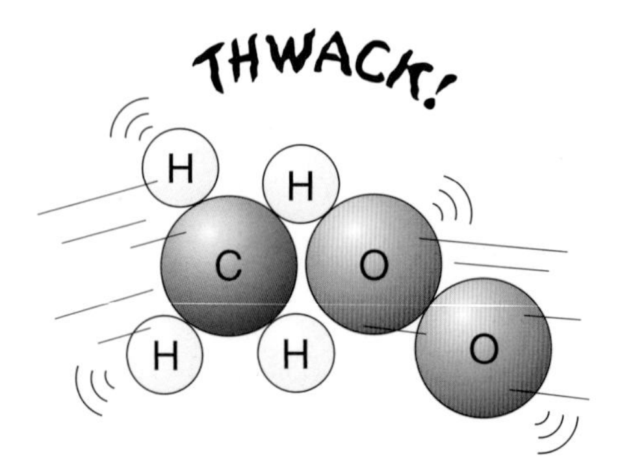

Figure 7.10 When two molecules meet, they only react if the collision transfers a sufficient amount of energy.

Figure 7.11 Reaction time is very important in cooking. There are several factors that change the rate of cooking reactions. The cook has to control the reactions so the parts of the meal are all ready at the same time.

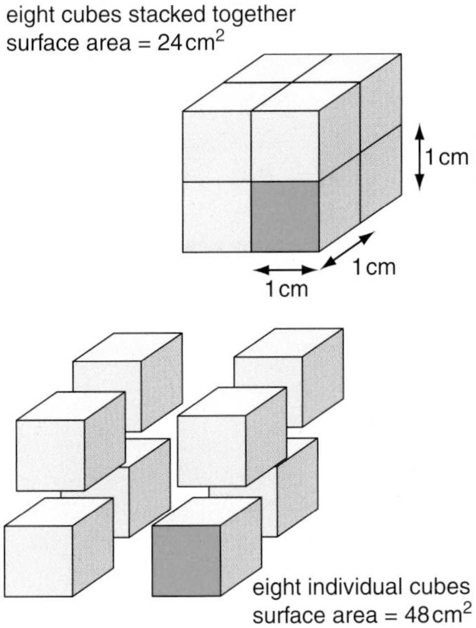

eight cubes stacked together
surface area = 24 cm^2

1 cm

1 cm

1 cm

eight individual cubes
surface area = 48 cm^2

Figure 7.12 Making the pieces of reactant smaller increases the surface area and the number of particles exposed on the surface.

❻ Use the collision model to explain why temperature has a large effect on rates of reaction.

❺ Explain why some collisions between reactants do not result in a chemical change.

Making reactions go faster

There are two ways to make a reaction go faster (increase the rate):

- increase the number of collisions;
- make the collisions have more energy.

You already know that increased temperature causes a faster reaction (increases the rate). Other factors that can change the number of collisions or change the number of effective collisions are concentration, pressure, particle size and presence of a catalyst.

The collision model explains how these factors can increase the number or energy of collisions.

- Increasing the temperature of the reactants makes the particles move about faster so they collide more often. The particles also collide with greater kinetic energy. Both increase the number of effective collisions.
- Increasing the concentration means there are more particles in a certain volume of a solution. The particles are crowded more closely together, so there are more collisions.
- Increasing the pressure of a reacting gas is like increasing the concentration of a solution; there are more collisions.
- If a solid reactant is cut up into more pieces then there are more reactant particles exposed on the surface and able to react.

Predicting changes in rates of reaction

Figures 7.13–7.15 show how the rate of formation of products in a reaction is related to concentration, temperature and the particle size of the reactants.

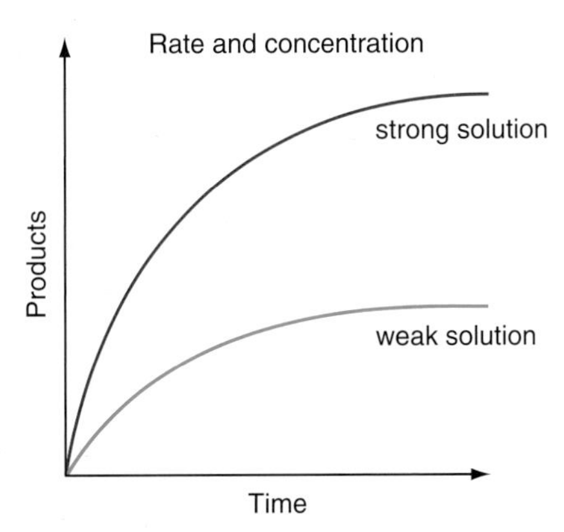

Figure 7.13 If a reactant in a solution is diluted to half the concentration this would halve the number of collisions in a certain time.

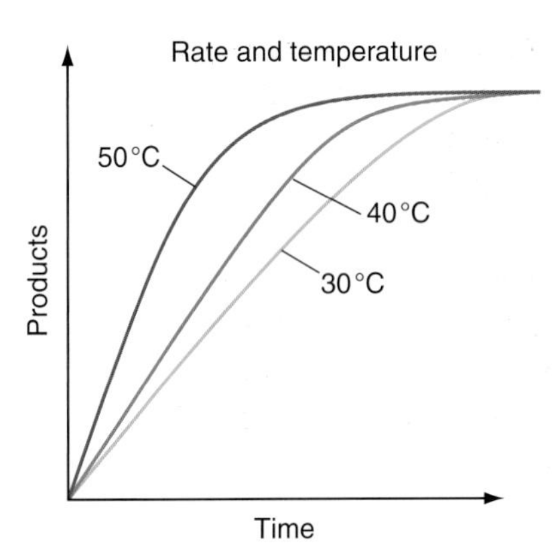

Figure 7.14 Each time the solution becomes 10 °C hotter the rate roughly doubles (this is a very approximate rule of thumb).

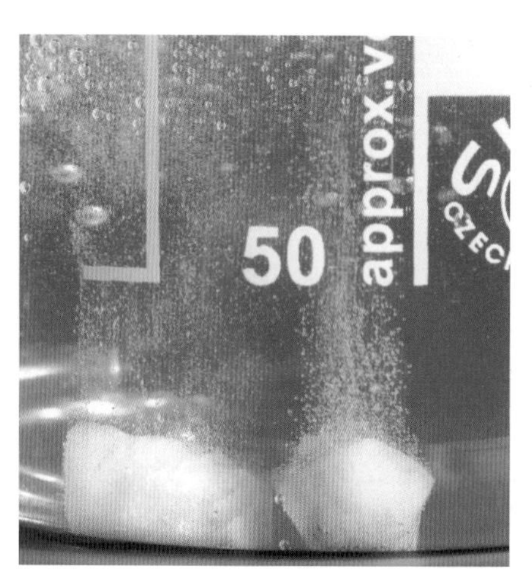

Figure 7.16 In acid at 20 °C not all the particles have the same energy – some are moving faster than others. Only the faster acid particles are moving fast enough when they collide with the marble chip to react and produce carbon dioxide gas. At a higher temperature many more of the acid particles have enough energy.

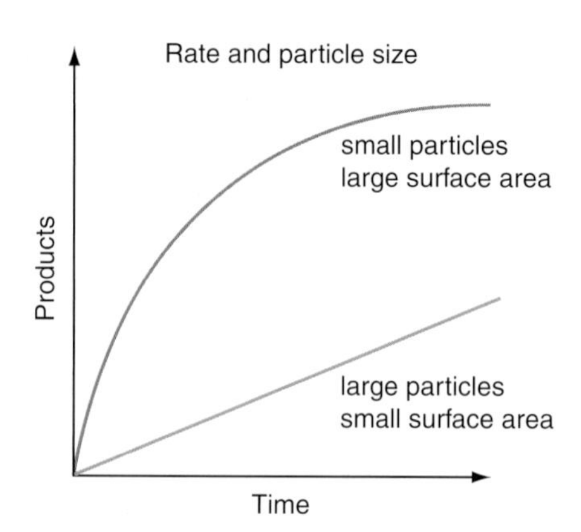

Figure 7.15 If more surface area is exposed then the reaction rate increases.

A thinking model for collision theory – blind mice and rice crispy bar

This is a thinking model for magnesium reacting with hydrochloric acid. The acid particles (the mice) react with the magnesium (rice crispy bar) and gradually the magnesium is used up (the rice crispy bar is eaten).

Figure 7.17 The acid particles (mice) will react with the magnesium (bar) only if they collide directly with it.

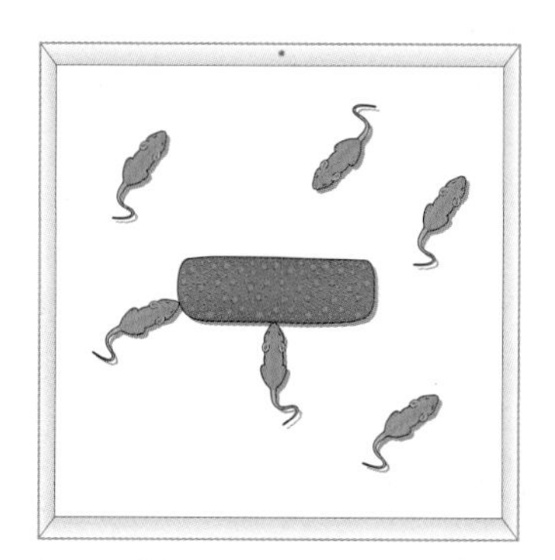

Figure 7.18 Slow-moving mice (low temperature) take all day to find and eat the rice crispy bar (slow reaction).

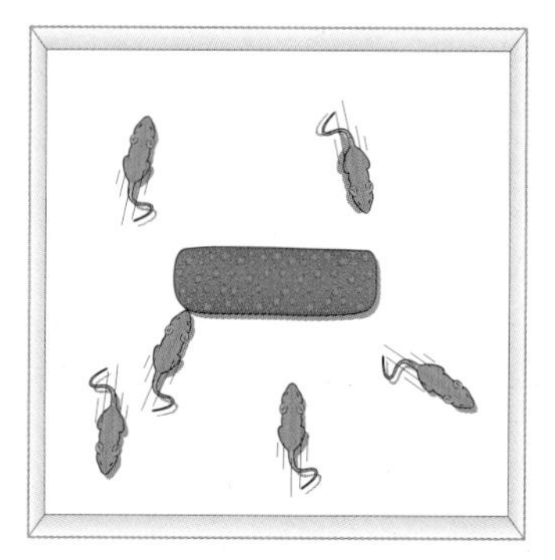

Figure 7.19 Fast scurrying mice (higher temperature) eat the bar up much more quickly (faster reaction).

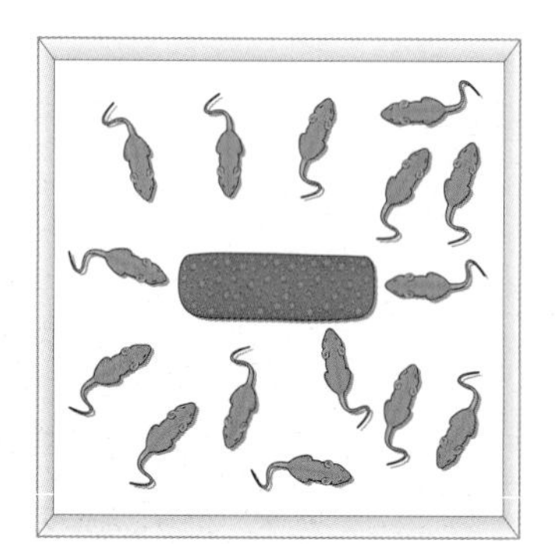

Figure 7.20 Lots more mice in the same space (higher concentration) eat the bar more quickly than a small number of mice.

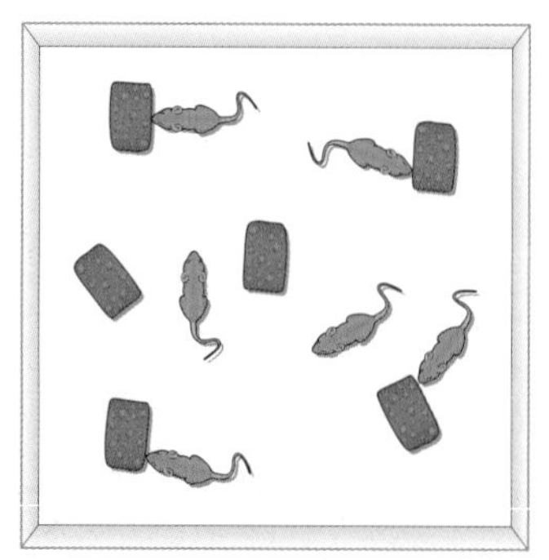

Figure 7.21 Even with a few mice, if the rice crispy bar is cut up into pieces it's more likely the mice will bump into a piece.

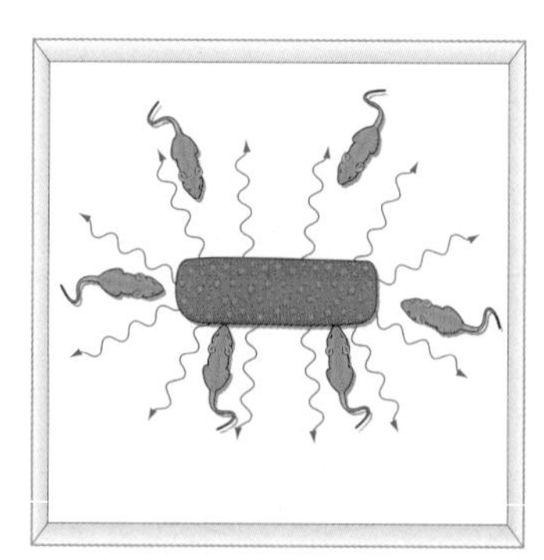

Figure 7.22 Explaining catalysts with this thinking model is difficult. But if a vibration attracted the mice to the rice crispy bar, then it would get eaten much more quickly (faster reaction).

The mice are not clever. For a mouse to take a bite out of the rice crispy bar it has to bump its nose into the bar first (this is like an effective collision needing the correct orientation of particles). If the mouse just brushes past the bar (unsuccessful collision) then no rice crispies get eaten.

7. Explain why chips that are deep-fried in oil cook in 8 minutes, but boiling potatoes takes about 20 minutes.
8. If you cut the potatoes up smaller they cook faster. Crisps only take 20–30 seconds to deep fry. Explain why this is so.
9. Explain why crinkle cut chips cook faster than straight chips.

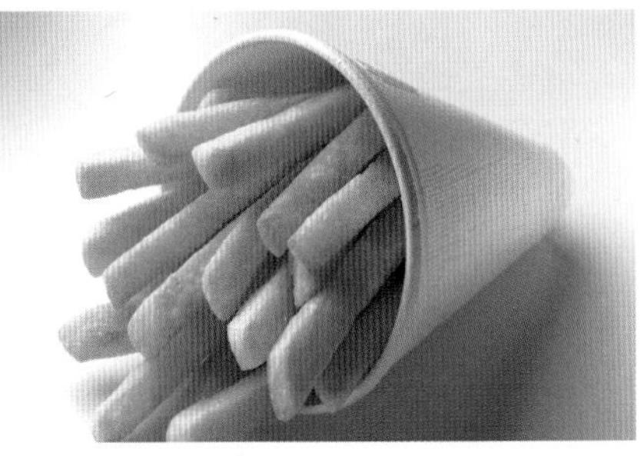

Figure 7.23 Cooking potatoes is a chemical change that sometimes takes ages and sometimes only a minute or two. The cooking process breaks down the strong cell walls, making the food easier to digest. Cooking also releases the starch from inside the cells.

Catalysts

Catalysts speed up or catalyse the rate of chemical reactions but are not used up in the reaction. This is why the formula for the catalyst does not appear in the equation for the reaction.

Enzymes are called biological catalysts. The enzyme peroxidase catalyses the decomposition of hydrogen peroxide to water and oxygen. The reaction is also catalysed by manganese(IV) oxide.

A **catalyst** is a substance that speeds up the rate of a chemical change, without being used up itself. The same amount of catalyst will be present after the reaction has finished.

10. Name two catalysts that make hydrogen peroxide decompose more quickly.
11. Write a word equation for the decomposition of hydrogen peroxide.
12. Suggest how you could measure the rate of decomposition of hydrogen peroxide with a manganese(IV) oxide catalyst.

a)

b)

c)

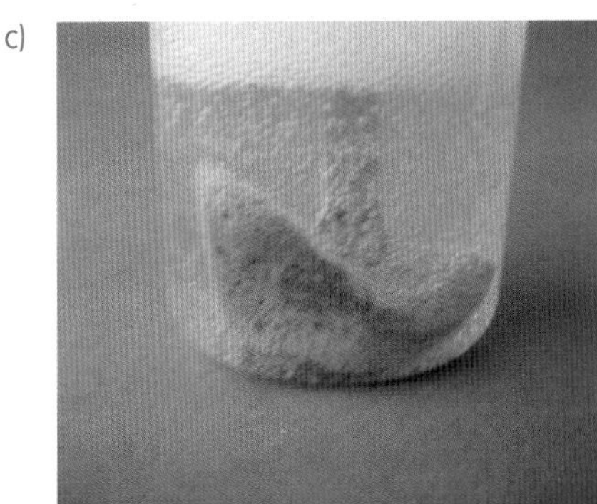

Figure 7.24 a) Hydrogen peroxide (H_2O_2) decomposes very slowly at room temperature to form water and oxygen. Adding a catalyst such as b) manganese(IV) oxide or c) peroxidase (an enzyme found in raw liver) makes the decomposition reaction much faster.

Figure 7.25 Hydrogen peroxide is found in highlight kits and permanent hair dyes. It doesn't just bleach – the oxygen it releases combines with the dye molecules to produce different coloured products.

13. Sketch the graph that you would plot of your results for the decomposition of hydrogen peroxide with manganese (IV) oxide catalyst. Indicate the point at which the reaction stops.

14. When 1.0 moles of hydrogen peroxide (H_2O_2) decomposes, it forms 0.5 moles of oxygen.
    a) What is the mass of 1.0 moles of hydrogen peroxide?
    b) What is the mass of 0.5 moles of oxygen molecules?

15. a) Write a balanced symbol equation for the decomposition of hydrogen peroxide.
    b) What mass of oxygen should be released from 1.0 moles of hydrogen peroxide by the end of the experiment? Suggest one reason why less gas than this might be collected.

Catalysts are used to increase the rate of reaction of some important industrial processes. Without the catalysts, the reactions would happen only very slowly, or would have to be carried out at a higher temperature. One disadvantage of using catalysts is that the catalyst needs to be separated from the product after the reaction. This is easy if the catalyst is a solid and the reactants are gases or solutions, but could be difficult if both the catalyst and reactants are liquids.

Different catalysts speed up different reactions, so each catalyst has only one use. Researchers are continually testing new catalysts for new industrial processes.

Reactants	Catalyst	Product	Notes
Nitrogen and hydrogen	Iron mesh to give large surface area	Ammonia	Haber process, manufacture of ammonia for fertilisers
Sulfur dioxide and oxygen	Granules of vanadium pentoxide	Sulfur trioxide	Contact process, manufacture of sulfuric acid
Hydrogen and unsaturated fats or oils	Fine nickel particles	Margarine	Nickel catalyses breakdown of carbon double bonds to make saturated fat, which is solid.
Carbon monoxide and nitrogen oxides	Platinum and rhodium on a honeycomb metal support to increase surface area	Carbon dioxide and nitrogen (less polluting gases)	Used in catalytic converters in vehicle exhaust systems. Catalyst is ruined if lead in the fuel gets on its surface
Ethene and steam	Phosphoric acid absorbed on to silica (sand)	Ethanol	Very low yield reaction; gases are recycled to avoid waste
Cracking of hydrocarbons	Zeolite crystals	Shorter chain hydrocarbons	The rate of cracking and the formation of diesel and petrol strongly depend on the presence of a catalyst
Ammonia and oxygen	Platinum mesh	Nitrogen pentoxide	Used to make nitric acid for fertilisers and explosives

Table 7.1 Industrial reactions and catalysts

16 Use Table 7.1 to describe three examples that show the importance of a large surface area when using a catalyst.

17 Describe a problem that can reduce the effectiveness of catalysts. Explain how you would overcome this problem.

18 Catalysts are used in industrial processes to reduce costs.
a) Name two industrial processes where catalysts are used.
b) Suggest how catalysts can reduce energy costs in industrial processes, and explain how this might work.
c) Suggest how catalysts can increase the rate of industrial chemical reactions and explain how this might work.

19 Some catalysts are supported on an inert substance so they stay in one place. Other catalysts are mixed with the reacting mixture.
a) Give one example of a catalyst from Table 7.1 that is supported on an inert substance.
b) Margarine has nickel particles in it when it is manufactured. What has to happen to these nickel particles before the margarine can be sold?
c) What does this processing do to the cost of making margarine?

Activity – Is there a cost-effective amount of a catalyst?

Lyubov investigated the question above as part of her GCSE practical work. She chose the decomposition of hydrogen peroxide (H_2O_2) using different masses of manganese(IV) oxide catalyst.

$$2H_2O_2(aq) \xrightarrow{MnO_2} 2H_2O(l) + O_2(g)$$

She used 60 cm^3 of 2 volume hydrogen peroxide solution in all her tests. All the tests were carried out at room temperature and atmospheric pressure. She used the same apparatus each time.

Lyubov eventually produced the results in Table 7.2.

She needs to analyse the data to reach a conclusion.

Hydrogen peroxide solutions do not have their concentration shown on them as molarity (molar concentration). They have an indication of how much oxygen you can get from a certain volume of the solution.

For example 100 cm^3 of '20 volume' hydrogen peroxide produces 100 cm^3 $\times$ 20 = 2000 cm^3 of oxygen gas.

Grams of catalyst used	0 g	5 g	10 g	15 g	20 g	25 g
Time in seconds	Total volume of oxygen gas released in cm^3					
0	0	0	0	0	0	0
10	0	5	12	20	22	24
20	0	10	20	37	40	43
30	0	15	30	55	57	60
40	1	20	39	72	74	76
50	1	25	47	86	87	88
70	1	35	65	110	110	110
100	2	48	90	115	115	115
150	2	70	118	120	120	120
200	2	92	120	120	120	120

Table 7.2

1. Materials such as platinum are used as catalysts in a number of chemical processes. Explain why it is important to be cost-effective in the use of catalysts.
2. Draw the apparatus that Lyubov could have used to carry out this investigation.
3. Copy and complete Table 7.3 to summarise how Lyubov dealt with the possible factors that could affect this investigation.
4. a) What is the maximum volume of oxygen produced in this experiment?
   b) What is the longest time measured?
   c) Use these answers to produce an x-axis scale (time) and a y-axis scale (oxygen gas given off) for a set of graphs.
5. Plot a line graph for each set of results. Plot them all on the same axes on one piece of graph paper.
6. Write a conclusion answering the question in the title. Suggest a quantitative answer for this set of reaction conditions.

Factor	Type of variable	Range chosen	Controlled by Lyubov
Type of catalyst used		Manganese(IV) oxide	
Temperature and pressure			
Concentration of peroxide solution		2.0 volume solution	yes
Amount of catalyst			
Volume of oxygen given off			

Table 7.3

7.4 Calculating concentrations of reactants in solution

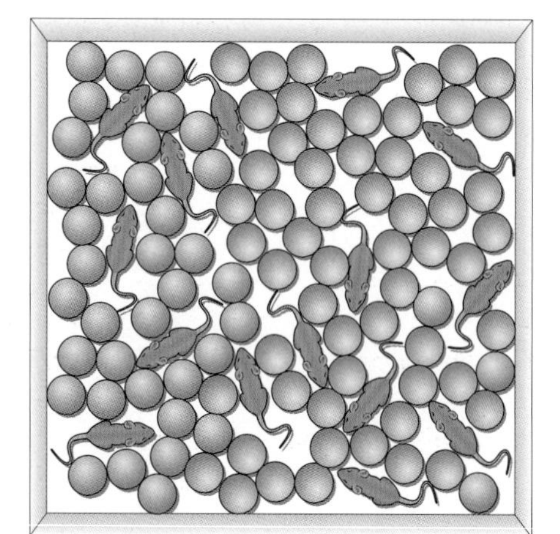

Figure 7.26 The mice represent the acid particles. The balls represent the water molecules. In a very concentrated solution, there would be many more mice (acid particles) in the same volume of solution.

In a well-mixed solution the solute particles are spread out evenly through the whole mixture. The solvent chemists use most often is water. We refer to solutions in water as aqueous solutions.

Chemicals are easy to handle in solutions. They are easy to measure, pour, heat and mix.

Using the mice model

As you can be see from Figures 7.17–7.20, the **concentration** of a solution indicates how many reactant particles there are in each unit volume of solution. Figure 7.26 shows the 'mice thinking model' for solutions.

Concentrations in moles per cubic decimetre

The unit of solute concentration in solutions is moles per cubic decimetre (mol/dm^3). Remember that 1 mole (the relative formula mass in grams) of a compound always contains 6×10^{23} particles (see Section 5.2).

The **concentration** of a solution or a gas is the amount of solute or gas in each unit volume. Gas and solute concentrations are expressed in moles per cubic decimetre (mol/dm^3).

When making a solution, always dissolve the solute in less solvent than is needed, then make up to the correct total volume.

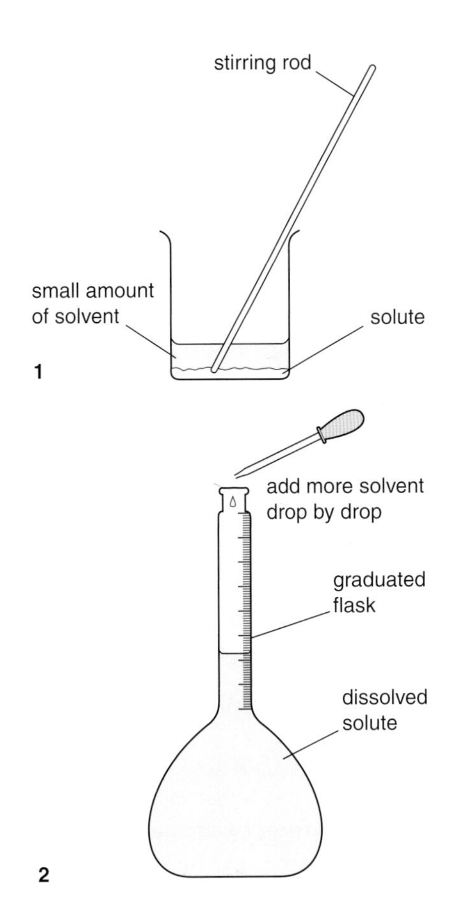

Figure 7.27 Making a solution with an accurately known concentration

A decimetre (dm) is 10 cm, so 1 dm^3 is 1000 cm^3. If you dissolve 1 mole of a solute and make the solution up to 1000 cm^3 (1 dm^3) in volume, then you will have 1000 cm^3 (1 dm^3) of 1.0 mol/dm^3 solution. Sometimes this is called a 1.0 molar solution.

Molarity and calculations

Example

100 cm^3 of waste water from a bottling factory was analysed and found to have 0.4 g of sodium hydroxide in it. Sodium hydroxide is an alkali used in cleaning solutions. The alkali in the waste water needs to be neutralised with hydrochloric acid.

NaOH(aq)	+	HCl(aq)	→	NaCl(aq)	+	$H_2O(l)$
sodium hydroxide		hydrochloric acid		sodium chloride		water

From the equation, 1 mole of hydrochloric acid neutralises 1 mole of sodium hydroxide.

relative formula mass of sodium hydroxide (NaOH)
$= 23 + 16 + 1 = 40$

so there is 40 g of sodium hydroxide in 1 mole.

Therefore in a 100 cm^3 sample of waste water there is:

$$\frac{0.4\text{ g}}{40\text{ g/mol}} = 0.01 \text{ moles of sodium hydroxide}$$

A chemist at the bottling factory therefore needs 0.01 moles of hydrochloric acid to neutralise this sample.

You could use:

- a small amount of concentrated acid; or
- a large amount of dilute acid.

It is important to use exactly the right amount of hydrochloric acid to neutralise the alkaline sodium hydroxide. So it is very important to be able to measure the volumes and concentrations of solutions accurately.

When the waste water has been treated it is neither acid nor alkali. The neutral, slightly salty water can be released into the environment if there are no other pollutants in it.

Equal volumes of solutions of the same molar concentration contain the same number of moles of solute, so they contain the same number of particles. So 25 cm^3 of 0.5 mol/dm^3 glucose ($C_6H_{12}O_6$) solution contains the same number of molecules as 25 cm^3 of 0.5 mol/dm^3 ammonia (NH_3) solution. Because each glucose molecule is big and heavy and each ammonia molecule is small and light, the glucose solution contains a greater mass of solute, but the same number of particles.

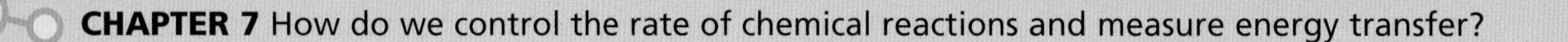

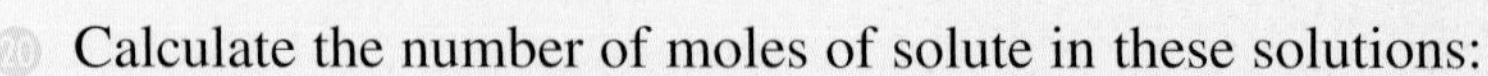

20 Calculate the number of moles of solute in these solutions:

a) 1000 cm^3 of 0.5 mol/dm^3 potassium hydroxide solution;
b) 25 cm^3 of 1.0 mol/dm^3 potassium hydroxide solution;
c) 100 cm^3 of 0.1 mol/dm^3 sodium chloride solution;
d) 10 cm^3 of 6 mol/dm^3 copper sulfate solution;
e) 3 dm^3 of 0.01 mol/dm^3 copper sulfate solution.

21 Sulfuric acid reacts with solid magnesium oxide and is neutralised. Excess magnesium oxide remains undissolved and sinks to the bottom.

MgO	+	H_2SO_4	→	$MgSO_4$	+	H_2O
magnesium oxide		sulfuric acid		magnesium sulfate		water

How much magnesium oxide would exactly neutralise 250 cm^3 of 0.5 mol/dm^3 sulfuric acid?

22 Potassium hydroxide reacts with sulfuric acid.

KOH	+	H_2SO_4	→	K_2SO_4	+	H_2O
potassium hydroxide		sulfuric acid		potassium sulfate		water

How much 1.0 mol/dm^3 potassium hydroxide solution is required to exactly neutralise 250 cm^3 of 0.5 mol/dm^3 sulfuric acid?

23 a) A detergent company releases quantities of sulfuric acid in its waste water. Explain how it could treat this water to make it environmentally safe.
b) Explain why magnesium oxide is a better chemical to use for this purpose than potassium hydroxide solution.

The **molar volume of a gas** is the volume of 1 mole of the gas at a fixed temperature and pressure. At normal temperature and pressure this volume is 24.0 dm^3 (24 000 cm^3).

Particles, moles and gases

Liquids are easy to handle and the volumes of aqueous solutions do not change on heating. Gases are another matter. They are easy to compress and expand a lot on heating. One fact makes handling gases easy: No matter what gas you have, one mole of the gas takes up the same volume at normal temperature and pressure. This is the **molar volume of a gas.**

For example, in the combustion of methane:

CH_4	+	$2O_2$	→	CO_2	+	$2H_2O$
methane		oxygen		carbon dioxide		water vapour

One molecule of methane reacts completely with two molecules of oxygen. Since the same volume of gases at the same temperature and pressure contain the same number of molecules, to get twice as many oxygen molecules the volume of oxygen that reacts must be twice the volume of methane. Therefore 100 cm^3 of methane reacts completely with 200 cm^3 of oxygen.

Figure 7.28 A gas phase reaction

Convention

'Normal temperature and pressure' for measuring gas volumes is 20 °C and one atmosphere. A standard is needed because gases expand when they are heated and when pressure is reduced.

24 a) Draw a particle picture to represent six methane molecules in a box.
b) Draw a particle picture to represent the number of oxygen molecules needed to react with this methane. Draw a box twice the size and put the right number of oxygen molecules in it.
c) What do you notice about the spacing between the molecules in the two boxes you have drawn?

7.5 Energy change in reactions

Many chemical changes transfer energy to the surroundings. Usually the energy heats up the surroundings. Sometimes the energy is transferred as light (glow sticks) or sound (explosions). These energy changes are called **exothermic reactions**. Combustion is an example of an exothermic reaction between a fuel and oxygen in the air.

Here is an example of an exothermic change you can try in the laboratory. During neutralisation, energy is given out when the new bonds are formed and this causes the solution to be heated.

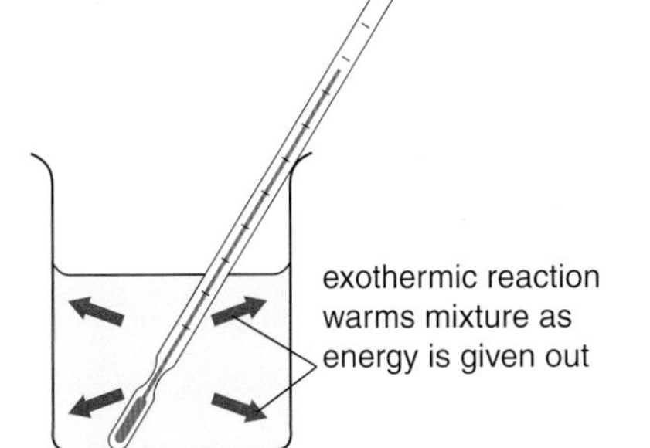

Figure 7.29 An exothermic reaction transfers energy to the surroundings.

⚠ Wear eye protection.

Mix 25 cm^3 of 2.0 mol/dm^3 hydrochloric acid (irritant) with 25 cm^3 of 2.0 mol/dm^3 sodium hydroxide (corrosive) in a polystyrene cup supported in a 250 ml beaker.

Temperature before mixing: 20 °C Temperature after mixing: 29 °C

Not all chemical reactions cause a temperature increase. Some chemical changes cool down the reactants and their surroundings. To do this the reaction takes in energy from the surroundings and transfers it to the reactants. This is called an **endothermic reaction**. Thermal decomposition is an endothermic reaction, as energy must be transferred to the reactant before it breaks down.

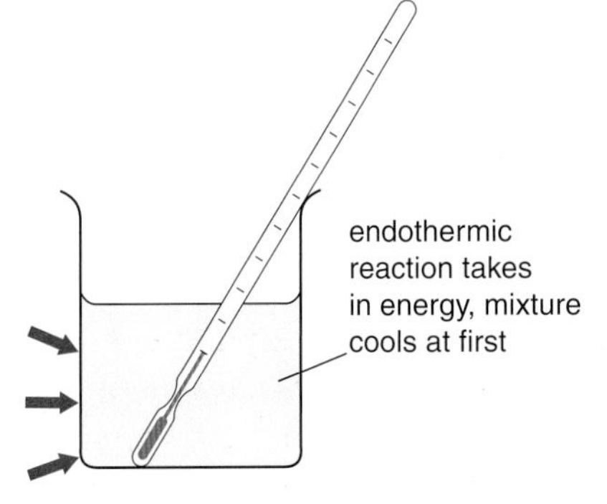

Figure 7.30 An endothermic reaction absorbs energy from the surroundings.

An **exothermic reaction** is a chemical change where energy is transferred to the surroundings, often causing heating.

An **endothermic reaction** is a chemical change where energy has to be absorbed from the surroundings, often as a heat transfer that causes cooling of the surroundings.

Figure 7.31 A self-warming can of coffee heats the coffee with an exothermic reaction. Press the bulb on the bottom of the can and wait 2 minutes. Water reacts with anhydrous calcium oxide in a sealed compartment inside the can. The reaction gives out energy. This energy is transferred to the coffee, causing it to heat up.

Here is an example of an endothermic change you can try in the laboratory. When sodium thiocyanate and barium hydroxide dissolve in water, this causes cooling.

⚠ Wear eye protection.

Mix 4 g sodium thiocyanate and 8 g barium hydroxide (corrosive) in 100 cm^3 of water.

Temperature before mixing: 20 °C Temperature after mixing: 6 °C

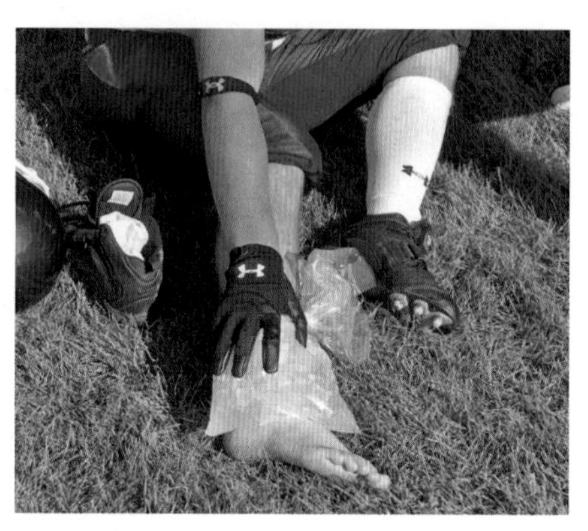

Figure 7.32 Twist a cold pack and it gets cold enough to numb the pain of a sports injury. Injuries such as sprains or bruises hurt because of the swelling; cold helps to reduce swelling. The chemicals combine to take in energy, so the pack gets cold.

Figure 7.33 Glow sticks contain luminol, a chemical that gives out light energy when it reacts with another liquid in the glow stick.

25 When would 'self-heating coffee' be useful?

26 The can is heated by the reaction between calcium oxide and water. The only product of this reaction is calcium hydroxide. Write a word equation for the reaction.

27 What type of injuries are treated with cold packs?

28 Complete Table 7.4.

Chemical change	Endothermic or exothermic?
Acid–alkali reaction	
Methane burning in a Bunsen burner	
Sodium reacting with water	
Calcium carbonate decomposing to calcium oxide and carbon dioxide	
Making aluminium metal	
Luminol reacting	

Table 7.4

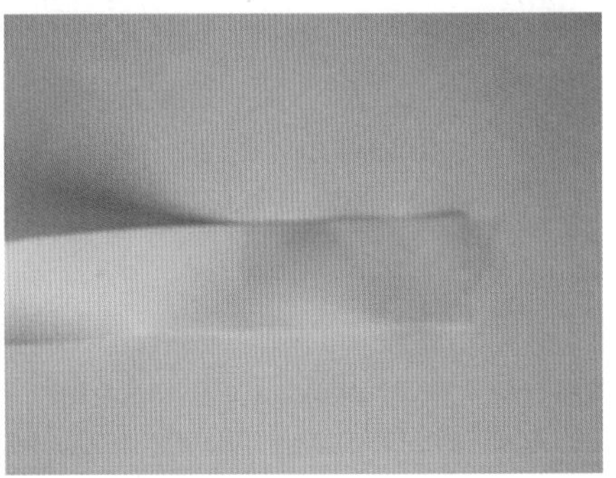

Figure 7.34 Testing for water. Filter paper is soaked in copper sulfate solution and dried in a hot oven. When the paper comes in contact with water, the anhydrous copper sulfate turns blue.

29 Describe how you would test a few drops of a mystery colourless liquid, to see if it is water.

Energy changes in a reaction that goes both ways

Figure 7.35 Hydrated (blue) copper sulfate crystals (left) have a formula of $CuSO_4.5H_2O$. They can be turned into anhydrous copper sulfate (right) on heating. Anhydrous copper sulfate ($CuSO_4$) has a slight tinge of colour due to water absorbed from the air.

When blue crystals of hydrated copper sulfate are heated, water vapour is given off and anhydrous copper sulfate is formed. The reaction is reversible, depending on how much water is present. The blue copper sulfate reforms when water is added to anhydrous copper sulfate. This colour change can be used as a test for water (Figure 7.34).

$$CuSO_4.5H_2O(s) \underset{\text{exothermic}}{\overset{\text{endothermic}}{\rightleftharpoons}} CuSO_4(s) + 5H_2O(g)$$

hydrated copper sulfate (blue) — anhydrous copper sulfate (white) — water

The thermal decomposition of hydrated copper sulfate is endothermic as heat is absorbed. The reverse reaction is exothermic. When 5 g of anhydrous copper sulfate are dissolved in 50 g of water the temperature rises by 10 °C.

In a reversible reaction, if the forward reaction is endothermic, then the reverse reaction is exothermic. Exactly the same amount of energy is transferred in each case.

7.6 Equilibrium in reversible reactions

When a reversible reaction occurs in a closed system, the reactants and products cannot escape. The amount of reactants and products reaches a balance, or **equilibrium**, in which the amounts of reactants and products do not change. At equilibrium, the rates of the forward and reverse reactions are the same, so as much product is being made in the forward direction as there are reactants being used up in the reverse direction.

Equilibrium is a state of balance in a reversible reaction. The amounts of each chemical do not change, because the reactions occur at exactly the same rate in each direction. **Dynamic equilibrium** means that the reactions are still continuing.

The reactions don't stop – the particles keep swapping from reactants to products and back, but these effects balance out at equilibrium. This is called a **dynamic equilibrium**.

The relative amounts of each reacting substance at equilibrium depend on the reaction conditions, such as temperature and pressure or the concentration of reactants. If you change the reaction conditions then you will change the equilibrium mixture. This affects the yield of products. If the forward reaction is faster this will result in more products being formed. A faster reverse reaction turns more of the products back into the original reactants.

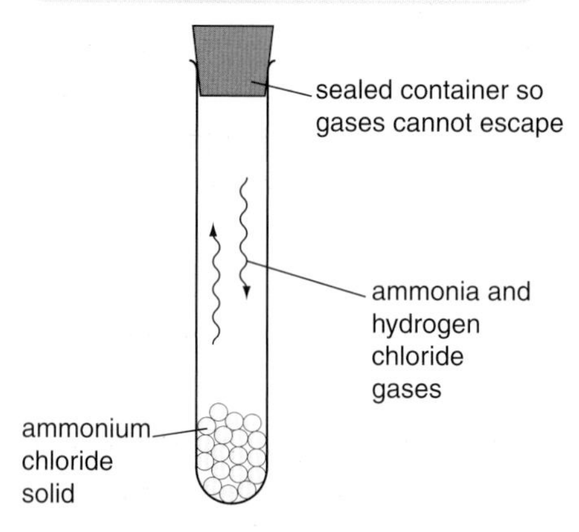

Figure 7.36 A reversible reaction in a closed system soon reaches a fixed amount of reactants and products at equilibrium.

Example

In a sealed container of solid ammonium chloride, some of the space above the solid contains molecules of ammonia and hydrogen chloride gases.

$$NH_3(g) + HCl(g) \rightleftharpoons NH_4Cl(s)$$

At room temperature, the equilibrium mixture is mainly ammonium chloride. Heating the container increases the rate of decomposition of ammonium chloride, so there will be a greater concentration of ammonia and hydrogen chloride.

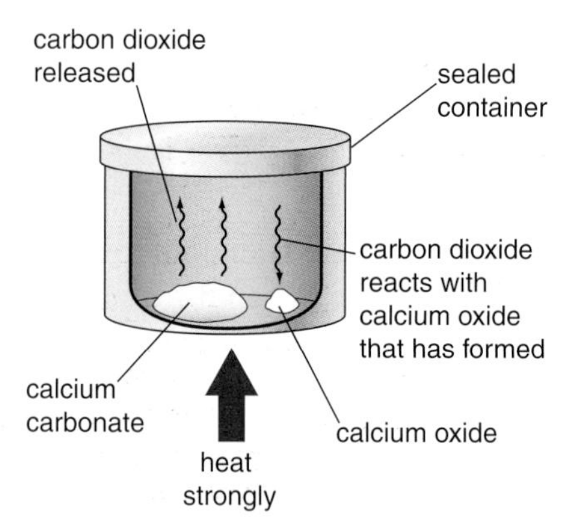

Figure 7.37 When calcium carbonate is heated strongly in a closed system, both decomposition and formation of calcium carbonate can occur.

Temperature in °C	Pressure of CO_2 in Pa × 10^2
700	19
800	250
850	474
900	790

Table 7.5

The decomposition of calcium carbonate is endothermic.

$$CaCO_3 \rightleftharpoons CaO + CO_2$$

The amount of carbon dioxide in the mixture can be shown by the pressure exerted by the carbon dioxide inside the container (Figure 7.37). The greater the pressure, the more molecules of carbon dioxide are present (Table 7.5).

30. Draw simple particle pictures of the equilibrium mixture at 700 °C and 900 °C.
31. What is the effect of increasing the temperature on the equilibrium mixture?
32. If the closed system were heated to 900 °C and allowed to cool, describe what would happen inside the container as the mixture cooled.
33. If the reactants were heated to 900 °C, with the bung removed so that gases could escape, describe what the difference would be as the mixture cooled.
34. When calcium oxide and carbon dioxide react to form calcium carbonate, is this reaction endothermic or exothermic?

7.7 The Haber process

Compounds made from ammonia (ammonium compounds) are used to make fertilisers. In the first part of the twentieth century, population growth and the demand for more fertilisers meant the world demand for ammonia exceeded the existing supply. The largest natural source of ammonia was a huge guano (bird dung) deposit in Chile. This 380 km long deposit was fast disappearing.

Just before the First World War, Fritz Haber invented a process to manufacture ammonia without using guano. Ammonia is also used to make the explosives for shells and bullets. Haber's invention allowed Germany to fight the First World War even though it was cut off from Chile. Haber was also involved in directing the first poison gas (chlorine) attacks against Allied troops in the First World War, resulting in 10 000 deaths.

Although ammonia and its exploitation had the ability both to sustain life and destroy it, Haber did not have either reason in mind when performing his research.

Figure 7.38 Fritz Haber is known for making ammonia from its elements. He was born in Prussia in 1868. Haber won a Nobel prize for his work.

In 1909 Karl Bosch developed a large-scale catalytic process for making ammonia from the process Haber had invented. The reactants (the elements, hydrogen and nitrogen gas) are easy to obtain and inexpensive:

- hydrogen from natural gas (methane, CH_4) or from cracking other hydrocarbons;
- nitrogen obtained from the air.

In the 1930s when Jewish academics were persecuted, Haber realised his strong patriotism could not outweigh his Jewish heritage. Because of his service to his country in the First World War, Haber's life was not actually threatened, but he realised that it was time to emigrate. He was offered a position at Cambridge University and he left Germany in 1933.

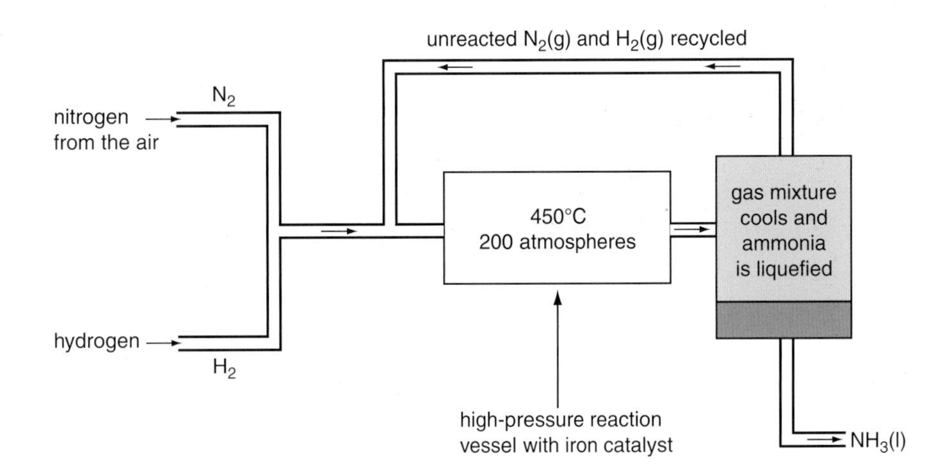

Figure 7.39 Flow diagram for the Haber process. Not all the hydrogen and nitrogen react. The remaining gases are recycled.

An iron catalyst is used to speed up the reaction, so that ammonia can be produced more quickly.

$$\underset{\text{nitrogen}}{N_2(g)} + \underset{\text{hydrogen}}{3H_2(g)} \rightleftharpoons \underset{\text{ammonia}}{2NH_3(g)}$$

The forward reaction is exothermic and the reverse reaction is endothermic.

This is an important industrial process, and chemists want the yield to be reasonable.

Equilibrium conditions – maximising the yield

As mentioned in Section 7.6, different reaction conditions shift the equilibrium mixture (the proportions of reactants and products) one way or another. A rule to predict the change in an equilibrium mixture is **Le Chatelier's principle**.

For a reversible reaction in equilibrium conditions, **Le Chatelier's principle** states that: 'The equilibrium position will respond to oppose a change in the reaction conditions.'

For the Haber process, this means that if you increase the pressure, the equilibrium mixture changes to oppose the change, i.e. to reduce the pressure.

In the Haber process when you turn hydrogen and nitrogen molecules into ammonia molecules, four molecules (one of nitrogen and three of hydrogen) are replaced by two molecules of ammonia. When 2 moles of gas are produced from 4 moles of gas the pressure reduces. This helps to reduce the pressure if the mixture has been exposed to increased pressure, and more ammonia is produced.

- If there are more molecules on the reactants side of the balanced symbol equation, an increase in pressure produces more product.
- If there are more molecules on the product side of the balanced symbol equation, an increase in pressure produces less product.

Pressures of 200 times normal atmospheric pressure are used in the Haber process.

Changing the temperature conditions also changes the equilibrium. The forward reaction in the Haber process is exothermic. If chemists increase the temperature, Le Chatelier's principle predicts that the reaction will move in the endothermic direction, so the yield of the forward reaction decreases. Less ammonia is produced.

- If the temperature is raised, the yield from the exothermic reaction decreases and the yield from the endothermic reaction increases.
- If the temperature is lowered, the yield from the endothermic reaction decreases and the yield from the exothermic reaction increases.

So, cooler temperatures would help to increase the yield in the Haber process.

Why are optimum conditions not used?

A graph of possible yield at different temperatures and pressures is shown in Figure 7.40. At very high pressures and lower temperatures the yield is higher. However, a temperature of 450 °C and a pressure of 200 atmospheres is used, giving a yield of about 30%. Even though these conditions do not give the highest yield, they ensure that less extra energy is used.

- Very high pressure reactors are expensive and require lots of energy input, which is a wasteful use of energy.
- A low temperature leads to a very slow forward reaction that could take days. This is wasteful as it is expensive to run the plant.

There is also a compromise between high yield and high rate of reaction.

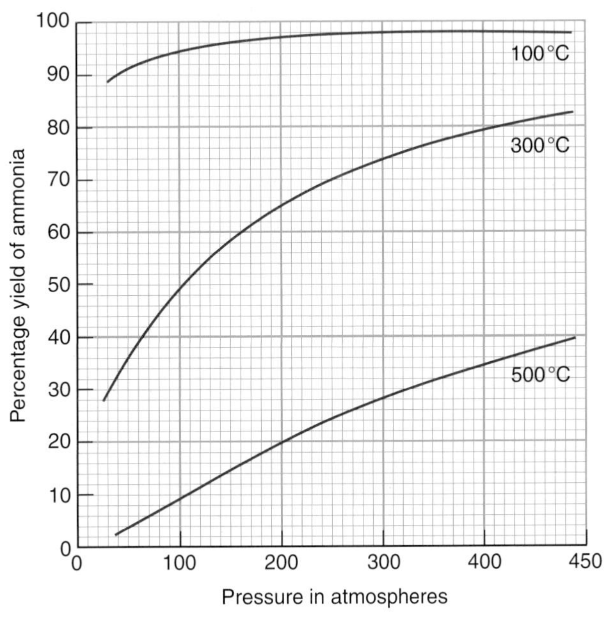

Figure 7.40 The percentage yield of ammonia at equilibrium for the Haber process at different temperatures and pressures

35 Look at Figure 7.40. Use this information to explain the effect of increasing the temperature on the yield of ammonia from the Haber process.

36 Explain the effect of increasing the pressure on the yield in the Haber process:
a) at 500 °C;
b) at 100 °C.

37 a) Why is the Haber process not carried out at 100 °C?
b) Why is the Haber process not carried out at the highest possible pressure?

Industrial chemistry and sustainable development

Scientific principles, such as that of Le Chatelier and the atom economy principles, are useful in determining the most efficient conditions. The atom economy of the Haber process is 100% since there are no by-products (see Section 7.1). In sustainable development terms, atom economy is more important than the yield since all parts of the starting material are used and there are no waste products.

It is important for sustainable development to run industrial processes at conditions of low energy waste. Running processes sustainably may also require:

- the use of catalysts to reduce energy costs by carrying out reactions at lower temperatures;
- new technology to recycle unreacted materials that could be reused;
- new technology to turn unwanted by-products (waste) into useful products;
- new reactions with higher atom economy to reduce volumes of waste and conserve resources.

The Haber process is run as a continuous reaction. This has benefits:

- The unused reactants can constantly be recycled. This is an efficient use of materials.
- The ammonia can be extracted easily from the mixture. It liquefies more easily than the other gases (nitrogen and hydrogen). This is an efficient use of energy and materials.

However, even with the use of a catalyst the Haber process has an enormous energy consumption, due to the high temperature of the reaction. Energy is also needed to extract nitrogen from the air.

38. The Haber process has a continuous flow of reactants through the reaction chamber rather than batches of reactants. Explain:
    a) why this has an advantage regarding the amount of reactants used;
    b) how the product (ammonia) is separated from the flow of reactants.

39. Describe the benefits for society from the Haber process. Explain why these benefits must be balanced against the need for a sustainable use of resources.

Summary

✓ Chemical reactions do not always turn all the reactants into useful products. Some of the material is wasted as by-products.

✓ In **reversible** reactions the products can react to form the original reactants, depending on the conditions.

✓ The amount of a product obtained is known as the **yield**. When this is compared with the maximum theoretical amount as a percentage, it is called the **percentage yield**.

$$\text{percentage yield} = \frac{\text{mass of product obtained}}{\text{theoretical mass possible}} \times 100\%$$

✓ **Atom economy** = the mass of atoms in the desired product divided by the mass of atoms in all the reactants, expressed as a percentage.

✓ A high atom economy helps the **sustainability** of chemical processes.

✓ The rate of a reaction can be measured as the amount of product formed divided by the time taken, or the amount of reactant used up divided by the time taken.

✓ The **concentration** of a solute in a solution is the quantity of solute in each unit volume of solution.

✓ Concentrations of solutions are given in moles per cubic decimetre (**mol/dm^3**).

✓ One mole of any gas occupies the same volume – 24 dm^3 – at normal temperature and pressure.

✓ Reactions occur when reacting particles collide with energy equal to or greater than the **activation energy**.

✓ The frequency and energy of collisions can be used to explain how factors affect reaction rate.

✓ **Catalysts** speed up the rate of a reaction but are not used up. Each reaction needs a specific catalyst.

✓ **Exothermic** reactions give out energy; **endothermic** reactions take in energy.

✓ In a closed system, a reversible reaction reaches a stable **equilibrium** between reactants and products and the amounts of each chemical do not change. In a **dynamic equilibrium** both the forward and backward reactions are still continuing at the same rate.

✓ **Le Chatelier's principle** can be used to make predictions about the effect of changing conditions on equilibrium.
- Low temperature favours (increases the yield of) exothermic reactions; high temperature favours endothermic reactions.
- High pressures favour gas phase reactions where the number of molecules reduces.
- The effect of these factors on yield, together with reaction rates, are considered when planning the conditions for industrial processes.

✓ The **Haber process** turns nitrogen and hydrogen into ammonia using an iron catalyst, high pressure and moderate temperatures.

✓ Industrial processes should choose sustainable methods that minimise energy requirements and material waste. Improved processes could also recycle unreacted materials, use renewable starting materials and discover ways of converting waste to useful by-products.

EXAMQUESTIONS

1 Two clear solutions, X and Y, are mixed in a conical flask. They gradually go cloudy until a cross under the conical flask can no longer be seen. The time for the cross to disappear is measured for different temperatures of the reacting solutions.

Temperature in °C	22	27	35	43	50	58
Time taken in seconds	327	260	198	194	118	100

Table 7.6

a) Draw a line graph of the data. Use temperature as the x-axis (from 20 °C to 60 °C) and time as the y-axis (from 0 to 360 seconds). *(6 marks)*

b) Identify the one anomalous result. *(1 mark)*

c) Use your graph to find how long it takes for the cross to become obscured at 30 °C. *(2 marks)*

d) Use collision theory to explain the shape of the graph. *(5 marks)*

2 In the Contact process, sulfur dioxide reacts with oxygen to form sulfur trioxide in an exothermic reaction. The sulfur trioxide is then reacted with water to make sulfuric acid. Vanadium pentoxide is used as the catalyst.

$$2SO_2(g) + O_2(g) \rightleftharpoons 2SO_3(g)$$

sulfur dioxide + oxygen ⇌ sulfur trioxide

a) What is the purpose of the catalyst? *(2 marks)*

b) Why is no extra catalyst added in the flow diagram in Figure 7.41? *(2 marks)*

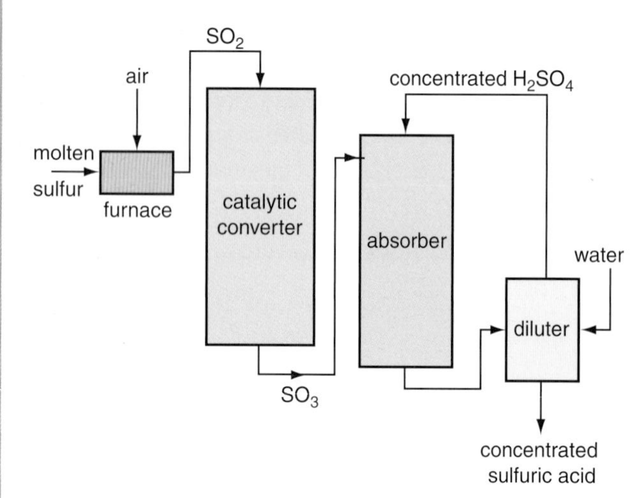

Figure 7.41 Flow diagram for making sulfuric acid

c) Explain how the best use is made of energy in the process. *(2 marks)*

d) Suggest why it is important to make the best use of energy and materials in an industrial process. *(2 marks)*

3 For the process described in question **2**, use Le Chatelier's principle to explain:

a) the effect of pressure on the equilibrium mixture; *(2 marks)*

b) the effect of temperature on the equilibrium mixture; *(2 marks)*

c) the effect of removing sulfur trioxide from the equilibrium mixture. *(2 marks)*

4 Ethanoic acid is used to make plastics and the synthetic fibre rayon. Before 1970, most ethanoic acid was made by heating butane in air with a catalyst:

$$2C_4H_{10} + 5O_2 \rightarrow 4CH_3COOH + 2H_2O$$

butane + oxygen → ethanoic acid + water

Several side products may also form, including butanone and formic acid. Formic acid is toxic.

a) Suggest one reason why the yield of ethanoic acid is never 100%. *(1 mark)*

Most ethanoic acid is now made by reacting methanol with carbon monoxide in the presence of a catalyst.

$$CH_3OH + CO \rightleftharpoons CH_3COOH$$

methanol + carbon monoxide ⇌ ethanoic acid

b) Calculate the atom economy of this reaction. *(3 marks)*

c) Explain why you think this would be greater than the atom economy of the first reaction. *(2 marks)*

Methanol and butane are both obtained from natural gas, but methanol is cheaper to obtain than butane. Methanol can also be obtained from wood or sewage.

d) Explain why making ethanoic acid from methanol is more sustainable than making ethanoic acid from butane. *(3 marks)*

Chapter 8
How can we use ions in solutions?

At the end of this chapter you should:

- ✓ know that when an ionic substance is melted or dissolved in water, the ions are free to move about within the liquid or solution;
- ✓ know that salt solutions can be crystallised to produce solid salt;
- ✓ know about the uses of precipitation reactions;
- ✓ understand about acids, bases and neutralisation;
- ✓ know how to predict what salt is produced in a reaction;
- ✓ know how ammonium salts are produced and their use as fertilisers;
- ✓ be able to suggest methods to make a named salt;
- ✓ be able to explain how the electrolysis of ionic liquids results in the decomposition of compounds into elements;
- ✓ be able to use half equations for the reactions occurring at electrodes;
- ✓ be able to explain electrolysis in terms of oxidation and reduction;
- ✓ understand that the electrolysis of sodium chloride solution produces important substances for the chemical industry;
- ✓ know how copper can be purified by electrolysis.

Figure 8.1 Common salt – it's used for more than chips. It's a raw material for many useful products. Salt is used to make bleach, soap, washing soda, caustic soda (sodium hydroxide), water softeners, chlorine, PVC, baking soda, the enema sodium phosphate, and glass. Salt is also used for preserving meat (and Egyptian mummies), making freezing mixtures, dyeing and removing traces of water from aviation fuel after it is purified. Every cell of your body contains salt, a total of 250 g per person. Salt in body fluids is essential for muscle action – too little and you get cramp. For many of these uses and products, a solution of salt is needed. If salt did not dissolve, none of these uses would be possible.

8.1 What are ions – and what does melting and dissolving do to ions?

Ions (charged particles) can form when atoms react. They can be made from a single atom (e.g. a magnesium ion, Mg^{2+}) or from a group of atoms (e.g. a nitrate ion, NO_3^-). Pure ionic substances are all solids at room temperature. There are strong electrostatic bonds holding the ions in a giant lattice. (Ionic bonds are discussed in Section 5.6 and giant ionic structures in Section 6.7.)

When an ionic substance melts, the ions are free to move in the liquid. Ions can also be freed from the lattice by dissolving in water. When soluble ionic solids are put into water, the water molecules can get between the ions and separate them.

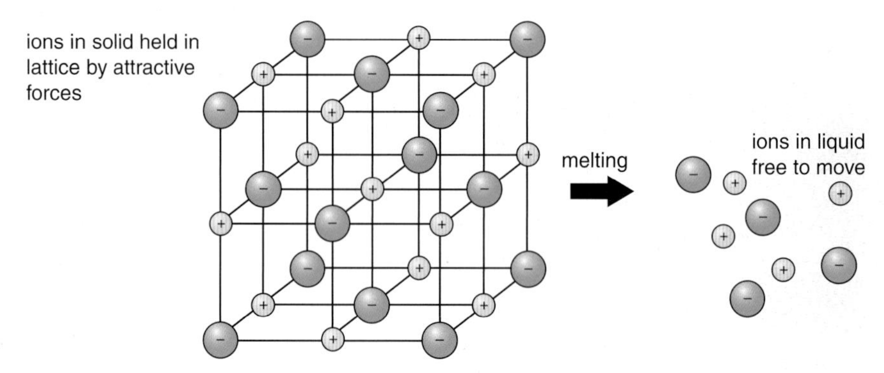

Figure 8.2 Melting an ionic compound means the ions are free to move.

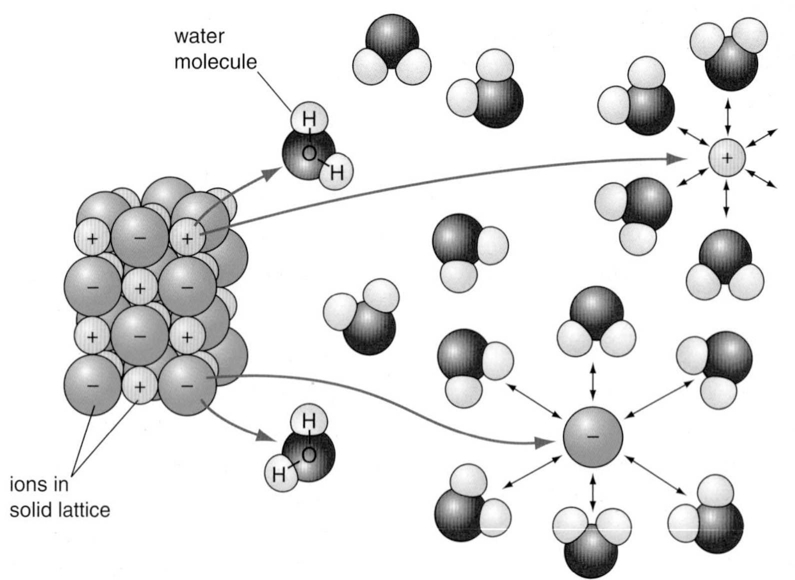

Figure 8.3 A model for dissolving an ionic substance. In a solution of an ionic compound the positive and negative ions are free to move about separately, kept apart by water molecules round the ions.

A solid, liquid or gas dissolved in a solvent is a **solution**.

A **solvent** is a substance (usually a liquid) that dissolves another substance to form a solution.

A **precipitation reaction is** a reaction that produces an insoluble solid when solutions are mixed. The solid produced is called a precipitate.

Crystallisation

If the water or other **solvent** in a **solution** is allowed to evaporate, the solution becomes more concentrated. When this happens there may not be enough water to hold the ions in solution. The positive and negative ions come together to form a solid. If the solvent evaporates slowly, the ionic solid forms as large regular crystals as each ion takes its place in an extended giant structure or lattice of ions (see Section 6.1).

Precipitation

If two solutions are mixed and one pair of ions reacts to form an insoluble substance as a solid, this is an example of **precipitation**. For example:

$BaCl_2(aq)$	+	$Na_2SO_4(aq)$	→	$BaSO_4(s)$	+	$2NaCl(aq)$
barium chloride solution		sodium sulfate solution		barium sulfate solid		sodium chloride solution

Figure 8.4 The colourless solutions in the pipette and test tube are two separate mixtures, one of barium and chloride ions and the other of sodium and sulfate ions. When the solutions are mixed, the barium and sulfate ions react together to form a white precipitate.

The solid barium sulfate precipitates out (Figure 8.4), leaving the sodium ions and chloride ions still in solution (remember, (aq) means aqueous or in solution). These ions are called 'spectator ions' because they do not take part in the reaction.

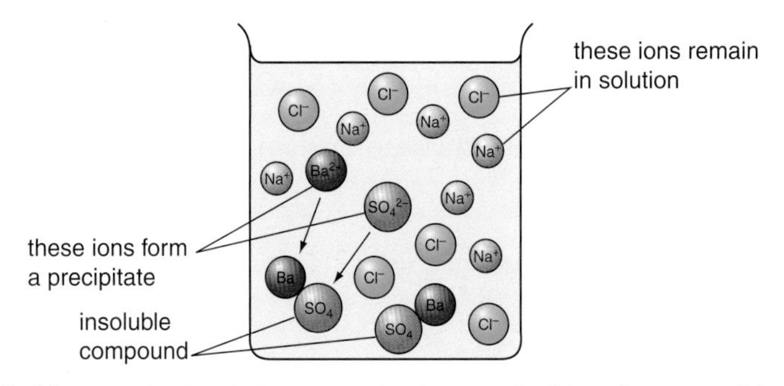

Figure 8.5 When two ionic solutions are mixed, one pair of ions forms a solid precipitate that can be removed from the solution, but the other ions remain in solution.

Uses of precipitation reactions

Precipitation can be used to remove selected ions from solution. The solution is mixed with another solution containing ions that form an insoluble precipitate with the unwanted ions.

Iron and some other metal ions in drinking water are not harmful to human health but can be a nuisance. Dissolved iron ions form brown stains on sinks, and darken cups when combined with tannin in tea. They also discolour vegetables during cooking. These ions can be removed in waste water treatment plants by adding calcium hydroxide solution. The hydroxide ions react with the iron ions in solution to form insoluble iron hydroxide. Filters remove the precipitated particles.

Precipitation reactions are used in forensic analysis to test whether a sample contains particular ions, for example those from poisons or explosives residues. Barium chloride solution is used as a test for dissolved sulfate ions.

A **salt** is an ionic compound. Most salts consist of positively charged metal ions and negatively charged non-metal ions. Salts form when an acid neutralises a base.

The same precipitation reaction can be used to remove soluble metal ions from mine discharges and industrial effluent. Dissolved heavy metal ions such as cadmium and lead are precipitated out as metal hydroxides when the effluent is mixed with calcium hydroxide.

Phosphate fertilisers can contaminate rivers and groundwater and cause eutrophication by excess weed growth. Adding a metal **salt** (such as aluminium sulfate) to the waste water precipitates the phosphate ions as aluminium phosphate.

1. Use the data from page 257 to write the formulae of:
   a) sodium nitrate;
   b) aluminium chloride;
   c) magnesium sulfate;
   d) ammonium carbonate;
   e) calcium hydroxide.
2. Explain why a dissolved metal ion can form a solid when you mix a colourless sample solution with another colourless solution.
3. Draw a diagram to represent ions and molecules in a solution of magnesium chloride ($MgCl_2$).
4. Use the idea of ions and lattices to explain why large crystals of sodium chloride take a long time to dissolve.
5. Table 8.1 gives some data about solutions and precipitates.

Solutions mixed	Colour of precipitate
Silver nitrate and sodium chloride	White precipitate
Silver nitrate and calcium chloride	White precipitate

Table 8.1

What is the white precipitate in each case? Explain your answers.

8.2 How do acids and alkalis react?

An **acid** is a substance that releases H^+ ions to make an acidic solution with a pH less than 7.

An **alkali** is a substance that releases OH^- ions to make an alkaline solution with a pH greater than 7.

Oxygen is very reactive. It oxidises both metal and non-metal elements. Some non-metal oxides form acids when they dissolve in water. Metal oxides form alkalis when they dissolve in water. That's where many acid and alkali solutions come from.

Acid solutions all contain hydrogen ions. Hydrogen ions (H^+) are the acid ions. **Alkali** solutions all contain hydroxide ions (OH^-). These are the alkali ions.

When an acid or alkali dissolves in water, the substance splits up to make hydrogen ions or hydroxide ions. The other dissolved ion is a 'spectator ion'.

$$HCl(aq) \rightarrow H^+(aq) + Cl^-(aq)$$

$$NaOH(aq) \rightarrow Na^+(aq) + OH^-(aq)$$

All acids react in similar ways. This is because only the hydrogen ion in the acid reacts. The alkali reactions of the hydroxide ion are all the same for a similar reason. The spectator ions take no part in acid or alkali reactions.

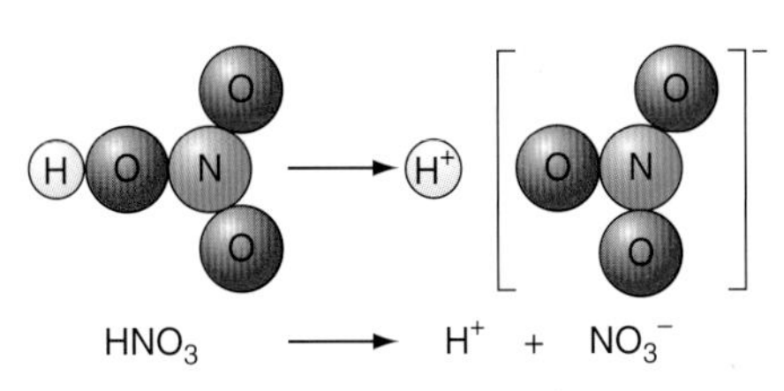

Figure 8.6 Nitric acid HNO_3 splits into H^+ (hydrogen ions) and NO_3^- (nitrate ions).

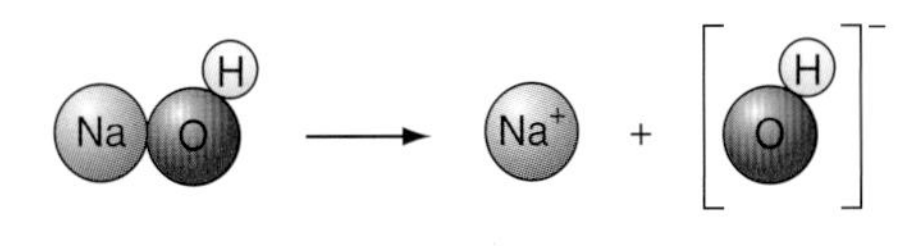

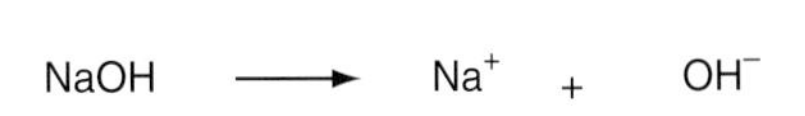

Figure 8.7 Sodium hydroxide NaOH splits into Na^+ (sodium ions) and OH^- (hydroxide ions).

Explaining neutralisation

In a neutralisation reaction between an acid and an alkali, hydrogen ions react with hydroxide ions to produce water:

$$H^+(aq) + OH^-(aq) \rightarrow H_2O(l)$$

The spectator ions stay dissolved in the solution.

Example

$$\underbrace{H^+(aq) + NO_3^-(aq)}_{\text{nitric acid}} + \underbrace{Na^+(aq) + OH^-(aq)}_{\text{sodium hydroxide}} \rightarrow \underbrace{Na^+(aq) + NO_3^-(aq)}_{\text{sodium nitrate}} + \underbrace{H_2O(l)}_{\text{water}}$$

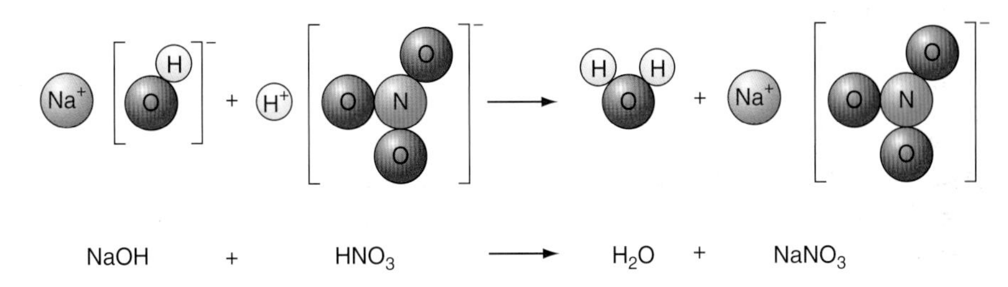

Figure 8.8 A neutralisation reaction. When sodium hydroxide (an alkali) reacts with nitric acid, the H+ and OH^- ions react to form water, but the Na^+ and NO_3^- ions stay dissolved.

Note that:

- If we remove the spectator ions from the picture, then all neutralisation equations become the same:

  $$H^+(aq) + OH^-(aq) = H_2O(l)$$

  This means there is really only one change happening in any acid/alkali reaction – hydrogen ions joining with hydroxide ions to make water molecules.
- The spectator ions are left over. They form the salt in the reaction.

pH scale

The pH scale, used to measure the acidity or alkalinity of a solution, is a measure of hydrogen ion concentration.

- pH 7 is the value for neutral solutions (1 in every 10^7 water molecules has split to make an H^+ ion);
- pH 1 is strong acid (1 water molecule in 10 has split to make an H^+ ion)
- pH 14 is a strong alkali (very low concentration of H^+ ions).

8.3 How can we make different salts by neutralisation?

When acids are neutralised the spectator ions are left as ions in solution. If all the water is evaporated, a solid salt is left behind. An acid plus an alkali forms a salt plus water. The acid and alkali used determine which salt is made (Table 8.2).

Figure 8.9 Universal indicator is a mixture of substances that responds to the pH of a solution by turning different colours. This range of colours matches the pH scale.

Using indicators to prepare a salt

As acid and alkali solutions are usually both colourless, when we are making a salt we need to determine when the solution is exactly neutral. In a neutral solution only the dissolved salt will be present with no excess acid or alkali.

You could use a pH meter to test when the solution is neutral. You could also add the acid or alkali solution a little at a time and check the pH after each addition by dropping a sample onto indicator paper (Figure 8.10). Another method is to leave the indicator in the reaction flask as the solutions are mixed. The indicator is removed before crystallising the salt, by adding two spatulas of charcoal and heating. The charcoal absorbs the indicator and can be filtered off, leaving a colourless solution.

Acid or alkali	Salt made
Nitric acid	Nitrates
Hydrochloric acid	Chlorides
Sulfuric acid	Sulfates
Ethanoic acid	Ethanoates
Sodium hydroxide	Sodium salts
Potassium hydroxide	Potassium salts
Ammonia solution (alkali)	Ammonium salts

Table 8.2

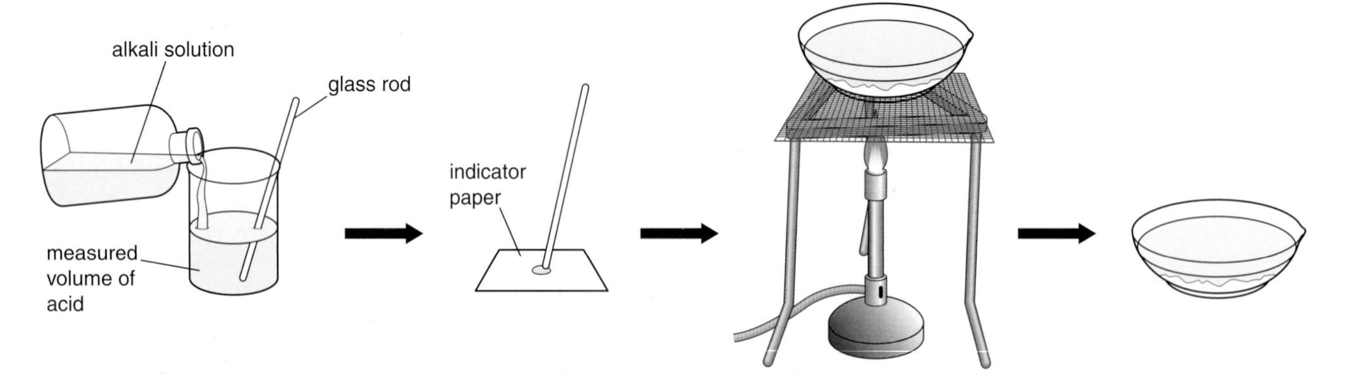

Figure 8.10 Preparing a pure soluble salt by neutralisation

Figure 8.11 Measuring the volume of acid accurately for a neutralisation reaction

Ammonia

Ammonia (NH_3) is a very smelly gas. It is very soluble in water – 1 dm^3 of water dissolves 1500 dm^3 of ammonia gas. When ammonia dissolves in water, some ammonia molecules react with water to form an ammonium ion (NH_4^+) and a hydroxide ion to give an alkaline solution.

$$NH_3(aq) + H_2O(l) \rightarrow NH_4^+(aq) + OH^-(aq)$$

Ammonium salts are important constituents of fertilisers. The ammonia provides the source of nitrogen for plants to make proteins.

6 a) What substances dissolve in water to produce acids?
b) What substances dissolve in water to produce alkalis?

7 a) 'Only the hydrogen ion matters in an acid reaction.' Explain what is meant by this statement.
b) What is the corresponding ion in an alkali reaction?
c) What new molecule is produced in neutralisations?
d) What is meant by a 'spectator ion'?

8 Copy and complete Table 8.3.

Name of acid	Name of alkali	Name of salt	Formula of salt
Sulfuric acid	Sodium hydroxide		
Hydrochloric acid		Lithium chloride	
		Calcium nitrate	
			$MgCl_2$
			NH_4NO_3

Table 8.3

9 Draw a flow chart for a method to produce clean, dry salt crystals by neutralising sulfuric acid with potassium hydroxide solution, using Universal indicator.

8.4 What other ways are there of making salts?

A **base** is a compound that will neutralise an acid by forming water molecules as one of the products.

Alkalis are soluble **bases**. Don't confuse the chemical word 'base' with other meanings of the word base, for example in sports, in apparatus, in mountain climbing, in 'back to base', and so on. In chemistry, a base is a metal oxide or hydroxide. If a base dissolves in water, it is an alkali, but most bases are insoluble.

Figure 8.12 Copper oxide dissolves in hydrochloric acid to give a blue solution of copper chloride. Unreacted copper oxide has settled to the bottom.

Acid plus insoluble base forms a soluble salt

$$\underbrace{2(H^+ + Cl^-)(aq)}_{\text{hydrochloric acid}} + \underbrace{CuO(s)}_{\text{copper oxide}} \rightarrow \underbrace{(Cu^{2+} + 2Cl^-)(aq)}_{\text{copper chloride}} + \underbrace{H_2O(l)}_{\text{water}}$$

To prepare copper chloride, add copper oxide to hydrochloric acid until no more will react. The oxygen in the oxide reacts with hydrogen ions in the acid to form water. The 'spectator ions' are left in solution to make copper chloride. Once all the acid is neutralised, filter off the remaining (excess) copper oxide. Then heat the mixture gently for a few minutes. This evaporates the copper chloride solution until it starts to crystallise.

Many metal carbonates react in a similar way to metal oxides. Carbon dioxide is an additional product as the carbonate reacts with the hydrogen ions from the acid.

$$\underbrace{2(H^+ + Cl^-)(aq)}_{\text{hydrochloric acid}} + \underbrace{MgCO_3(s)}_{\text{magnesium carbonate}} \rightarrow \underbrace{(Mg^{2+} + 2Cl^-)(aq)}_{\text{magnesium chloride}} + \underbrace{H_2O(l)}_{\text{water}} + \underbrace{CO_2(g)}_{\text{carbon dioxide}}$$

A **displacement reaction** is a reaction in which a more reactive element displaces a less reactive one from its compound.

Acid plus metal forms a soluble salt plus hydrogen

Many reactive metals form salts when added to acids. The reaction is a **displacement reaction**, not a neutralisation reaction. The reaction does not work for the less reactive metals such as copper or gold.

Reactivity series for metals
Potassium
Sodium
Lithium
Calcium
Magnesium
Aluminium
Zinc
Iron
Lead
Hydrogen
Copper
Silver
Gold

Only the metals above hydrogen in the reactivity series react with an acid to make a salt. Aluminium does not make salts with acids readily since a hard aluminium oxide layer forms on the surface of aluminium and this resists acid attack.

To make a salt from a metal plus an acid, add an excess of metal. Leave the reaction for some time, but don't heat, since hydrogen gas forms an explosive mixture with air. Once the reaction has finished, filter off the unreacted metal to leave a salt solution.

For example, you can make zinc sulfate by adding zinc to sulfuric acid:

$$\underset{\text{zinc}}{Zn(s)} + \underset{\text{sulfuric acid}}{(H_2SO_4)(aq)} \rightarrow \underset{\text{zinc sulfate}}{ZnSO_4(aq)} + \underset{\text{hydrogen}}{H_2(g)}$$

Figure 8.13 Soluble zinc sulfate is formed when zinc reacts with sulfuric acid. The fizz is hydrogen being released.

Insoluble salts

In Section 8.1 you saw that when two ionic solutions are mixed a precipitate can sometimes form. The precipitate is an insoluble salt and can be filtered off. The precipitate can be washed with pure water to remove any residual salt solution, and dried to give a pure dry insoluble salt.

10. What is the difference between a base and an alkali?
11. Magnesium chloride could be made by reacting magnesium with hydrochloric acid.
    a) What safety precautions do you need to take?
    b) Why would this method not work for copper chloride?
    c) Write a step-by-step report for making magnesium chloride crystals by this method, detailing what you would expect to see at each stage.
12. Write a step-by-step plan for making salt crystals from zinc oxide and nitric acid.
13. a) Explain how you would know that a certain volume of sulfuric acid had been completely neutralised by copper carbonate.
    b) What salt is made in this reaction?
14. Write a balanced symbol equation for the reaction between sulfuric acid and copper carbonate in question **13**.

8.5 Electrolysis

When an ionic substance is molten or dissolved in water, the ions are free to move about. In **electrolysis** an electric circuit is connected to the liquid and the ions move towards the **electrodes** that have the opposite charge. Chemical changes take place at the electrodes as the ions gain electrons at the cathode and lose electrons at the anode. The energy transferred by the electricity can re-form elements from ions and break a compound down into its elements.

Figure 8.14 Electrolysis does many jobs in science. These batteries are being used to electrolyse brine.

Electrolysis is the process of splitting a chemical compound into its elements by passing an electric current through a molten liquid or aqueous solution containing ions (the **electrolyte**).

An **electrode** is the physical connection between an electric circuit and an electrolyte. It is usually made from a material that won't corrode such as a carbon rod or platinum foil.

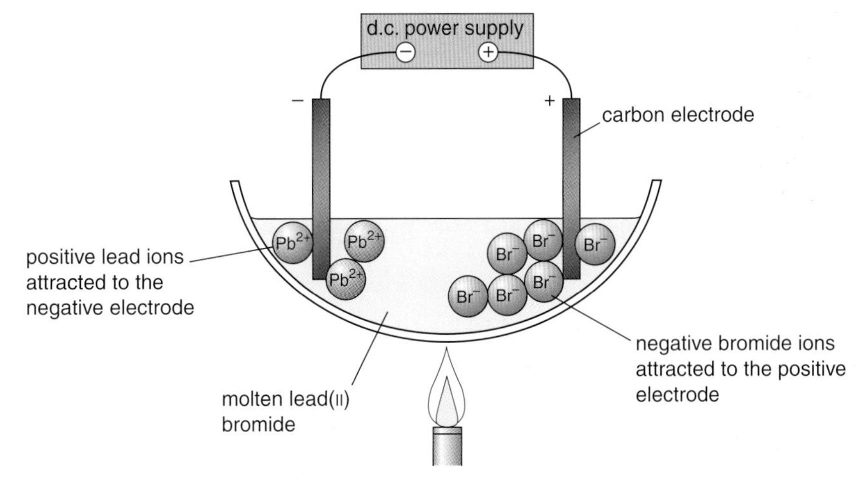

Figure 8.15 Lead bromide can be spilt into its elements, lead and bromine, by electrolysis.

Let us look at the electrolysis of molten lead(II) bromide (Figure 8.15).

Lead ions (Pb^{2+}) are attracted to the **cathode** (negative electrode). The lead ions gain electrons so they become lead metal:

lead ions + electrons → lead metal

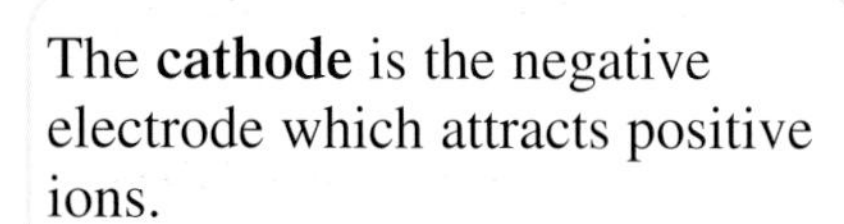

The **cathode** is the negative electrode which attracts positive ions.

The **anode** is the positive electrode which attracts negative ions.

Bromide ions (Br^-) are attracted to the **anode** (positive electrode). Electrons are pulled away from the bromide ions forming bromine.

$$\text{bromide ions} - \text{electrons} \rightarrow \text{bromine}$$

If a salt can exist in a molten state, then it can be decomposed into its elements like this. Electrolysis forms the basis of the production of most reactive metals, such as sodium and aluminium, from their molten metal ores.

Electrolysis equations

A **half equation** describes the reaction that takes place at one of the electrodes in an electrolysis cell. This is only half the total process because another, different, reaction takes place at the other electrode.

The reactions at the electrodes can be represented by **half equations**. For example, for the electrolysis of lead bromide described above, we have:

$$\text{cathode:} \quad Pb^{2+} + 2e^- \rightarrow Pb$$

$$\text{anode:} \quad 2Br^- - 2e^- \rightarrow Br_2$$

These are called half equations because each equation represents only half of the overall electrolysis reaction. The pairs of half equations must obey the same rules. Each half equation must be balanced in terms of atoms and charges, and the loss and gain of electrons in each pair must match.

Oxidation is the loss of electrons. In electrolysis this happens at the positive electrode when a negatively charged ion loses electrons.

Reduction is the gain of electrons. In electrolysis this happens at the negative electrode when a positively charged ion gains electrons.

When oxygen reacts with a metal and oxidises it, the oxygen atoms remove electrons from the metal to make negatively charged oxide ions. The metal becomes a positively charged ion. When any other non-metal reacts with a metal, electrons are also transferred from the metal to the non-metal. This process is also called **oxidation**.

Reduction of metal oxides with carbon or hydrogen is the reverse process. The carbon pulls the oxygen away from the less reactive metal ion and gives the electrons back, so the ion becomes a metal atom again.

We can define oxidation and reduction as follows:

Oxidation Is Loss (of electrons)
Reduction Is Gain (of electrons)

This gives us the mnemonic OIL RIG.

In electrolysis, the anode takes electrons away – so the anode oxidises materials. The cathode gives electrons to materials – so the cathode reduces them.

Electrolysis of solutions

Salt solutions can be electrolysed in a similar way to molten ionic compounds. The ions in solution are attracted to the electrode with the opposite charge, where they are turned into atoms of the element. Negatively charged ions (e.g. Cl^-) lose electrons (oxidation) and positively charged ions (e.g. Cu^{2+}) gain electrons (reduction). However, the presence of water in a solution means that a different substance can sometimes be produced at the electrodes.

With some solutions, hydrogen ions (H^+) in the water gain electrons at the cathode to form molecules of hydrogen gas (H_2). For example, when sodium chloride solution is electrolysed, hydrogen is formed at the cathode. The negative ion from the salt (Cl^-) loses electrons at the anode to form molecules of chlorine gas (Cl_2). The positive ion from the salt (Na^+) stays in solution. Hydroxide ions (OH^-) from the water also stay in solution. This means the original sodium chloride solution forms three products: hydrogen, chlorine and sodium hydroxide solution.

The half equations for the reactions at the electrodes are:

cathode: $2H^+ + 2e^- \rightarrow H_2$

anode: $2Cl^- \rightarrow Cl_2 + 2e^-$

Mixtures of ions in solution

In solutions containing a mixture of ions, the products formed depend on the reactivity of the elements. If a mixture of positive metal ions is present, then the ion of the element with the lowest reactivity will form an atom first (the reactivity series for metals is given in Section 8.4, page 160). For example, in a mixture of magnesium chloride and hydrochloric acid, hydrogen is produced, not magnesium. In a mixture of iron sulfate and copper sulfate, copper is produced, not iron.

The negative ions in salt solutions are often simple halide ions (e.g. Cl^-, Br^-). The reactivity series for the halogens determines what product forms in electrolysis of solutions with a mixture of halide ions. For example, in a mixed solution of sodium bromide and sodium chloride, bromine is less reactive so bromine gas is produced at the anode.

15 Explain what these terms mean.
a) electrolyte;
b) electrolysis;
c) anode;
d) cathode.

16 a) Write a plan for a procedure to turn copper chloride crystals into copper metal and chlorine using electrolysis.
b) What safety precautions should you take?
c) Draw a diagram of the apparatus you would use.

17 Copy and complete Table 8.4.

Liquid being electrolysed	Product at cathode	Product at anode
Molten lead bromide		
Molten sodium oxide		
Copper chloride solution		
Potassium iodide solution		
A mixture of sodium iodide solution and copper chloride solution		

Table 8.4

8.6 Uses of electrolysis

Electrolysis of water

Pure water is a very poor conductor. If a little acid is added to the water then it will conduct electricity – acidified water can be electrolysed. This reaction breaks down water into hydrogen and oxygen. It's the basis of making hydrogen for use in fuel cells for electric vehicles.

cathode: $4H^+(aq) + 4e^- \rightarrow 2H_2(g)$

anode: $4OH^-(aq) - 4e^- \rightarrow O_2(g) + 2H_2O(l)$

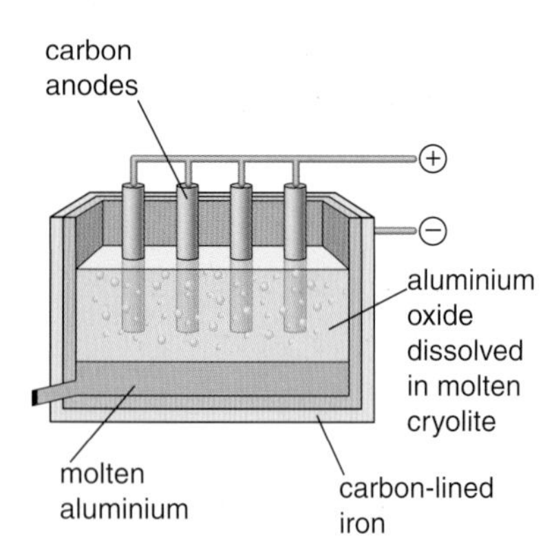

Figure 8.16 Aluminium oxide is decomposed by electrolysis in this industrial cell. The molten aluminium is siphoned from the bottom of the cell.

Making aluminium metal

Aluminium is too high in the reactivity series to extract it from its ore by reduction with carbon. Most reactive metals are produced by electrolysis of a molten compound.

Aluminium metal is produced by the electrolysis of molten aluminium oxide (obtained from bauxite ore). The aluminium oxide has a very high melting point and is dissolved in molten sodium aluminium fluoride (cryolite) to make it melt at a lower temperature. Only aluminium and no sodium is produced during electrolysis because sodium is higher in the reactivity series.

cathode: $4Al^{3+} + 12e^- \rightarrow 4Al(l)$

anode: $6O^{2-} - 12e^- \rightarrow 3O_2(g)$

Purification of copper metal by electrolysis

Copper ore is reduced to copper metal by smelting with carbon. However, the metal obtained contains several impurities (lead, nickel, zinc, gold, silver) that reduce its electrical conductivity. The impure copper is purified by electrolysis.

Large impure slabs of copper are used as anodes and pure sheets of copper are used as cathodes. The electrolyte is acidified copper sulfate solution, which remains unchanged overall. During electrolysis copper from the impure anode is reduced to copper ions and these ions are transferred into solution. The positive copper ions are attracted to the cathode, where they are reduced to copper atoms and deposited on the cathode. As the impure anode crumbles away, all the impurities collect at the bottom of the cell as sludge. This anode sludge can be reprocessed to recover precious metals.

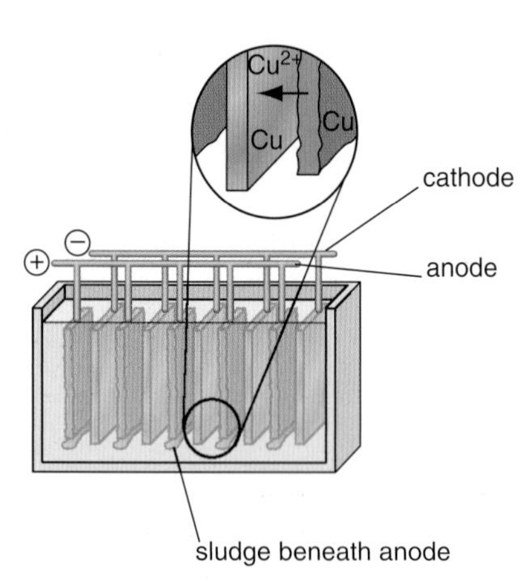

Figure 8.17 The impure copper anode dissolves away and pure copper is deposited on the cathode

cathode: $Cu^{2+}(aq) + 2e^- \rightarrow Cu(s)$

anode: $Cu(s) - 2e^- \rightarrow Cu^{2+}(aq)$

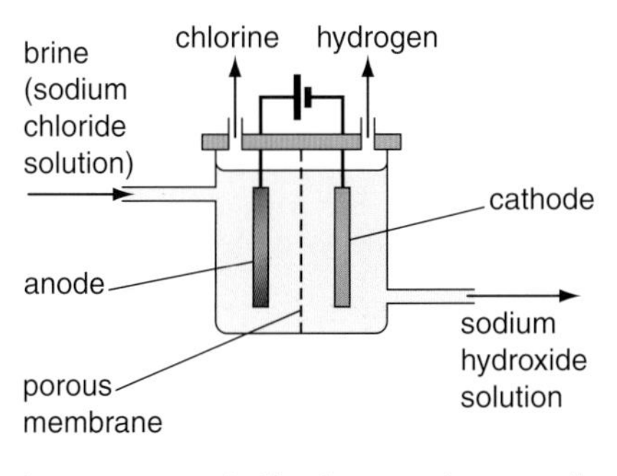

Figure 8.18 Cells like this membrane cell are used in a chloralkali plant.

Electrolysis of sodium chloride solution

The electrolysis of concentrated brine (salt solution) is used industrially for the production of chlorine and sodium hydroxide, an alkali known as caustic soda. Hydrogen gas is a by-product of the process (see page 163). Some of the many industrial uses of salt and its products are described on page 153.

Sodium chloride solution is electrolysed in a membrane cell that separates the anode chamber from the cathode chamber. This method is called the chloralkali process. A porous membrane divides the electrolytic cell and prevents the mixing of brine on one side with the sodium hydroxide which is left on the other side.

18 a) What is the name of the aluminium ore from which aluminium oxide is obtained for electrolysis?
b) Why is cryolite used in the process?
c) Explain why no sodium metal is produced, even though sodium ions are present.
d) Use your knowledge of the reactivity series for metals to predict three metals other than aluminium that cannot be reduced by carbon and so must be made by electrolysis.

19 Name three products that can be made from the electrolysis of sodium chloride solution.

20 Write half ionic equations for the reactions at the anode and cathode of the membrane cell shown in Figure 8.18.

21 a) Calculate the atom economy for the manufacture of aluminium by electrolysis of molten aluminium oxide.
b) Electrical energy is needed for electrolysis. What other costs are involved in the manufacture of aluminium?
c) Explain why producing aluminium by electrolysis is not as sustainable as recycling aluminium.

Summary

- ✓ Ions are free to move within a molten ionic compound or an ionic solution.
- ✓ Evaporation of a salt solution **crystallises** the solid salt.
- ✓ Ions dissolved in ionic solutions can form an insoluble **precipitate** when two solutions are mixed. Precipitation can be used to remove unwanted ions from a solution.
- ✓ Acid solutions contain **hydrogen ions** (H^+); alkalis contain **hydroxide ions** (OH^-).
- ✓ Metal oxides and hydroxides are bases. Soluble bases are called alkalis.
- ✓ **Neutralisation** of an acid with an alkali produces water molecules.
- ✓ There are several types of substances that will react with acids and form a **salt**:

acid + alkali → salt + water
acid + metal oxide (base) → salt + water
acid + reactive metal → salt + hydrogen

- ✓ Ammonium salts produced by adding acid to ammonia solution are important as fertilisers.
- ✓ Liquids containing ions conduct electricity.

✓ When electricity passes through liquids there are chemical changes at the electrodes. The ions are either **oxidised** (lose electrons) or **reduced** (gain electrons).

✓ Positive ions are attracted to the **cathode**, while negative ions are attracted to the **anode**.

✓ The reactions at the electrodes can be described using **half equations.**

✓ Electrolysis is an important industrial process. It can be used to:

- produce hydrogen for use as a fuel;
- extract aluminium and other reactive metals from their ores;
- purify copper metal for electrical wires and cables;
- obtain chlorine and sodium hydroxide from the electrolysis of rock salt (sodium chloride).

EXAMQUESTIONS

1. Briefly suggest the chemicals needed for making each of the salts in Table 8.5 below.

Name of salt	Function	Formula
a) Washing soda	Prevents scum from soap	Na_2CO_3
b) Epsom salts	Helps digestive system	$MgSO_4$
c) Moss killer for lawns	Gardening aid	$FeSO_4$
d) Road salt	Prevents water freezing	NaCl
e) Fish tank cleaner	Stops growth of fungi	$CuCl_2$

Table 8.5

(10 marks)

2. Making salts is about neutralising acids. How would you make sure that all the acid had been exactly neutralised in each of the following reactions?
   a) an acid plus insoluble metal carbonate *(2 marks)*
   b) an acid plus metal oxide *(2 marks)*
   c) an acid plus alkali. *(2 marks)*

3. Write balanced half equations for the electrolysis of these liquids:
   a) acidified water (H_2O) *(2 marks)*
   b) molten zinc chloride ($ZnCl_2$) *(2 marks)*
   c) concentrated sodium chloride solution (NaCl(aq)) *(2 marks)*
   d) dilute copper sulfate solution ($CuSO_4$(aq)). *(2 marks)*

4. Oxidation is loss. Reduction is gain.
   a) Explain how the oxidation of a metal fits in with the definition of oxidation above. *(4 marks)*
   b) How can this principle for defining oxidation and reduction be applied to electrolysis? *(4 marks)*

5. The apparatus in Figure 8.19 is used in industry to purify copper. The anode is impure copper and the cathode is a sheet of pure copper. The electrolyte is copper sulfate solution.

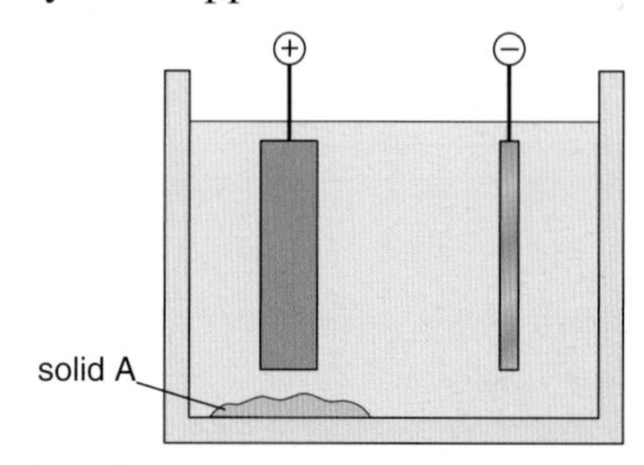

Figure 8.19

   a) Explain why the anode is made of a large block of impure copper. *(2 marks)*
   b) Explain why the cathode is made of a thin sheet of pure copper. *(2 marks)*
   c) What is the difference in properties of pure and impure copper? *(1 mark)*
   d) What is the solid A that accumulates below the anode? *(2 marks)*
   e) What are the main differences between this process and the electrolysis of copper sulfate solution with carbon electrodes? *(4 marks)*

Chapter 9
How can the Periodic Table help us to understand reactions?

At the end of this chapter you should:

- ✓ know how Newlands and then Mendeléev attempted to classify the elements by arranging them in order of their atomic weights (relative atomic masses);
- ✓ be able to explain why scientists regarded a periodic table as a curiosity at first, then as a useful tool and finally as an important summary of the structure and reactions of elements;
- ✓ appreciate how Mendeléev overcame some of the deficiencies in early periodic tables;
- ✓ know that in the modern Periodic Table, all the elements are arranged in order of atomic numbers with elements in the same group having the same number of electrons in their outer shell (highest occupied energy level);
- ✓ know about the properties and reactions of Group 1 elements (alkali metals) and that they become more reactive down the group;
- ✓ know about the properties and reactions of Group 7 elements (halogens) and that they become less reactive down the group;
- ✓ know about the similar properties and special properties of transition elements (transition metals),
- ✓ be able to compare the properties and reactions of transition metals with the alkali metals in Group 1.
- ✓ be able to explain the trends in reactivity within groups in the Periodic Table;
- ✓ appreciate that the similar properties and special properties of transition elements arise because a lower energy level is being filled in the atoms of these elements.

Figure 9.1 In a library, the books are classified and arranged so that similar types of books are near one another. The books are classified into sections such as fiction, travel, science and so on. In a similar way, chemists classify elements by grouping together those with similar properties.

9.1 Families of elements: Dobereiner's Triads

Lithium Li 6.9	Chlorine Cl 35.5
Sodium Na 23.0	Bromine Br 79.9
Potassium K 39.1	Iodine I 126.9

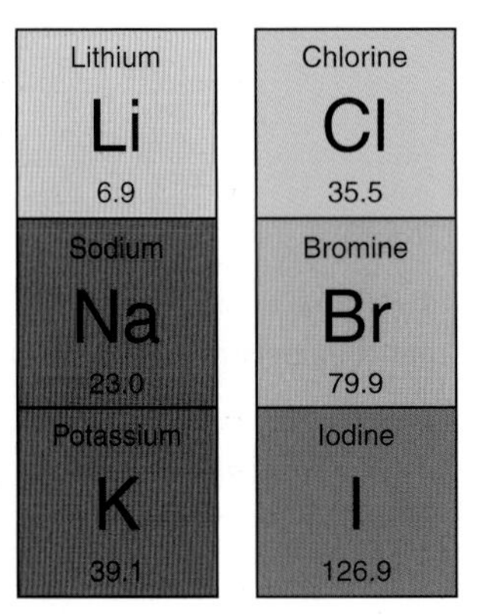

Figure 9.2 Two of Dobereiner's Triads – an early attempt to group chemical elements.

Early in the nineteenth century, a German chemist called Johann Wolfgang Dobereiner noticed that several elements could be arranged in groups or families of three. These groups or families became known as Dobereiner's Triads.

Two of Dobereiner's Triads are shown in Figure 9.2. One of the triads contains the metals lithium, sodium and potassium. The other contains the non-metals chlorine, bromine and iodine.

Dobereiner also noticed an interesting point regarding the atomic weights of the elements in each of his triads. (Atomic weight was the original name that scientists used for relative atomic masses.) When Dobereiner arranged the elements in a triad in order of their atomic weights (relative atomic masses), the properties of the middle element were half way between those of the other two elements. For example, the relative atomic mass of sodium (23.0) is the average of (half way between) the relative atomic masses of lithium (Li) and potassium (K):

$$\text{average relative atomic mass of Li and K} = \frac{6.9 + 39.1}{2} = \frac{46.0}{2} = 23.0$$

In addition, other properties of the middle element (such as melting point and boiling point) also had a value close to the average of the other two elements in the triad. The discovery of these triads and the links between the properties of elements and their relative atomic masses led other chemists to search for patterns in the properties of all the known elements.

9.2 Patterns in the properties of elements

In 1864, John Newlands, a British chemist, arranged all the known elements in order of their relative atomic masses. He found that one element often had similar physical and chemical properties to those of the element eight places after it in his list. Newlands called this the 'Law of Octaves'. He said that 'the eighth element is a kind of repetition of the first, like the eighth note of an octave in music'.

Figure 9.3 shows the first three of Newlands' Octaves. Notice that similar elements often occur eight places on and in the same column. For example, the first column contains the similar elements fluorine and chlorine and the second column contains lithium, sodium and potassium.

H	Li	Be	B	C	N	O
F	Na	Mg	Al	Si	P	S
Cl	K	Ca	Cr	Ti	Mn	Fe

Figure 9.3 The first three of Newlands' Octaves

A **periodic table** is a table in which the chemical elements are carefully arranged so that elements with similar properties occur at regular intervals and fall in the same vertical column.

The repetition of elements with similar properties at regular intervals in Newlands' table led to the name **periodic table.**

Unfortunately, Newlands' classification grouped together some elements which were very different from others in the same column. For example, iron (Fe) was placed in the same column as oxygen (O) and sulfur (S). Because of this, Newlands' Law of Octaves was criticised and rejected.

Activity – John Newlands presents his Law of Octaves

Figure 9.4 shows the report of a meeting in March 1866 at which Newlands presented his ideas to the Chemical Society.

Read the report in Figure 9.4 carefully and then answer the following questions.

1. What does Newlands mean by the term 'elements analogous in their properties' which is mentioned in the report?
2. Why did Newlands point out that 'chlorine, bromine, iodine and fluorine are thus brought into the same line' in his table?
3. Why does the report make the point that 'Nitrogen and phosphorus, oxygen and sulfur are considered as forming true octaves.'?
4. Which elements in Newlands' first octave now have different symbols from those that he used?
5. a) On what grounds did Dr Gladstone object to Newlands' ideas?
   b) What evidence did he quote to support his objection?
6. Suggest one other criticism that could be made of Newlands' proposal.
7. In Newlands' time, scientists hoped to get their ideas accepted by other scientists through articles in journals or by presentations at meetings like that reported in Figure 9.4. These methods are still used today, but how do you think modern scientists operate differently from Newlands in reporting their results and ideas?

PROCEEDINGS OF SOCIETIES.

CHEMICAL SOCIETY.

Thursday, March 1.

Professor A. W. WILLIAMSON, *Ph.D., F.R.S., Vice-President, in the Chair.*

Mr. JOHN A. R. NEWLANDS read a paper entitled "*The Law of Octaves, and the Causes of Numerical Relations among the Atomic Weights.*" The author claims the discovery of a law according to which the elements analogous in their properties exhibit peculiar relationships, similar to those subsisting in music between a note and its octave. Starting from the atomic weights on Cannizzarro's system, the author arranges the known elements in order of succession, beginning with the lowest atomic weight (hydrogen) and ending with thorium (= 231·5); placing, however, nickel and cobalt, platinum and iridium, cerium and lanthanum, &c., in positions of absolute equality or in the same line. The fifty-six elements so arranged are said to form the compass of eight octaves, and the author finds that chlorine, bromine, iodine, and fluorine are thus brought into the same line, or occupy corresponding places in his scale. Nitrogen and phosphorus, oxygen and sulphur, &c., are also considered as forming true octaves. The author's supposition will be exemplified in Table II., shown to the meeting, and here subjoined :—

Table II.—Elements arranged in Octaves.

	No.		No.		No.		No.		No.		No.		No.		No.
H	1	F	8	Cl	15	Co & Ni	22	Br	29	Pd	36	I	42	Pt & Ir	50
Li	2	Na	9	K	16	Cu	23	Rb	30	Ag	37	Cs	44	Os	51
G	3	Mg	10	Ca	17	Zn	24	Sr	31	Cd	38	Ba & V	45	Hg	52
Bo	4	Al	11	Cr	19	Y	25	Ce & La	33	U	40	Ta	46	Tl	53
C	5	Si	12	Ti	18	In	26	Zr	32	Sn	39	W	47	Pb	54
N	6	P	13	Mn	20	As	27	Di & Mo	34	Sb	41	Nb	48	Bi	55
O	7	S	14	Fe	21	Se	28	Ro & Ru	35	Te	43	Au	49	Th	56

Dr. GLADSTONE made objection on the score of its having been assumed that no elements remain to be discovered. The last few years had brought forth thallium, indium, cæsium, and rubidium, and now the finding of one more would throw out the whole system. The speaker believed there was as close an analogy subsisting between the metals named in the last vertical column as in any of the elements standing on the same horizontal line.

Professor G. F. FOSTER humorously inquired of Mr. Newlands whether he had ever examined the elements according to the order of their initial letters? For he believed that any arrangement would present occasional coincidences, but he condemned one which placed so far apart manganese and chromium, or iron from nickel and cobalt.

Mr. NEWLANDS said that he had tried several other schemes before arriving at that now proposed. One founded upon the specific gravity of the elements had altogether failed, and no relation could be worked out of the atomic weights under any other system than that of Cannizzarro.

Figure 9.4 The report of Newlands' presentation to the Chemical Society

Mendeléev's periodic table

In spite of the rejection of Newlands' Octaves, chemists carried on searching for a pattern linking the properties of elements and their relative atomic masses (atomic weights).

In 1869, the Russian chemist Dmitri Mendeléev produced new ideas to support the ideas which Newlands had suggested 5 years earlier.

Figure 9.5 shows part of Mendeléev's periodic table, published in 1869. Notice that elements with similar properties, such as lithium, sodium and potassium, fall in the same vertical column. These vertical columns of similar elements are called groups and the horizontal rows of elements are called periods.

	Group							
	1	2	3	4	5	6	7	8
Period 1	H							
Period 2	Li	Be	B	C	N	O	F	
Period 3	Na	Mg	Al	Si	P	S	Cl	
Period 4	K Cu	Ca Zn	* *	Ti *	V As	Cr Se	Mn Br	Fe Co Ni
Period 5	Rb Ag	Sr Cd	Y In	Zr Sn	Nb Sb	Mo Te	* I	Ru Rh Pd

Figure 9.5 Part of Mendeléev's periodic table published in 1869.

Most scientists did not accept Mendeléev's periodic table at first. They simply treated it as a curiosity. However, Mendeléev was more successful than Newlands because of three steps that he took.

- He left gaps in his table so that similar elements fell in the same vertical columns (groups). Four of these gaps are labelled with asterisks in Figure 9.5.
- He said that elements would be discovered in time to fill the gaps.
- He predicted the properties of undiscovered elements which would fill the gaps in his table from the properties of the elements above and below the gaps.

Figure 9.6 The element germanium was predicted by Mendeléev shortly after he published his periodic table in 1869. It was discovered in 1886 and today it is used as an important semiconductor in transistors and other electronic components.

Within 15 years of Mendeléev's predictions, three of the missing elements in his table had been discovered. They were named scandium, gallium and germanium, and their properties were very similar to Mendeléev's predictions.

The success of Mendeléev's predictions persuaded other scientists that his ideas were correct and sparked off much more research. Scientists used his periodic table as a working tool to predict the properties of missing elements and fill the gaps in the table.

❶ Early periodic tables, like the one proposed by Newlands, were incomplete. In addition, some elements were obviously placed in the wrong group when a strict order of relative atomic masses (atomic weights) was followed.

a) How did Mendeléev overcome these problems in his periodic table?
b) How were Newlands' proposals similar to those of Mendeléev?

Moseley relates atomic structure to the periodic table

Early in the twentieth century, after the discovery of electrons and protons, a British chemist called Henry Moseley studied the atomic structure of elements using X-rays. Moseley showed that the atomic number of an element, which was equal to the number of its electrons, could explain the pattern of elements with similar properties in the periodic table. Because of Moseley's work, elements in the modern Periodic Table (Figure 9.9) are arranged in order of their atomic (proton) numbers and not their relative atomic masses (atomic weights).

Moseley's work also led to the use of the Periodic Table as an important summary for understanding the properties and reactions of all elements.

From the Periodic Table, it is possible to predict:

- the atomic structure of an element (Section 5.5);
- the electron structure of an element (Section 5.5);
- whether an element is a metal or a non-metal (Section 1.3);
- the physical properties of an element;
- the reactions between elements;
- the relative reactivities of elements (Sections 9.4 and 9.5);
- the charges on ions and formulae of compounds (Sections 5.5 and 5.6).

❷ Chlorine, bromine and iodine formed one of Dobereiner's Triads. Using values for the relative atomic mass, melting point and boiling point of chlorine, bromine and iodine in Table 9.2 on page 178, answer the following questions.

a) For chlorine and iodine, calculate the average of:
   i) their relative atomic masses;
   ii) their melting points;
   iii) their boiling points.
b) How close is the average of the relative atomic masses of chlorine and iodine to the relative atomic mass of bromine?
c) How close is the average of the melting points of chlorine and iodine to the melting point of bromine?
d) How close is the average of the boiling points of chlorine and iodine to the boiling point of bromine?
e) Does bromine appear to have properties halfway between those of chlorine and iodine as Dobereiner claimed?

❸ Figure 9.7 shows Group 3 from Mendeléev's periodic table with the atomic weights which Mendeléev used in brackets below the symbols of the elements.

a) Predict the atomic weights of the two missing elements shown by asterisks in Figure 9.7. Explain how you obtained your answers.
b) i) Which group is missing from Mendeléev's periodic table?
   ii) Why is this group missing?

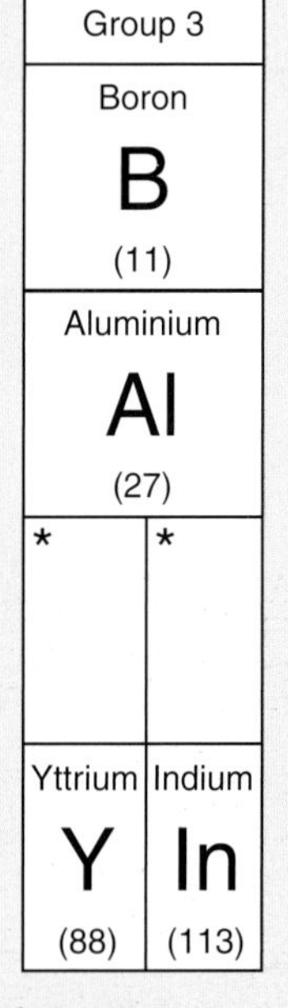

Figure 9.7 Group 3 from Mendeléev's periodic table

4 a) Why do you think most scientists regarded Mendeléev's periodic table as a curiosity when it was first proposed?
b) What happened to make scientists realise that Mendeléev's periodic table was a useful scientific tool?
c) During the twentieth century, scientists began to realise that the periodic table was an important summary for the atomic structure and reactivity of all elements. Why was this?

Activity – Following in Mendeléev's footsteps

The year is 1869. Suppose you are Mendeléev. You are just about to announce the discovery of your periodic table at the Russian Academy of Sciences. Write down what you will say.

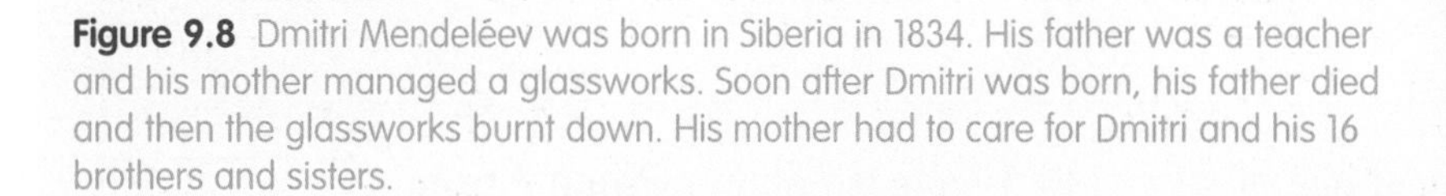

Figure 9.8 Dmitri Mendeléev was born in Siberia in 1834. His father was a teacher and his mother managed a glassworks. Soon after Dmitri was born, his father died and then the glassworks burnt down. His mother had to care for Dmitri and his 16 brothers and sisters.

9.3 Modern Periodic Tables

All modern Periodic Tables are based on the one proposed by Mendeléev in 1869. A modern Periodic Table is shown in Figure 9.9. From Chapters 1 and 5, you should already know that elements are arranged in order of their atomic numbers in the modern Periodic Table starting with period 1, then period 2, and so on.

The most obvious difference between modern Periodic Tables (Figure 9.9) and Mendeléev's (Figure 9.5) is the position of the transition elements. These have been taken out of the numbered groups and placed between Group 2 and Group 3. The first series of transition elements occurs in period 4. It includes titanium, chromium, iron, copper and zinc. The other major difference in modern Periodic Tables is the addition of Group 0, the noble gases. None of the noble gases were known until 1890, 31 years after Mendeléev produced his periodic table.

In modern Periodic Tables, non-metals are clearly separated from metals above the thick stepped line in Figure 9.9. A few elements close to the step have some properties like metals and some properties like non-metals. These elements are called metalloids.

Apart from the noble gases, the most reactive elements are near the left and right hand edges of the Periodic Table. Sodium and potassium, two very reactive metals, are on the left hand edge. The next most reactive

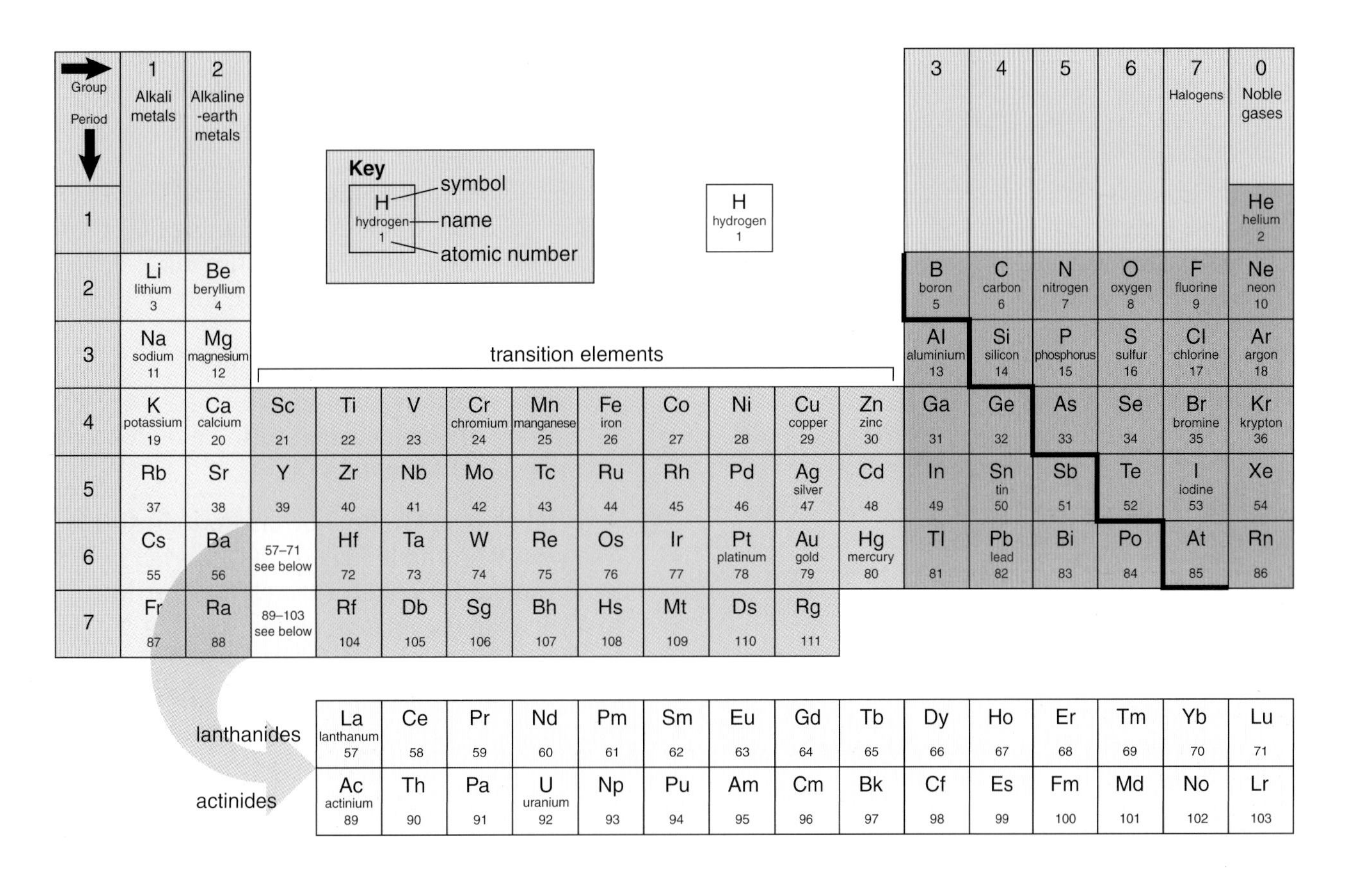

Figure 9.9 The modern Periodic Table

metals, including magnesium and calcium, are in Group 2. Less reactive metals, like iron and copper, are in the centre of the table. Carbon and silicon, unreactive non-metals, are also near the centre of the Periodic Table. Sulfur and oxygen, which are nearer the right hand edge, are more reactive. Fluorine and chlorine, the most reactive non-metals, are very close to the right hand edge.

You should recall from Section 5.5 that the arrangement of elements in modern Periodic Tables is closely related to their electron structures. All the elements in Group 1, which contains sodium and potassium, have just one electron in their outermost shell. This electron is lost fairly easily, so these elements form ions with one positive charge.

At the other extreme, all the halogens have seven electrons in their outermost shell, just one less than the next noble gas. So, halogens gain one electron very readily to form ions with one negative charge.

These points help to illustrate the fact that elements in the same group have similar properties and react in a similar way because they have the same number of electrons in their outer shell (highest occupied energy level).

5 Modern Periodic Tables can be divided into five blocks of elements. Elements within each block have similar properties. These blocks are shown in different colours in Figure 9.10.

a) The five blocks of similar elements are called noble gases, non-metals, poor metals, reactive metals and transition metals. Which name belongs to which coloured block in Figure 9.10?

b) Which groups of the Periodic Table make up the reactive metals?

c) Some non-metals have properties like poor metals. Give one example of this.

d) The noble gases are very unreactive. The first noble gas compound was not made until 1962. Today, several compounds of them are known.

i) Why were the noble gases once called 'inert gases'?

ii) Why do you think their name was changed to 'noble gases'?

iii) Why are there no noble gases in Mendeléev's periodic table?

6 Draw an outline of the Periodic Table like the one in Figure 9.10. On your outline, indicate the position of:

a) all non-metals;

b) the element with atomic number 37;

c) the element with atoms containing 20 protons;

d) atoms with an electron structure of 2, 8, 3;

e) a group of elements that form ions with a charge of 2−;

f) the most reactive alkali metal;

g) the most reactive non-metal;

h) a group of elements that form chlorides with the general formula XCl_2.

7 A metal M is in Group 3 of the Periodic Table. The relative formula mass of its chloride is 176.5.

a) Write the formula for the chloride of M.

b) Calculate the relative atomic mass of M. (Cl = 35.5)

c) Which element is M?

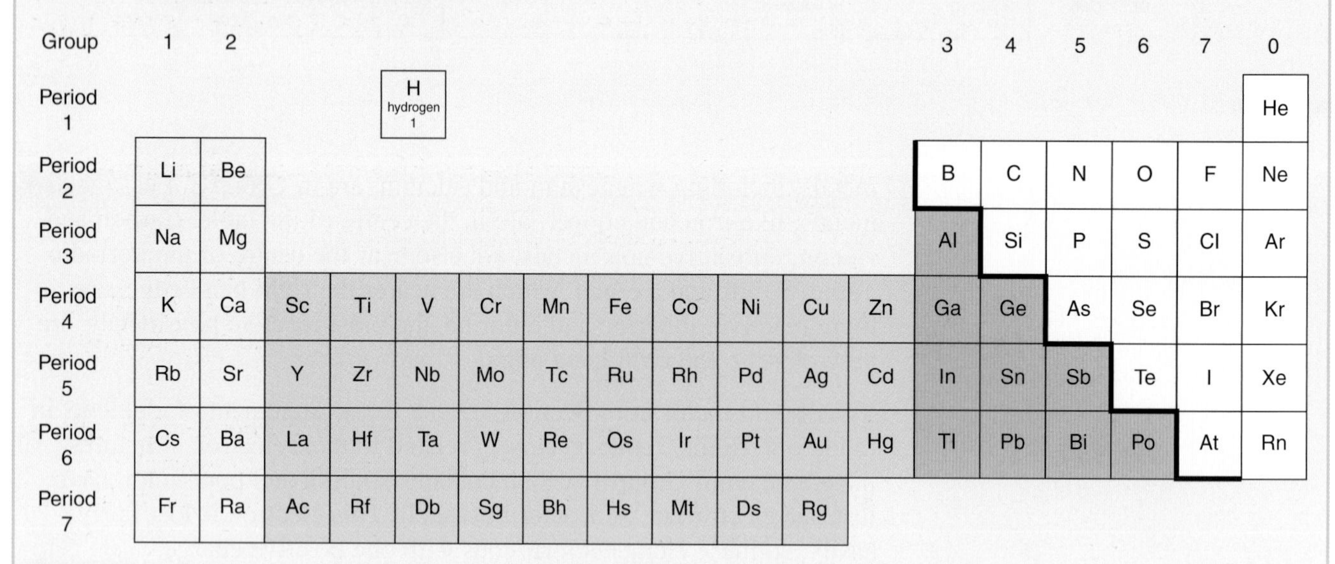

Figure 9.10 The five blocks of similar elements in the Periodic Table

9.4 The alkali metals

You should remember from Section 5.5 that the elements in Group 1 of the Periodic Table are called alkali metals (Figure 9.11). They are called alkali metals because they react with water to form alkaline solutions. The first three elements in the group, lithium, sodium and potassium, are the best known. The other elements in Group 1 are rubidium, caesium and francium.

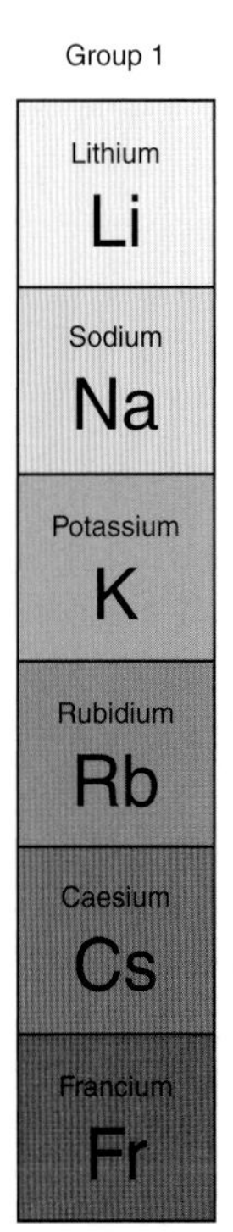

Figure 9.11 The alkali metals in Group 1

Some properties of the alkali metals are summarised in Table 9.1.

Property	Character or behaviour
Appearance	Shiny grey surface that quickly forms a dull layer of oxide
Hardness	Easily cut with a knife – not as hard as typical metals
Density	Low density compared with other metals – lithium, sodium and potassium are less dense than water (1.0 g/cm^3)
Melting and boiling points	Low compared with typical metals
Reaction with non-metals	React vigorously forming ionic compounds. For example: sodium + oxygen → sodium oxide, $(Na^+)_2O^{2-}$ $4Na + O_2 \rightarrow 2Na_2O$
Reaction with water	React with water producing hydrogen and metal hydroxides which dissolve in water to give an alkaline solution. For example: sodium + water → sodium hydroxide + hydrogen $2Na + 2H_2O \rightarrow 2NaOH + H_2$ (Lithium reacts steadily, sodium vigorously, potassium violently.)
Type of compound and ion	Compounds are white ionic solids which dissolve in water to form colourless solutions (unless the anion is coloured). Ions have a charge of 1+ (see Section 5.5), so their compounds have similar formulae. For example, oxides are Li_2O, Na_2O, K_2O and so on.

Table 9.1 Some properties of the alkali metals

Notice in Table 9.1 that alkali metals have some unusual properties for metals.

- They are soft enough to cut with a knife.
- Their melting points and boiling points are unusually low.
- Their densities are much lower than other metals. In fact, lithium, sodium and potassium have such low densities that they float on water as they react with it.

Notice also that sodium reacts with water more vigorously than lithium, and potassium reacts more vigorously than sodium. In general, the elements in Group 1 are more reactive moving down the group from lithium to francium.

The alkali metals become more reactive as their relative atomic mass increases down the group. This trend in reactivity can be explained by the increasing size of their atoms. As their atoms get larger, the outer electrons are in higher energy levels further from the nucleus.

As their outer electrons get further from the nucleus down Group 1, they are less strongly attracted by the positive protons in the nucleus. The outer electrons are also shielded from the attractions of the positive nucleus by more inner shells of electrons as the atoms get larger. For both these reasons, the outer electrons are more easily lost during reactions as the atoms get larger and this makes them more reactive.

8 a) The elements lithium, sodium and potassium are called alkali metals. How are these three metals similar in:
   i) appearance;
   ii) strength (hardness);
   iii) melting point and boiling point compared with other metals;
   iv) density compared with other metals;
   v) reaction with oxygen in the air?
   b) Why are these metals called alkali metals?

9 Write word equations and then balanced chemical equations for the reaction of:
   a) lithium with oxygen;
   b) potassium with water;
   c) sodium with chlorine.

10 Rubidium (Rb) is in Group 1 below potassium. Use the information in Table 9.1 to predict:
   a) the symbol for a rubidium ion;
   b) the formula of rubidium chloride;
   c) the colour of rubidium oxide;
   d) how rubidium reacts with water.

11 The metals in Group 2 are similar to those in Group 1, but not so reactive. Magnesium and calcium are the most common metals from Group 2.
   a) Write the name and formula of the products which form when:
      i) magnesium reacts with oxygen;
      ii) calcium reacts with water.
   b) Which is the more reactive – magnesium or calcium?

Activity – Looking at the properties of alkali metals

The graph in Figure 9.12 shows the melting points of the first four alkali metals plotted against their relative atomic mass.

1. Describe the change in melting points with increasing relative atomic mass shown in Figure 9.12.
2. Use Figure 9.12 to predict the melting point of caesium (Cs = 132.9).
3. Use a data book or the website www.chemicalelements.com to compare your prediction in question **2** with the actual melting point of caesium.
4. Do the melting points for Group 1 elements change in the way you expected? Say 'Yes' or 'No' and explain your answer.
5. Now use a data book or the website to find the boiling points of the elements in Group 1.
6. Plot a graph of the boiling points of the elements in Group 1 against relative atomic mass.
7. Describe the change in boiling points with relative atomic mass for the elements in Group 1.

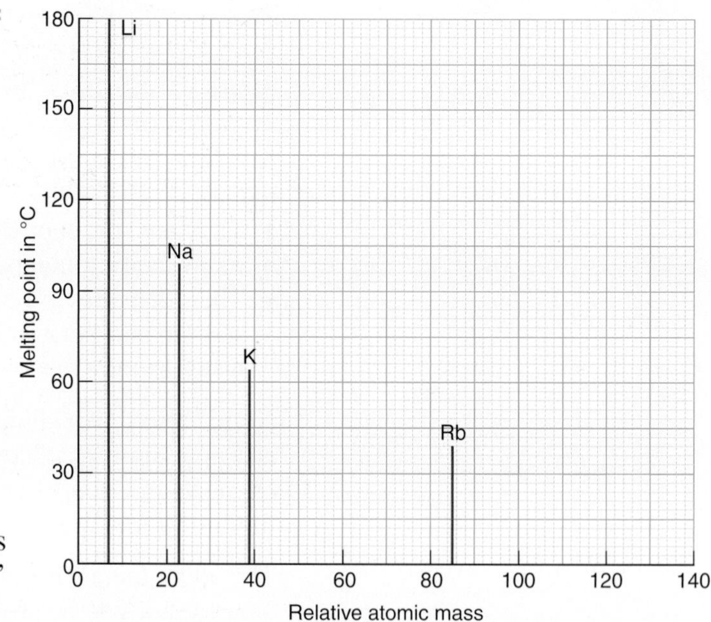

Figure 9.12 The melting points of the first four alkali metals

8. Do the boiling points for elements in Group 1 change in the same way as their melting points?

The halogens

You should recall from Section 5.5 that the elements in Group 7 of the Periodic Table are called halogens (Figure 9.13). The halogens form a group of reactive non-metals. The common elements in the group are chlorine (Cl), bromine (Br) and iodine (I). The other elements in the group are fluorine (F), which is so reactive it is difficult to handle, and astatine (At), which is radioactive and does not occur naturally.

Group 6	Group 7	Group 0
		He
O	**F**	Ne
S	**Cl**	Ar
	Br	Kr
	I	Xe
	At	Rn

Figure 9.13 The halogens

The halogens are so reactive that they never occur naturally as the elements. They are usually found combined with metals in simple ionic compounds called salts, such as sodium chloride (common salt, NaCl) and magnesium bromide ($MgBr_2$). This gives rise to the name 'halogens' which means 'salt formers'.

Figure 9.14 Sodium and most sodium compounds are made from sodium chloride (common salt). In hot countries, impure sodium chloride is left when seawater evaporates. Sodium chloride also occurs as rock salt in salt beds below the ground.

The commonest chlorine compound is, of course, sodium chloride (NaCl) which occurs in seawater and below the ground as rock salt (Figure 9.14). Every kilogram of seawater contains about 30 grams of sodium chloride. Seawater also contains small amounts of bromide and traces of iodides. Certain seaweeds concentrate iodine from seawater.

Patterns in the properties of halogens

The elements in a group of the Periodic Table are similar to each other, but their properties change gradually as you go down the group from one element to the next. Group 7 shows this behaviour very well. The three halogens in Figure 9.16 may look quite different, but a closer look at their properties shows they have a lot in common.

- They have coloured vapours.
- They are all poisonous and smelly.
- They form diatomic molecules – molecules made up of pairs of atoms (Cl_2, Br_2, I_2) – see Figure 9.15.

- They react with metals to form ionic salts with similar formulae (sodium chloride, NaCl; sodium bromide, NaBr; sodium iodide, NaI) in which the halide ions (chloride, Cl^-; bromide, Br^- and iodide, I^-) each carry a charge of −1.
- They react with other non-metals to form simple molecular compounds.

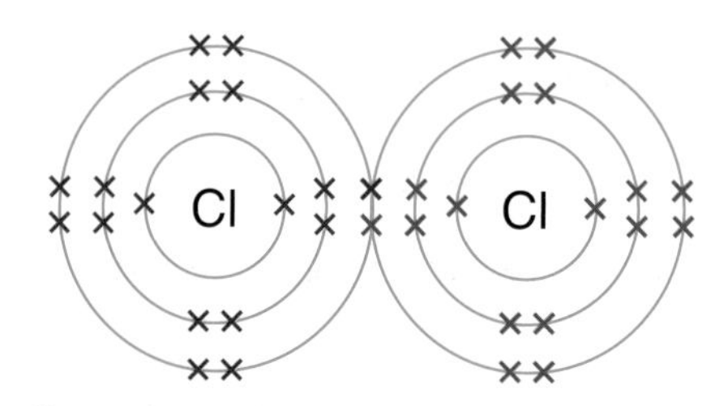

Figure 9.15 All the halogen elements have seven electrons in their outer shell. So, pairs of atoms form stable molecules by sharing one pair of electrons in a covalent bond. This gives each atom an electron structure similar to the next noble gas in the Periodic Table.

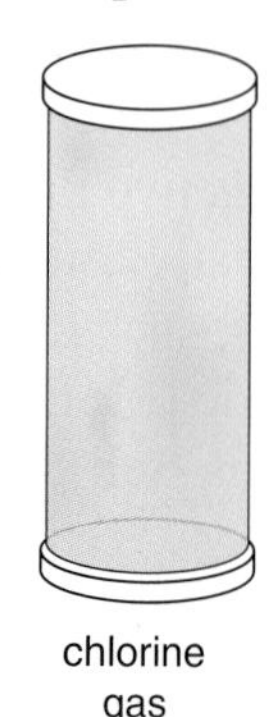

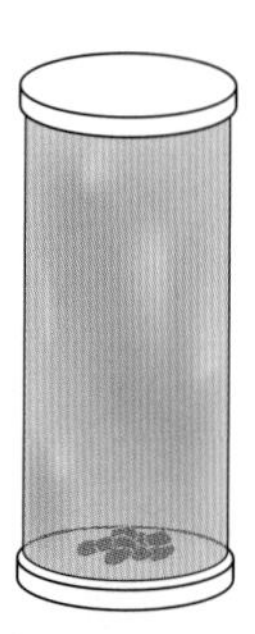

Figure 9.16 At room temperature, chlorine is a gas, bromine is a liquid with a dense vapour and iodine is a solid with vapour.

Activity – Looking at the properties of the halogens

Look at the information in Table 9.2 and then answer the questions below.

1. How do the following properties of the halogens change as their relative atomic mass increases?
   a) state at room temperature
   b) colour of vapour
   c) melting point
   d) boiling point.
2. Use the data in Table 9.2 to plot a graph of the melting points of the halogens against their relative atomic mass.
   a) Why are the melting points of the halogens relatively low?
   b) Explain the change in melting point for the halogens as their relative atomic mass increases.
3. Do you think there will be a clear relationship between the melting points and boiling points of the halogens? Say 'Yes' or 'No' and explain your answer.
4. a) Plot an appropriate graph to check your prediction in question **3**.
   b) What conclusion can you make from your graph?
5. Use the data in Table 9.2 and the graphs you have just drawn to predict the following properties for astatine (A_r(At) = 210):
   a) colour
   b) state at room temperature
   c) melting point
   d) boiling point.

Element	Relative atomic mass	Colour and state at room temperature	Colour of vapour	Melting point in °C	Boiling point in °C
Fluorine	19.0	Pale yellow gas	Pale yellow	−220	−188
Chlorine	35.5	Pale green gas	Pale green	−101	−35
Bromine	79.9	Red-brown liquid	Orange	−7	59
Iodine	126.9	Dark grey solid	Purple	114	184

Table 9.2

9.6 Reactions of the halogens

The halogens are a reactive group of elements. In fact, fluorine is the most reactive of all the non-metals. As you would expect, there is a gradual change as you go down the group from one element to the next.

Table 9.3 summarises some reactions of the halogens.

Reaction	Chlorine	Bromine	Iodine
With damp litmus paper	Bleaches the litmus paper very quickly. This is a test for chlorine	Bleaches the litmus paper slowly	Bleaches the litmus paper very slowly
With iron	Iron wool reacts vigorously to form iron(III) chloride once the reaction has started $2Fe + 3Cl_2 \rightarrow 2FeCl_3$ (Figure 9.17)	Iron wool reacts steadily to form iron(III) bromide provided it is heated all the time $2Fe + 3Br_2 \rightarrow 2FeBr_3$	Iron wool reacts very slowly to form iron(III) iodide even if heated all the time $2Fe + 3I_2 \rightarrow 2FeI_3$
With sodium chloride solution	No reaction	No reaction	No reaction
With sodium bromide solution	Chlorine displaces yellow-orange bromine $Cl_2 + 2NaBr \rightarrow Br_2 + 2NaCl$	No reaction	No reaction
With sodium iodide solution	Chlorine displaces brown iodine $Cl_2 + 2NaI \rightarrow I_2 + 2NaCl$	Bromine displaces brown iodine $Br_2 + 2NaI \rightarrow I_2 + 2NaBr$	No reaction

Table 9.3 Some reactions of the halogens chlorine, bromine and iodine

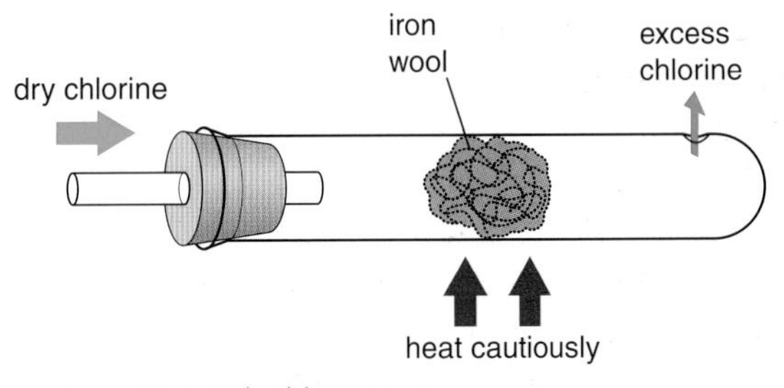

Figure 9.17 Iron wool reacting with chlorine

Notice in Table 9.3 that the reactions of the halogens are very similar with gradual changes down the group. Chlorine is the most reactive, then bromine and then iodine.

These results show that the halogens get less reactive down the group as their relative atomic mass increases. This is opposite to the trend in Group 1 where the alkali metals get more reactive down the group.

Displacement reactions

The **displacement reactions** in Table 9.3 confirm the relative reactivity of the halogens. Displacement reactions sometimes occur when we try to react one halogen with compounds of other halogens.

Displacement reactions occur when a more reactive element displaces a less reactive element from one of its compounds.

For example, when chlorine water (water containing dissolved chlorine) is added to a solution of sodium bromide, the chlorine displaces bromine. The chlorine is more reactive than bromine, so it pushes out the less reactive bromine and forms a solution of sodium chloride.

chlorine + sodium bromide → bromine + sodium chloride

$$Cl_2(aq) + 2NaBr(aq) \rightarrow Br_2(aq) + 2NaCl(aq)$$

In the same way, bromine is more reactive than iodine. So, bromine reacts with sodium iodide, displacing iodine and forming sodium bromide.

bromine + sodium iodide → iodine + sodium bromide

$$Br_2(aq) + 2NaI(aq) \rightarrow I_2(aq) + 2NaBr(aq)$$

These results illustrate the general rule that a more reactive halogen can displace a less reactive halogen from an aqueous solution of its salt.

Explaining the trend in reactivity of halogens

In Group 7, the halogens get less reactive down the group from fluorine to astatine. All the halogens have seven electrons in their outer shell. When the halogens react, they gain one electron to obtain a stable electron structure like the next noble gas in the Periodic Table. This additional electron is gained less easily as the atoms get larger because the attraction of the positive nucleus is much less. Also, as the atoms get larger, the nucleus is surrounded by more shells of electrons. The extra shells provide an increasingly large repelling effect to any electron that might be added.

12. Arrange the following pairs of elements in order of decreasing vigour of reaction.
    A lithium and iodine
    B potassium and chlorine
    C rubidium and fluorine
    D sodium and chlorine.

13. Describe experiments which show the relative reactivity of bromine, chlorine and iodine. Write equations for the reactions you describe.

14. When chlorine is bubbled into water, some chlorine dissolves in the water and some reacts with water to form a mixture of hydrochloric acid ($HCl(aq)$) and chloric acid ($HClO(aq)$). When blue litmus paper is added to the final solution, it turns red at first and then goes white.
    a) What does (aq) mean in the formula $HCl(aq)$?
    b) Write a word equation for the reaction of chlorine with water.
    c) Write a balanced chemical equation with state symbols for the reaction of chlorine with water.
    d) Why does blue litmus paper turn red at first with the final solution above?
    e) Why does the blue litmus paper finally go white?

9.7 The transition metals

In the Periodic Table between Groups 2 and 3, there is a block of elements known as the transition elements (Figure 1.38, Section 1.7). The elements are all metals and are often called the transition metals.

The transition elements have very similar properties to one another. Unlike other parts of the Periodic Table, there are similarities throughout the whole block of the transition metals, across the periods as well as down the groups.

Two of the most important transition metals are iron and copper. Iron is the most widely used metal. More than 700 million tonnes of iron are manufactured every year throughout the world (Section 1.7). Almost all of this is converted to steel, which is hard, strong and relatively cheap. It is used in supports for bridges and buildings, and in vehicles, engines and tools.

Figure 9.18 The roof and arches of the new Wembley Stadium are supported by massive amounts of iron in steel girders.

Figure 9.19 This statue has been made from copper. Copper is used because it is easy to shape and it gets an attractive green coating as it weathers. The green coating is a thin layer of copper hydroxide and copper carbonate on the surface of the copper.

Copper is the third most widely used metal after iron and aluminium. About 12 million tonnes are manufactured each year. Copper is a good conductor of heat and electricity. It is also malleable and ductile, which means that it can be made into different shapes and drawn into wires. Because of these properties, copper is used in electrical wires and cables and in hot water pipes.

Activity – Comparing the transition metals with the alkali metals in Group 1

Some properties of iron and copper are shown in Table 9.4. The information for iron and copper in the table is fairly typical of transition metals. In this activity you will be comparing these properties of iron and copper in Table 9.4 on page 182 with the properties of alkali metals in Table 9.1 and Section 9.4.

1. How do the melting points and boiling points of transition metals compare with those of alkali metals? (Remember that the properties of iron and copper are typical of those for other transition metals.)
2. One transition metal has a very anomalous melting point, much lower than all the others. It is even liquid at room temperatures.
   a) Which element is this?
   b) What is meant by anomalous in this context?
3. How do the densities of transition metals compare with those of the alkali metals?

4. From your own experience of iron (steel) and copper, how does the strength and hardness of transition metals compare with that of the alkali metals?
5. Compare the reactions of transition metals and alkali metals with water and oxygen (air). Would you say the transition metals were much more reactive / more reactive / less reactive / much less reactive than alkali metals?
6. The salts of transition metals are usually coloured. What is the usual colour of the salts of alkali metals?
7. Compare the transition metals and alkali metals in terms of the number of ions that they form.

Element	Melting point in °C	Boiling point in °C	Density in g/cm^3	Reaction with water	Symbols of ions	Colour of salts
Iron	1540	3000	7.9	Does not react with pure water. Reacts slowly with steam	Fe^{2+} Fe^{3+}	Fe^{2+} salts are green Fe^{3+} salts are yellow or brown
Copper	1080	2600	8.9	No reaction with water or steam	Cu^{+} Cu^{2+}	Cu^{2+} salts are blue or green

Table 9.4 Some properties of iron and copper

Special properties of the transition elements

The transition elements have some special properties unlike those of any other elements in the Periodic Table.

Ions with different charges

Most of the transition elements form stable ions with different charges. For example, iron forms Fe^{2+} and Fe^{3+} ions. This results in two sets of compounds, iron(II) compounds and iron(III) compounds. So, iron forms both iron(II) chloride, $FeCl_2$ and iron(III) chloride, $FeCl_3$; iron(II) sulfate, $FeSO_4$ and iron(III) sulfate, $Fe_2(SO_4)_3$. Copper forms Cu^+ and Cu^{2+} salts resulting in copper(I) and copper(II) compounds (Figure 9.20).

Figure 9.20 Copper has two oxides and two chlorides.

In contrast, other elements form only one stable ion. For example, the alkali metals in Group 1 form only one stable ion with a charge of 1+ (Na^+, K^+, etc.) and the halogens in Group 7 form only one stable ion with a charge of 1− (Cl^-, Br^-, etc.).

Coloured compounds

Transition metals usually have coloured compounds which produce coloured solutions if they dissolve in water. Most iron(II) compounds are green, most iron(III) compounds are yellow or brown, and most copper(II) compounds are blue or green. In contrast, alkali metals have white salts with colourless solutions, unless their anions are coloured. The coloured compounds of transition elements produce the vivid colours we see in oil paintings, pottery glazes and precious stones (Figure 9.21).

Figure 9.21 Artists produce vivid colours on their paintings using pigments which often contain the compounds of transition metals.

Catalytic properties

Transition elements and their compounds can act as catalysts. For example, iron or iron(III) oxide is used as a catalyst in the Haber process to manufacture ammonia, NH_3 (Section 7.7). Nickel is used as a catalyst in the production of margarine from vegetable oil (Section 3.4). Platinum is used as a catalyst in the manufacture of sulfuric acid and in catalytic converters in vehicles.

Why do transition elements have these special properties?

The transition elements have similar properties to one another, and some special properties. The reason for their similarities is that, instead of adding extra electrons to the outer shell as their atomic number increases, an inner shell (at a lower energy level) is being filled.

This extra space for more electrons appears for the first time in the third energy level, once the fourth energy level contains two electrons at calcium. Once the extra space for electrons in the third energy level has become available, the third shell can hold a maximum of 18 electrons even though all the elements already have two electrons in the fourth shell.

As electrons go into a lower energy level between Groups 2 and 3, all the transition elements have two electrons in their outer shell. This similarity in outer shell electrons gives all transition metals similar chemical properties.

The addition of electrons to a lower energy level also results in atoms of the same element having two or more stable ions with different charges. This is because the elements can lose a variable number of electrons from the lower energy level when they form ions. The different ions produce sets of compounds with different colours.

15 a) What are the transition elements?
b) Make a list of the six most characteristic properties of the transition elements.

16 Write balanced chemical equations for the following word equations:
a) iron + oxygen → iron(III) oxide
b) iron + chlorine → iron(III) chloride
c) iron + hydrogen chloride → iron(II) chloride + hydrogen

17 a) Explain why iron is manufactured in much larger quantities than any other metal.
b) Explain why copper is used more than any other metal in the production of electrical wires and cables.
c) Look at the photo in Figure 9.19, page 181.
i) What compounds in the atmosphere does copper react with to form copper hydroxide and copper carbonate?
ii) Write the symbol and charge for the copper ion in these compounds.

Summary

✓ Early in the nineteenth century, Dobereiner first tried to classify elements in a systematic way by identifying triads of similar elements. This was followed by Newlands' Octaves. Soon afterwards, Mendeléev's early periodic table attempted to classify elements by arranging them in order of their atomic weights (relative atomic masses).

✓ A **periodic table** is a table in which the chemical elements are carefully arranged so that elements with similar properties occur at regular intervals and fall in the same vertical column.

✓ In a periodic table, the vertical columns of similar elements are called groups and the horizontal rows of elements are called periods.

✓ Mendeléev overcame deficiencies in earlier periodic tables by leaving gaps for elements that had not been discovered. He went on to predict accurately the properties of some of the missing elements. Because of this, his periodic table was soon accepted as an important working tool.

✓ The most obvious differences between modern Periodic Tables and that proposed by Mendeléev are:
- the removal of the transition elements from the numbered groups to their placement between Groups 2 and 3; and
- the addition of Group 0, the noble gases.

✓ With the rapid growth of chemical knowledge in the early part of the twentieth century, chemists realised that the Periodic Table provided a brilliant summary not only of the properties of elements, but also of their atomic structures.

✓ The elements in Group 1, the alkali metals:
- are metals with low densities, low melting points and low boiling points;
- react with non-metals to form ionic compounds;
- react with water, producing hydrogen and alkaline solutions of hydroxides;
- become more reactive down the group.

✓ The elements in Group 7, the halogens:
- are reactive non-metals;
- exist as diatomic molecules;
- have coloured vapours;
- react with metals forming ionic solids;
- react with other non-metals to form simple molecular compounds;
- become less reactive down the group.

✓ **Displacement reactions** occur when a more reactive element (e.g. a more reactive halogen) displaces a less reactive element (e.g. a less reactive halogen) from one of its compounds.

✓ Transition elements (transition metals) are a block of elements between Group 2 and Group 3 in the Periodic Table.

✓ Compared with the alkali metals in Group 1, transition metals:
- have higher densities, melting points and boiling points.
- are stronger and harder;
- are much less reactive.

✓ Transition metals also:
- form coloured compounds;
- have ions with different charges;
- are useful as catalysts.

EXAMQUESTIONS

1. Look at Mendeléev's periodic table in Figure 9.5.
   a) Name two elements in Group 1 of Mendeléev's periodic table which are not in Group 1 of the modern Periodic Table. *(2 marks)*
   b) Which group in the modern Periodic Table is missing from Mendeléev's table? *(1 mark)*
   c) Why is this group missing from Mendeléev's table? *(1 mark)*
   d) Why did Mendeléev leave gaps in his periodic table? *(1 mark)*
   e) On what basis did Mendeléev arrange the elements in his table? *(2 marks)*
   f) On what basis are elements arranged in the modern Periodic Table? *(1 mark)*

2. Figure 9.22 summarises some reactions of the element sodium which occurs in Group 1 of the Periodic Table. The products of the reactions are labelled A, B, C and D.

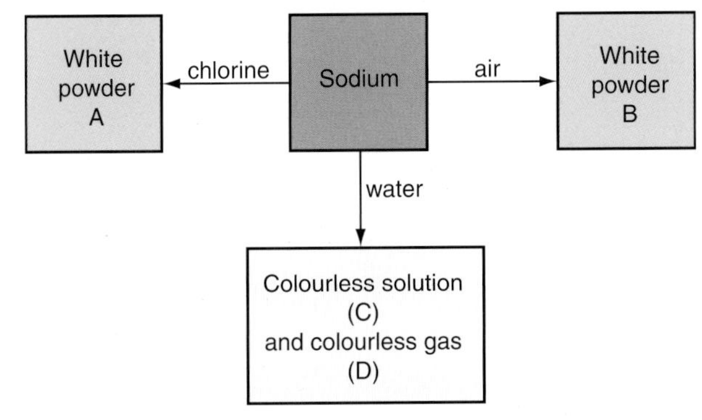

Figure 9.22

   a) Give the chemical names of substances labelled A, B, C and D. *(4 marks)*
   b) Copy, complete and balance the equation below for the reaction that occurs when sodium burns in air.

      ______ Na + ______ $O_2 \rightarrow$ ______ Na_2O *(3 marks)*

   c) i) Which reaction in Figure 9.22 shows that sodium is an alkali metal? *(1 mark)*
      ii) Why is sodium described as an alkali metal? *(1 mark)*

3. This question is about the halogen elements in Group 7 of the Periodic Table.
   a) Which element in Group 7:
      i) is most reactive; *(1 mark)*
      ii) occurs in the largest quantity in seawater; *(1 mark)*
      iii) is liquid at room temperature? *(1 mark)*
   b) Table 9.5 summarises the results of reactions when halogens are added to solutions of sodium halides.
      i) Copy and complete the table by adding a tick if a reaction takes place and a cross if there is no reaction.

Solution used	Halogen added: Bromine	Chlorine	Iodine
Sodium bromide	✗		
Sodium chloride	✗	✗	
Sodium iodide	✓		✗

Table 9.5

(4 marks)

      ii) The reactions which occur are examples of displacement reactions. Explain what is meant by a displacement reaction. *(3 marks)*
      iii) Write a word equation for the reaction which occurs when bromine is added to sodium iodide solution. *(2 marks)*
      iv) Write a balanced chemical equation for the reaction which occurs when bromine is added to a solution of sodium iodide (NaI). *(2 marks)*

4. Some of the properties of vanadium are given in the list below.
   - It has a high melting point.
   - It is a solid at room temperature.
   - It conducts electricity.
   - It forms coloured compounds.
   - It forms crystalline compounds.
   - It forms compounds that are catalysts.

   a) Which property suggests most strongly that vanadium is a metal? *(1 mark)*
   b) Which two properties suggest strongly that vanadium is a transition element? *(2 marks)*

EXAMQUESTIONS

c) Vanadium(V) reacts with chlorine to form vanadium(IV) chloride and it reacts with oxygen to form vanadium(V) oxide.
   i) Write the formulae of vanadium(IV) chloride and vanadium(V) oxide. *(2 marks)*
   ii) Write a balanced equation for the reaction of vanadium with oxygen to form vanadium(V) oxide. *(3 marks)*

5 a) State three properties in which alkali metals differ from typical transition metals. *(3 marks)*

b) How does the reactivity of the elements in Group 1 change with increasing atomic number? *(1 mark)*

c) Explain in terms of electron structure why lithium differs in reactivity from sodium. *(4 marks)*

Chapter 10
What are strong and weak acids and alkalis?

At the end of this chapter you should:

- ✓ know that acids produce H^+ ions and alkalis produce OH^- ions in aqueous solutions;
- ✓ know that the H^+ ion is a proton which is hydrated in water and represented as $H^+(aq)$;
- ✓ understand that strong acids and alkalis are completely ionised in water, but weak acids and alkalis are only partly ionised in water;
- ✓ understand that water is normally present when most substances act as acids and bases;
- ✓ know how the volumes of acid and alkali solutions that react with each other can be measured by titration;
- ✓ understand the choice of indicator for different combinations of strong or weak acids and alkalis;
- ✓ be able to calculate the volumes, amounts and concentrations of substances involved in titrations;
- ✓ be able to judge the contributions of Arrhenius, Brønsted and Lowry to our understanding of acid–base behaviour and the readiness with which their ideas were accepted;
- ✓ understand the definition of acids as proton donors and bases as proton acceptors.

Figure 10.1 Acids and alkalis are everywhere. They are present in our food, in our bodies, in our medicines and in the soil. Some plants, such as heathers, grow best in slightly acidic soils, whereas other plants, such as beech trees, grow best in slightly alkaline soils.

10.1 What do we already know about acids and alkalis?

Figure 10.2 All fruits contain acids. These acids give many fruits, such as oranges and lemons, a sharp taste. In other fruits, such as peaches and strawberries, there is less acid and the taste of the acid is masked by sugars and other constituents.

Acids and alkalis are soluble in water

Many drinks contain acids and other substances dissolved in water. Fruity drinks often contain citrus fruits such as oranges, lemons or grapefruit. These fruits contain citric acid. Alkalis are defined as soluble bases (Section 8.2). Alkalis such as calcium hydroxide and sodium hydroxide are important in industry. Large amounts of sodium hydroxide solution are used to make soaps, paper and ceramics.

Acids and alkalis give characteristic colours with indicators

Acids give a red colour with litmus and an orange or red colour with universal indicator. Alkalis give a blue colour with litmus and a green, blue or violet colour with universal indicator (Figure 10.3).

Indicators can be used to test for acids and alkalis. However, chemists use a scale of numbers called the pH scale (Section 8.2) when they want to describe accurately how acidic or how alkaline something is.

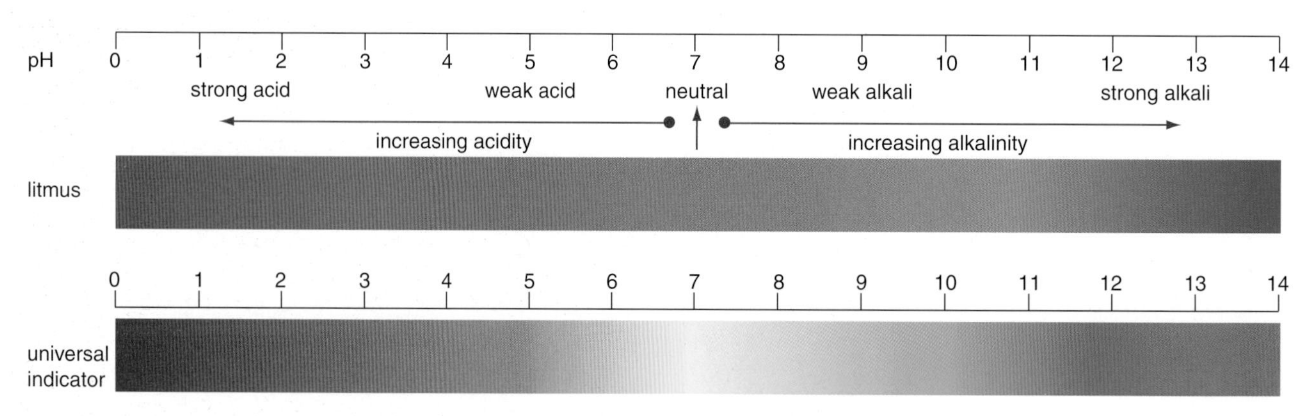

Figure 10.3 The colours of litmus and universal indicator with solutions of different pH.

Solutions of acids and alkalis contain ions

In Chapter 8 (Section 8.2) you learned that acid solutions contain H^+ ions and alkali solutions contain OH^- ions. This means that when acids and alkalis are dissolved in water they produce ions, and the solutions can conduct electricity. Table 10.1 shows the ions produced by some acids and alkalis.

Acids react with alkalis to form salts and water

You studied reactions between acids and alkalis, called neutralisation reactions, in Section 8.2. By choosing which acid and alkali to use, these reactions can be used to prepare particular salts.

Acid or alkali	Formula	Ions produced in solution
Hydrochloric acid	HCl	$H^+(aq) + Cl^-(aq)$
Sulfuric acid	H_2SO_4	$2H^+(aq) + SO_4^{2-}(aq)$
Ethanoic acid	CH_3COOH	$H^+(aq) + CH_3COO^-(aq)$
Sodium hydroxide	$NaOH$	$Na^+(aq) + OH^-(aq)$
Calcium hydroxide	$Ca(OH)_2$	$Ca^{2+}(aq) + 2OH^-(aq)$

Table 10.1 The ions produced by acids and alkalis in aqueous solution

10.2 What part does water play in acidity and alkalinity?

All these properties of acids and alkalis apply to solutions in water. But what happens when water is not present?

Dry hydrogen chloride and dry sulfuric acid have no effect on dry litmus paper or magnesium ribbon (Figure 10.4). So, substances which we call acids do not behave like acids in the absence of water. When water is added, they become acidic straight away. Blue litmus paper turns red and bubbles of hydrogen are produced with magnesium.

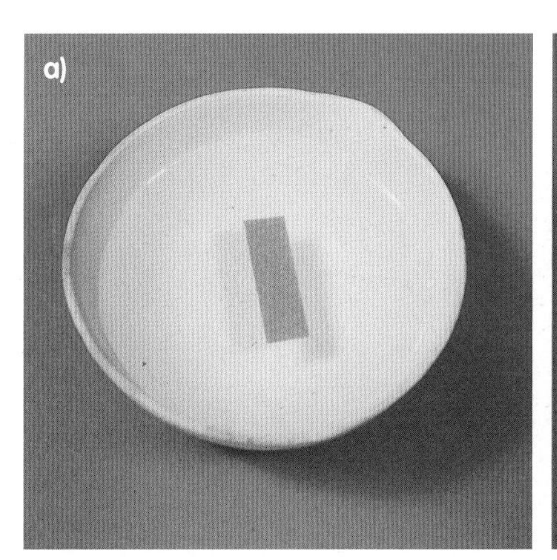

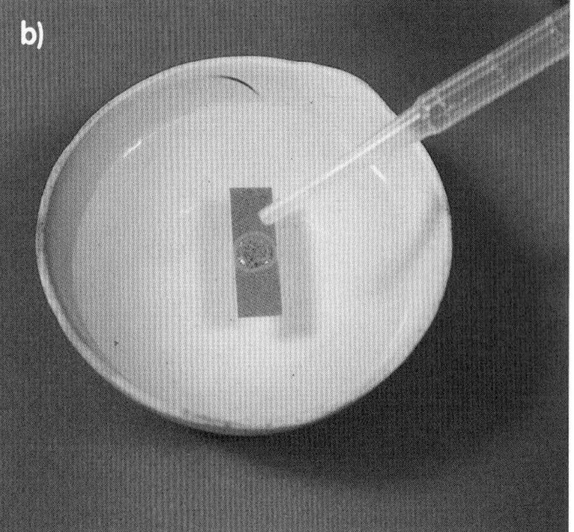

Figure 10.4 a) When blue litmus paper is placed on dry sulfuric acid, it remains blue. b) When one drop of water is added, the litmus paper turns red.

Similar experiments with dry calcium hydroxide show that water must be present before it behaves like an alkali.

As a general rule, water must be present for substances to act as acids and bases. When acids dissolve in water, H^+ ions are formed. When alkalis dissolve in water, OH^- ions are formed.

A hydrogen ion (H^+) is formed when a hydrogen atom loses its one electron. Hydrogen atoms contain just one proton and one electron, therefore the H^+ ion is a proton. In water, these protons are attached to water molecules. Because of this, the H^+ ions are described as hydrated and are represented as $H_3O^+(aq)$ (i.e. $H_2O + H^+ \rightarrow H_3O^+$), or more usually as $H^+(aq)$ for short.

There are no hydrogen ions in dry hydrogen chloride or dry sulfuric acid. If water is added, then these substances react with water to produce hydrated H^+ ions ($H_3O^+(aq)$), which makes them acidic.

$$HCl(g) + H_2O(l) \rightarrow H_3O^+(aq) + Cl^-(aq)$$

$$H_2SO_4(l) + 2H_2O(l) \rightarrow 2H_3O^+(aq) + SO_4^{2-}(aq)$$

In a similar way, alkalis like sodium hydroxide and calcium hydroxide cannot act as alkalis until hydroxide ions are present in aqueous solution.

$$NaOH(s) \xrightarrow{\text{water}} Na^+(aq) + OH^-(aq)$$

1. a) Describe how you would check the pH of a soil sample.
   b) Why is it important for gardeners to know the pH of their soil?
2. Make a list of six different foods which contain acids and state what acids they contain.
3. a) Write down the ions formed when the following dissolve in water:
   i) nitric acid, HNO_3;
   ii) potassium hydroxide, KOH.

   b) Copy and complete the following equation for the reaction between nitric acid and potassium hydroxide.

   $HNO_3(aq) + KOH(aq) \rightarrow$ ________ + ________
4. Acids produce hydrogen at the cathode during electrolysis. Why does this suggest that acids contain H^+ ions?
5. Pure ethanoic acid (CH_3COOH) is a liquid at room temperature. Describe how you would show that it does not act like an acid until water is added to it.

10.3 Strong and weak acids and alkalis

Strong acids and **strong alkalis** are completely ionised when they dissolve in water. Examples of strong acids are hydrochloric, sulfuric and nitric acids. Examples of strong alkalis are sodium and potassium hydroxide.

Weak acids and **weak alkalis** are only partially ionised in water. Examples of weak acids are ethanoic, citric and carbonic acids. An example of a weak alkali is ammonia.

When different acids with the same concentration in moles per cubic decimetre (mol/dm^3) are tested with universal indicator, they give different pHs (Figure 10.5). The different pH values mean that some acids produce H^+ ions more easily than others. The results in Figure 10.5 show that hydrochloric acid, nitric acid and sulfuric acid produce more H^+ ions in water than the other acids. We say that they ionise (split up into ions) more easily. They are described as **strong acids**. The other acids, including sulfurous acid, ethanoic acid and carbonic acid, are described as **weak acids**. Strong acids and weak acids also show a difference in their reactions with metals. Strong acids react much more quickly than weak acids.

We can show the difference between strong and weak acids in the way we write equations for their ionisation.

hydrochloric acid: $HCl(aq) \rightarrow H^+(aq) + Cl^-(aq)$

ethanoic acid: $CH_3COOH(aq) \rightleftharpoons H^+(aq) + CH_3COO^-(aq)$

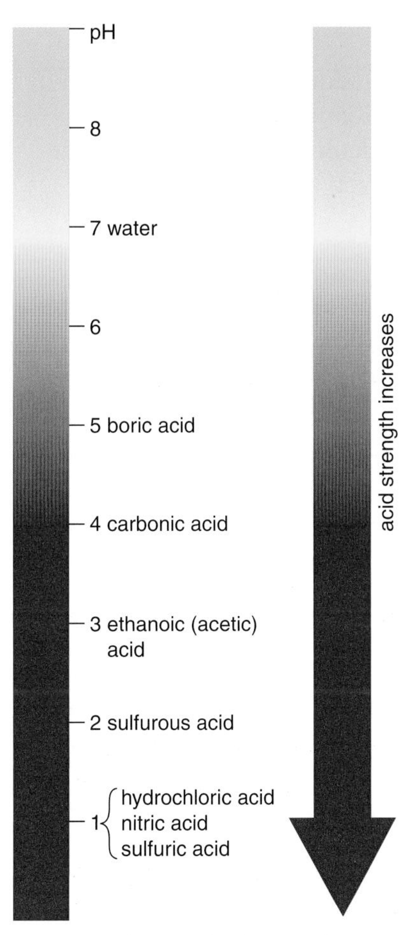

Figure 10.5 The pH of solutions of various acids. (All solutions have a concentration of 0.1 mol/dm^3.)

A single arrow in the hydrochloric acid equation shows that all the HCl molecules have formed H^+ and Cl^- ions in aqueous solution. The equilibrium arrows in opposite directions for ethanoic acid show that some CH_3COOH forms H^+ and CH_3COO^- ions, but part of it remains unchanged as CH_3COOH molecules.

Alkalis can also be classified by the extent of their ionisation in water. Strong alkalis, such as sodium hydroxide, are completely ionised in water.

$$NaOH(aq) \rightarrow Na^+(aq) + OH^-(aq)$$

Weak alkalis, such as ammonia, are only partially ionised in water.

$$NH_3(aq) + H_2O(l) \rightleftharpoons NH_4^+(aq) + OH^-(aq)$$

We use the terms 'concentration' and 'strength' when we describe acids and alkalis. These do not mean the same thing – a concentrated acid is not the same as a strong acid.

- Concentration tells you how much acid or alkali is dissolved in a solution, and we use the words 'concentrated' and 'dilute'. A concentrated solution has a lot of solute dissolved in it, while a dilute solution has only a small amount of solute.
- Strength tells you the extent to which an acid or alkali is ionised, and we use the words 'strong' and 'weak'.

6 a) What is the difference between a strong acid and a weak acid?
b) i) Why can vinegar (which contains ethanoic acid) be used to descale kettles?
ii) Why is sulfuric acid not used to descale kettles?

7 a) Calcium hydroxide ($Ca(OH)_2$) is a strong alkali. What does this mean?
b) Write an equation for the ionisation of calcium hydroxide in water. (Hint: Look at Table 10.1.)
c) Do you think magnesium hydroxide is a strong or weak alkali? Write 'strong' or 'weak' and explain your answer.
d) How would you test to decide whether magnesium hydroxide is a strong or weak alkali?

10.4 Finding the volumes of acids and alkalis that react

When an alkali neutralises an acid, the pH of the solution changes. We can follow this pH change using an indicator. The indicator shows us how much alkali just reacts with all the acid. when there is no acid or alkali left over.

Activity – Comparing the acidity of different vinegars

Anna and Carl decided to compare the acidity of two different white wine vinegars. The acid in vinegar is ethanoic acid. They agreed to use 25 cm^3 of vinegar and measure how much sodium hydroxide solution just reacts with each vinegar in turn using phenolphthalein (pronounced 'feenolthayleen') indicator. Phenolphthalein is colourless in the acid and red in the alkali.

Anna measured 25 cm^3 of the first white wine vinegar into a conical flask using a pipette (Figure 10.6).

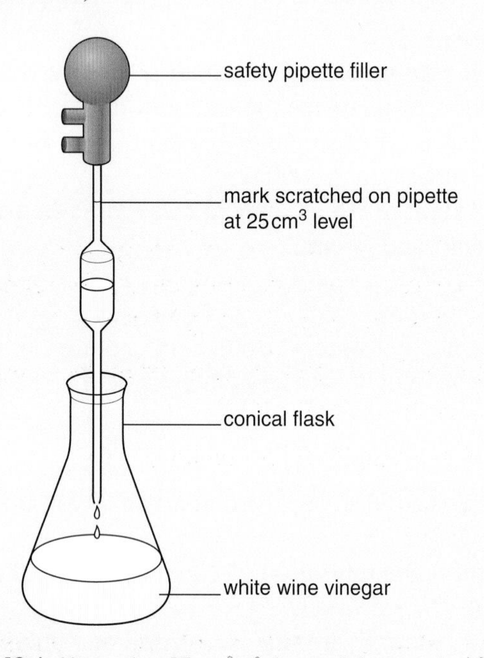

Figure 10.6 Measuring 25 cm^3 of vinegar into a conical flask using a pipette

After Anna had added 5 drops of phenolphthalein indicator, Carl added the sodium hydroxide solution slowly from a burette (Figure 10.7), keeping the contents of the conical flask well mixed. When the first tinge of red appeared in the mixture, Carl noted the volume of sodium hydroxide added.

This method of adding one solution to another to find out how much of the two solutions will just react is called a **titration**.

The titration which Anna and Carl have just done is only a rough titration. They now carry out three accurate titrations. To do this they add 1 cm^3 less sodium hydroxide solution than the amount they used in their rough titration. Then they add the sodium hydroxide solution one drop at a time until the first tinge of pink/red appears in the mixture.

A **titration** involves adding one solution from a burette to a measured amount of another solution in order to determine how much of the two solutions just react.

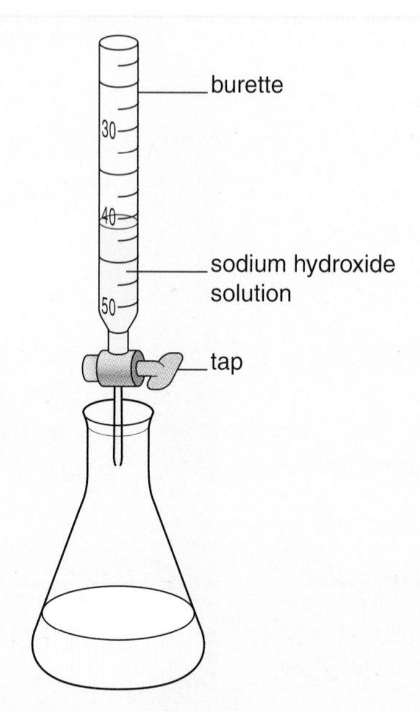

Figure 10.7 Adding sodium hydroxide solution from a burette to 25 cm^3 of white wine vinegar in a conical flask

Anna and Carl's results are shown in Table 10.2, together with those for the second white wine vinegar.

1. Why does the mixture in the conical flask suddenly change from colourless to pink/red during the addition of sodium hydroxide solution?
2. Why must the contents of the conical flask be well mixed during the titration?
3. Calculate the volumes of sodium hydroxide solution added in the rough and accurate titrations for both wine vinegars in Table 10.2.
4. Calculate an average value for the accurate titration with the first wine vinegar.
5. What did Anna and Carl do to increase the reliability of their results?

6 a) What value will you take for the accurate titration with the second wine vinegar?

b) Explain why you have taken this value.

7 Compare the acidity of the two vinegars.

8 State three factors Anna and Carl must control in their experiments in order to compare the two vinegars fairly.

Experiment	Rough titration	Accurate titrations		
First wine vinegar:				
final burette reading in cm^3	20.50	40.15	19.60	39.30
initial burette reading in cm^3	0.00	20.50	0.00	19.60
Volume of sodium hydroxide added in cm^3				
Second wine vinegar:				
final burette reading in cm^3	13.70	26.80	40.65	13.10
initial burette reading in cm^3	0.00	13.70	26.80	0.00
Volume of sodium hydroxide added in cm^3				

Table 10.2 The results of Anna and Carl's experiments

To get accurate results in experiments we must try to avoid errors in measurements.

Random errors cause readings to be different from the true value. These can be caused by faulty equipment, mistakes in reading scales or other human error. We can detect and compensate for random errors by taking many readings (e.g. by doing three accurate titrations in the Activity above).

Systematic errors affect all the results in an experiment. The results are shifted away from the true value because the apparatus has an error in it (e.g. if the scale on a burette is inaccurate).

A **zero error** is a type of systematic error where the instrument has a false zero reading.

Titrating acids and alkalis

As we found in the activity above, the volumes of acid and alkali that react with each other can be measured using indicators. When alkalis are added to acids, a point is reached when the pH changes very rapidly from acid (when there is more acid than alkali) through neutral to alkaline (when there is excess alkali). This point of rapid pH change is called the **end point** of the titration. However, the actual pH change varies depending on whether the acid and alkali are strong or weak (Figure 10.8).

Figure 10.8 shows that:

- the pH changes rapidly from pH 3 to pH 10 when a strong acid is titrated with a strong alkali;
- the pH changes rapidly from pH 3 to pH 7 when a strong acid is titrated with a weak alkali;
- the pH changes rapidly from pH 7 to pH 10 when a weak acid is titrated with a strong alkali;
- the pH changes slowly through pH 7 when a weak acid is titrated with a weak alkali.

The **end point** of a titration occurs when the two solutions just react and neither is in excess.

8 Look closely at the ranges of the indicators in Figure 10.8.
a) What is the pH of pure water?
b) What colour will the following indicators be in pure water?
i) methyl orange
ii) phenolphthalein
iii) bromothymol blue

9 a) Why was phenolphthalein used in Anna and Carl's experiments in the last Activity?
b) A chemist was asked to measure the concentration of ammonia in a commercial bathroom cleaner.
i) Suggest an acid that he could use for his titration and explain your choice.
ii) Suggest a suitable indicator.

10 Look closely at the pH change when a weak acid is titrated with a weak alkali in Figure 10.8.
a) How does the pH change around the end point?
b) How will the colour of bromothymol blue change around the end point?
c) Why is bromothymol blue unsuitable for detecting the end point?

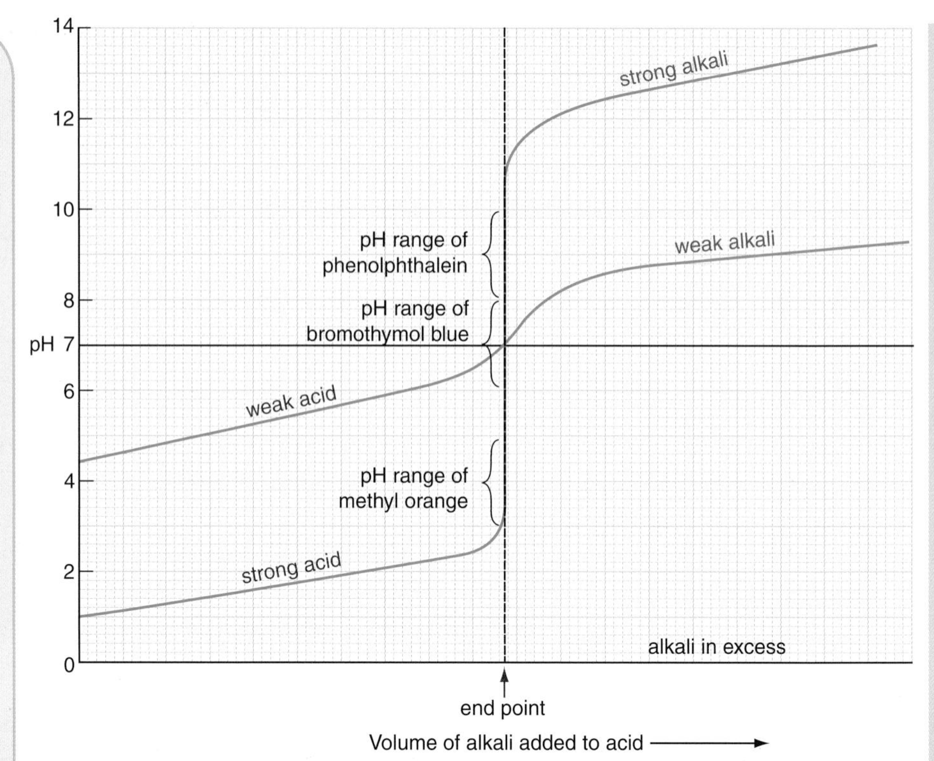

Figure 10.8 Graphs showing the pH changes during the titration of strong and weak acids with strong and weak alkalis.

Several indicators are available for titrations of acids with alkalis. However, litmus and universal indicator are not suitable because they contain a mixture of substances which change colour over a wide range of pH. The most useful indicators are:

- phenolphthalein, which changes from colourless at pH 8 to red at pH 10;
- methyl orange, which changes from red at pH 3 to yellow at pH 5;
- bromothymol blue, which changes from yellow at pH 6 to blue at pH 8.

So, if you look at Figure 10.8, you will see that the titration of:

- strong acid with strong alkali – can use any of the three indicators;
- strong acid with weak alkali – must use methyl orange;
- weak acid with strong alkali – must use phenolphthalein;
- no indicator is suitable for weak acid with weak alkali.

10.6 Concentrated and dilute acids and alkalis

Concentrated acids and alkalis contain a lot of acid or alkali in a small amount of water. Concentrated sulfuric acid has 98% sulfuric acid and only 2% water.

Dilute acids and alkalis have only a small amount of acid or alkali in lots of water. Dilute sulfuric acid has about 10% sulfuric acid and 90% water.

Figure 10.9 A car battery contains sulfuric acid.

Concentration tells us how much substance is dissolved in a certain volume of solution (see Section 7.4). It is usually given as the number of grams or the number of moles of solute per cubic decimetre (g/dm^3 or mol/dm^3) of solution. Dilute solutions usually have a concentration of 1.0 mol/dm^3 or less. For example, a solution of sodium hydroxide containing 1.0 mol/dm^3 has 1 mole of sodium hydroxide (40.0 g NaOH) in 1 dm^3 (1000 cm^3) of solution. This is sometimes written as 1.0 M NaOH.

The car battery in Figure 10.9 contains 2352 g of sulfuric acid (H_2SO_4) in 6 dm^3 of the battery liquid.

So, the concentration of sulfuric acid in the battery liquid is given by:

$$\text{concentration} = \frac{2352}{6} = 392\ \text{g/dm}^3$$

The concentration of sulfuric acid ($M_r(H_2SO_4) = 98$) in mol/dm^3 is therefore:

$$\text{concentration in mol/dm}^3 = \frac{392}{98} = 4\ \text{mol/dm}^3 \text{ or } 4.0\ \text{M}$$

11 Look at Figure 10.10. Identify two precautions that the student has taken to protect herself from the acid and alkali she is using.

12 What is the concentration in moles per dm^3 (mol/dm^3) of the following solutions? (Na = 23.0, O = 16.0, H = 1.0, S = 32.0)
a) 4.0 g sodium hydroxide in 1 dm^3
b) 8.0 g sodium hydroxide in 500 cm^3
c) 9.8 g sulfuric acid in 250 cm^3
d) 490 g sulfuric acid in 5 dm^3

13 How many moles of solute do the following solutions contain?
a) 250 cm^3 of 2.0 M sodium chloride
b) 400 cm^3 of 0.1 M sodium chloride
c) 25 cm^3 of 1.5 M sodium carbonate
d) 5 dm^3 of 0.2 M sodium carbonate

10.7 How much acid and alkali react in titrations?

Figure 10.10 This student is carrying out a titration.

We can calculate the quantities of different chemicals which react during titrations by doing calculations similar to those in the last section. Here is an example.

Example

25 cm^3 of a solution of sodium hydroxide (NaOH) which contains 0.2 mol/dm^3 reacts with 20 cm^3 of a solution of ethanoic acid (CH_3COOH) containing 0.25 mol/dm^3.

a) How many moles of sodium hydroxide reacted in the titration?

Number of moles of sodium hydroxide in 1 dm^3 (1000 cm^3) = 0.2

$$\therefore \text{Number of moles of sodium hydroxide in 25 cm}^3 = \frac{0.2}{1000} \times 25 = 5 \times 10^{-3}$$

$$\Rightarrow \text{Number of moles of sodium hydroxide reacting} = 5 \times 10^{-3}$$

b) How many moles of ethanoic acid reacted in the titration?

Number of moles of ethanoic acid in 1 dm^3 (1000 cm^3) = 0.25

$\therefore$ Number of moles of ethanoic acid in 20 cm^3 = $\frac{0.25}{1000} \times 20 = 5 \times 10^{-3}$

$\Rightarrow$ Number of moles of ethanoic acid reacting = 5×10^{-3}

c) How many moles of ethanoic acid react with 1 mole of sodium hydroxide?
From stages a) and b), we know that:

5×10^{-3} moles of ethanoic acid react with 5×10^{-3} moles of sodium hydroxide

$\therefore$ 1 mole of ethanoic acid reacts with 1 mole of sodium hydroxide.

14 Calculate the missing values in the statements below.
a) 25 cm^3 of 1.0 M HCl(aq) neutralise _____ cm^3 of 2.0 M NaOH(aq)
b) 20 cm^3 of 0.1 M HCl(aq) neutralise 10 cm^3 of _____ M KOH(aq)
c) 20 cm^3 of 0.5 M HCl(aq) neutralise _____ cm^3 of 0.5 M $Ca(OH)_2$(aq)
d) 10 cm^3 of 1.5 M H_2SO_4(aq) neutralise _____ cm^3 of 1.0 M NaOH(aq)

Finding the concentration of substances by titration

If we know the volumes of solutions which react and the concentration of one of the reactants, we can use the results of a titration to find the concentration of the other reactant.

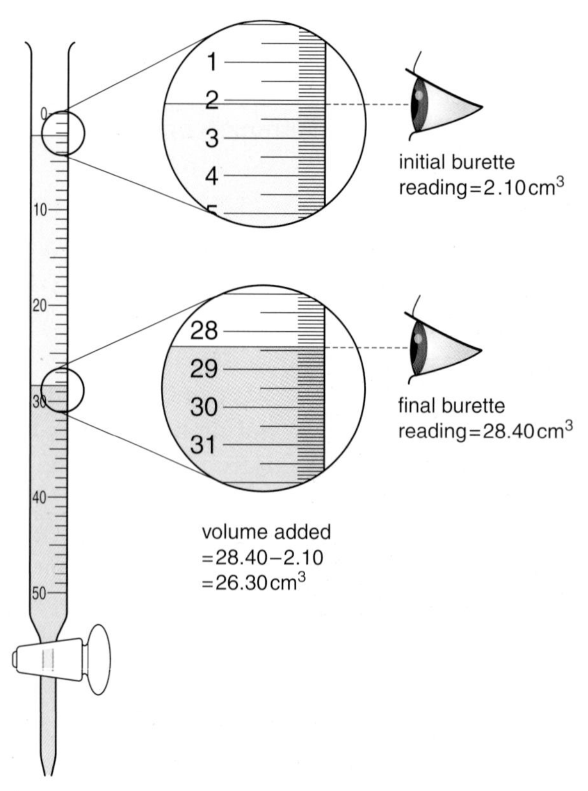

Figure 10.11 Make sure you read a burette correctly. Get your eye level with the meniscus and take the reading to the flat part of the meniscus.

Activity – Finding the concentration of sulfuric acid in battery acid

Katy and Zak were asked to find the concentration of sulfuric acid in battery acid which their teacher had diluted a hundred times.

After carrying out their titrations, they concluded that 31.5 cm^3 of the diluted battery acid just reacted with 25.0 cm^3 of 0.1 M sodium hydroxide.

1. Name a suitable indicator for the titration.
2. Write a word equation for the reaction of sodium hydroxide with sulfuric acid.
3. Copy and balance the following equation for the reaction.

   $$\_\_\_\_ \text{NaOH} + \text{H}_2\text{SO}_4 \rightarrow \_\_\_\_\_\_\_\_ + \_\_\_\_ \text{H}_2\text{O}$$

4. How many moles of NaOH react with 1 mole of H_2SO_4?
5. How many moles of sodium hydroxide reacted in their titrations?
6. How many moles of sulfuric acid reacted?
7. Copy and complete the following statements.

   31.5 cm^3 of the diluted battery acid contains _____ moles of H_2SO_4

   ∴ 1 cm^3 of the diluted battery acid contains _____ moles of H_2SO_4

   ∴ 1000 cm^3 of the diluted battery acid contains _____ moles of H_2SO_4

   ⇒ Concentration of sulfuric acid in diluted battery acid = _____ mol/dm^3
8. What was the concentration of the battery acid before it was diluted?
9. Why do you think their teacher diluted the battery acid?
10. How could Katy and Zak improve the reliability of their result?

10.8 How did our understanding of acids and alkalis develop?

Figure 10.12 Svante Arrhenius (1859–1927)

In 1887, the Swedish chemist Svante Arrhenius (Figure 10.12) put forward his theory of ionisation. He suggested that many substances, including acids and alkalis, ionise when they dissolve in water. Arrhenius defined acids as substances which produce hydrogen ions (H^+) and alkalis as substances which produce hydroxide ions (OH^-) when they dissolve in water.

Unfortunately for Arrhenius, his ideas were not readily accepted by other scientists. At that time, there was no model for the structure of atoms, electrons had not been discovered and it was difficult to understand how hydrogen ions (H^+) and hydroxide ions (OH^-) would form. In spite of these doubts, Arrhenius's theory was gradually accepted because it explained the behaviour of substances dissolved in water, and neutralisation in particular.

According to Arrhenius, the hydroxide (OH^-) ions in alkalis react with H^+ ions in acids to form water during neutralisations.

$$\text{H}^+(aq) + \text{OH}^-(aq) \rightarrow \text{H}_2\text{O}(l)$$

In 1923, when Johannes Brønsted and Thomas Lowry developed their theory of acids and bases, scientists understood how atoms contained protons and electrons. Brønsted and Lowry widened the scope of acid–base reactions by suggesting that:

Acids are proton (H^+ ion) donors and bases are proton acceptors.

Using these definitions, carbonates also act as bases when they react with acids to from carbonic acid, which then decomposes to water and carbon dioxide.

acid + carbonate → carbonic acid → water + carbon dioxide

$$2H^+ + CO_3^{2-} \rightarrow H_2CO_3 \rightarrow H_2O + CO_2$$

The reaction between the gases hydrogen chloride and ammonia, forming ammonium chloride, is also an acid–base reaction. Here, the HCl donates a proton (H^+) to NH_3 to form the NH_4^+ ion.

$$HCl(g) + NH_3(g) \rightarrow NH_4^+Cl^-(s)$$

In contrast to Arrhenius's theory, the ideas proposed by Brønsted and Lowry were readily accepted. Their ideas agreed with the theory of atomic structure and developed the ideas suggested by Arrhenius. They also explained acid–base reactions beyond the limits of aqueous solutions.

Despite the acceptance of Brønsted and Lowry's ideas of acids and bases, chemists continue to use Arrhenius's ideas of acids and alkalis as sources of H^+ and OH^- ions in many situations.

Figure 10.13 Johannes Brønsted (1879–1947) (left) and Thomas Lowry (1874–1936). Although Brønsted was working in Denmark and Lowry in England, they published separate, but almost identical, theories at the same time in 1923.

15 How did Arrhenius define:
a) acids; b) alkalis?

16 Why were Arrhenius's ideas not readily accepted by other scientists?

17 How did Brønsted and Lowry define:
a) acids; b) bases?

18 Why do you think the Brønsted and Lowry definitions were accepted more quickly than those of Arrhenius?

19 Which of the reactants acts as an acid and which acts as a base in the following equations?
a) $HCl(g) + H_2O(l) \rightarrow Cl^- + H_3O^+$
b) $NH_3(g) + HCl(g) \rightarrow NH_4^+Cl^-(s)$

20 Why do you think that Brønsted and Lowry defined acids and bases in terms of protons rather than H^+ ions?

Summary

- ✓ Acids and alkalis are soluble in water and give characteristic colours with indicators.
- ✓ An indicator is a substance which changes colour depending on how acidic or how alkaline a solution is.
- ✓ The pH scale measures how acid or how alkaline a solution is.
- ✓ The hydrogen ion is a proton. In water, H^+ ions are hydrated and represented as $H^+(aq)$.
- ✓ **Strong acids and alkalis** are completely ionised in water. **Weak acids and alkalis** are only partly ionised in water.
- ✓ The volumes of acids and alkalis which react can be measured by **titration** using an indicator.
- ✓ A titration involves adding one solution from a burette to another solution, in order to determine how much of the two solutions just react with each other.
- ✓ The **end point** of a titration occurs when the two solutions just react and neither is in excess.
- ✓ Concentration tells us how much solute is dissolved in a certain volume of solution. It is usually given as the number of grams or the number of moles of solute per cubic decimetre (g/dm^3 or mol/dm^3) of solution.
- ✓ Strength tells us the extent to which an acid or alkali is ionised in water.
- ✓ Arrhenius, Brønsted and Lowry developed our understanding of acids and bases in the nineteenth and twentieth centuries.
- ✓ Arrhenius defined acids as substances which ionise (split up into ions) producing H^+ ions, and alkalis as substances which ionise producing OH^- ions in aqueous solution.
- ✓ Brønsted and Lowry refined Arrhenius's ideas and defined acids as proton donors, and bases as proton acceptors.

EXAMQUESTIONS

1 One type of indigestion tablet contains magnesium hydroxide. Magnesium hydroxide is a base which is slightly soluble in water.

a) Is magnesium hydroxide an alkali? Explain your answer. *(2 marks)*

b) Copy and complete the equation below to show the ions formed by magnesium hydroxide ($Mg(OH)_2$) when it dissolves in water.

$$Mg(OH)_2(s) \xrightarrow{\text{water}} Mg^{2+}(aq) + \_\_\_\_\_\_$$

(2 marks)

c) Magnesium hydroxide neutralises acid in the stomach and relieves indigestion. Copy and complete the following equation for the reaction.

$$Mg(OH)_2 + \_\_\_\_ HCl \rightarrow \_\_\_\_\_\_ + \_\_\_\_ H_2O$$

(3 marks)

d) How does the pH in the stomach change after taking the tablets? *(1 mark)*

2 Pure ethanoic acid (CH_3COOH) forms a weak acid when added to water.

a) What is meant by a weak acid? *(2 marks)*

b) What ions are formed when ethanoic acid is added to water? *(2 marks)*

c) What is the colour of blue litmus paper with:

i) pure dry ethanoic acid; *(1 mark)*

ii) an aqueous solution of ethanoic acid? *(1 mark)*

d) Copy and complete the following equation for the reaction of ethanoic acid with sodium hydroxide solution.

$$CH_3COOH(aq) + NaOH(aq) \rightarrow \_\_\_\_\_\_ + \_\_\_\_\_\_$$

(2 marks)

e) A student titrated some ethanoic acid with sodium hydroxide. Name the two pieces of apparatus used to measure the volumes of solutions accurately in a titration. *(2 marks)*

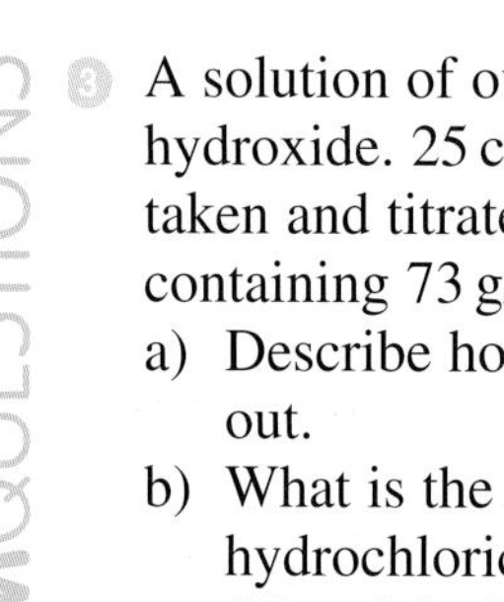

3 A solution of oven cleaner contains sodium hydroxide. 25 cm^3 of the oven cleaner was taken and titrated with hydrochloric acid containing 73 g/dm^3 of hydrogen chloride, HCl.

a) Describe how the titration should be carried out. *(5 marks)*

b) What is the concentration of the hydrochloric acid in mol/dm^3? (H = 1.0, Cl = 35.5) *(2 marks)*

c) 20 cm^3 of the hydrochloric acid just react with the 25 cm^3 of oven cleaner.

i) How many moles of hydrochloric acid react? *(2 marks)*

ii) How many moles of sodium hydroxide react? *(1 mark)*

iii) What is the concentration of sodium hydroxide in the oven cleaner in mol/dm^3? *(2 marks)*

Chapter 11
What's in the water we drink?

At the end of this chapter you should:

- ✓ understand how the water cycle works;
- ✓ know that most ionic substances and some molecular substances are soluble in water;
- ✓ be able to plot, use and interpret solubility curves;
- ✓ understand how temperature affects the solubility of substances, and that the effect on gases is different from the effect on solids;
- ✓ understand the effect of pressure on the solubility of gases;
- ✓ know how fizzy drinks become fizzy;
- ✓ understand the importance of dissolved oxygen for aquatic life;
- ✓ know that dissolved calcium and magnesium compounds cause hard water;
- ✓ know how hard water affects washing, water pipes and heating systems;
- ✓ understand the advantages and disadvantages of hard water;
- ✓ understand how hard water can be made 'soft';
- ✓ understand that water of the correct quality is essential for life;
- ✓ know how water is treated before it is supplied to homes and industry in the UK;
- ✓ understand how water quality can be improved through filtering and distillation;
- ✓ be able to consider and evaluate aspects of water quality and hardness.

Figure 11.1 The water tumbling down the mountain looks cool, clear and pure. Cool and clear it may be – but it is certainly not pure. Water is a very good solvent, and it dissolves many of the substances that it comes in contact with. Minerals and salts are dissolved out of the rocks and soil. Fish and plants in the stream rely on the dissolved oxygen and carbon dioxide.

11.1 The water cycle

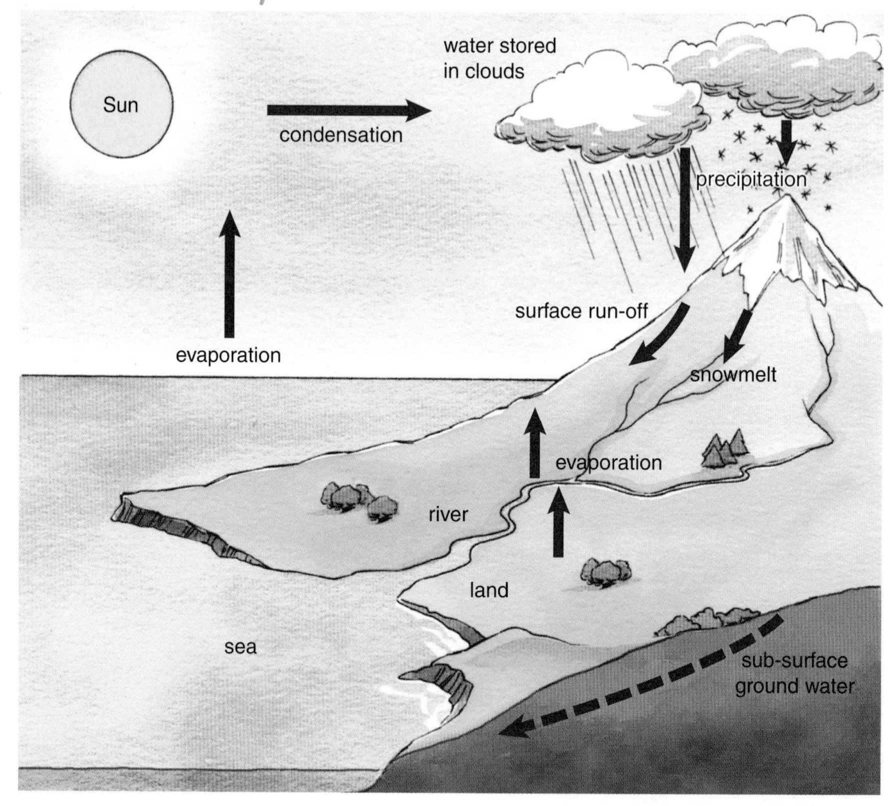

Figure 11.2 The water cycle

Water is found everywhere on Earth – in oceans, rivers and clouds, and in living things. But there is only a finite amount of water on Earth, which is continuously cycled from one location to another.

In the **water cycle**, water is cycled between the Earth's surface and the atmosphere. Water in rivers, lakes and oceans is evaporated by the heat of the Sun. The water vapour rises, cools and condenses to form clouds. The water droplets in the clouds join together to form rain and snow, which fall back to the Earth's surface.

The **water cycle** describes the continuous movement of water on, above, and below the surface of the Earth. Water can change state at various places in the water cycle, from vapour to liquid water or ice, and back again. These processes can occur very quickly (e.g. when water condenses to form rain), or over a much longer period (e.g. when glacial ice melts).

Although the amount of water on Earth remains fairly constant, individual water molecules move around all the time. The water in the salad you ate yesterday may have fallen as rain halfway around the world last month. It may also have been drunk by a dinosaur millions of years ago!

What causes the water cycle?

Seventy per cent of the Earth is covered by water – mainly in the oceans. Even at low temperatures, water will evaporate and turn into water vapour. That is why puddles on roads disappear even on cold days. Heat from the Sun and air movement caused by the wind help this evaporation.

The water vapour formed rises into the air. At 3000 metres above the ground, the air temperature may be many degrees below 0 °C. Here, the

Figure 11.3 Falling raindrops are often thought of as teardrop-shaped, with a narrow, elongated top part – but this is incorrect. Small raindrops are spherical, while larger ones are shaped like spheres flattened at the bottom.

vapour condenses, turning into liquid droplets or tiny ice crystals, which form the clouds. These microscopic droplets of water and crystals of ice stay suspended in the clouds.

The final stage in the water cycle is precipitation, when water falls from the clouds. Several factors can make raindrops form so that precipitation occurs.

- The clouds rise and cool further. More water vapour condenses from the colder air onto the tiny droplets, making them bigger.
- Convection currents in the clouds cause the droplets to clump together until they are too large to remain suspended.
- Dust particles act as starting points on which the water vapour condenses, forming large droplets.

1. Describe the general pattern of the water cycle. Why is it called a 'cycle'?
2. Name three types of precipitation.
3. Explain how both the Sun and wind contribute to evaporation of water from the surface of the Earth?
4. State and explain three ways in which raindrops can form.
5. Use the figures in Table 11.1 to show that the total amount of water on the Earth remains constant from year to year.

Water movement	Average movement rate in cubic km per year
Precipitation over land	107 000
Evaporation from land	71 000
Run-off and ground water from land into seas	36 000
Precipitation over oceans	398 000
Evaporation from oceans	434 000

Table 11.1 Average annual water movement

11.2 Solubility

You should remember from Chapter 8 that when one substance dissolves in another, a solution is formed. The dissolving substance is said to be soluble in the solvent (the substance that dissolves it). Water is a very effective solvent: it dissolves a huge variety of different substances. Many ionic compounds, such as sodium chloride, are soluble in water. When ionic solids dissolve in water, the water molecules get between the individual ions and surround them. Ionic solids will dissolve if the attractive forces between the ions and the solvent are greater than the forces between the ions in the solid ionic lattice. If the ionic lattice has greater attractive forces, the solid is insoluble and will not dissolve.

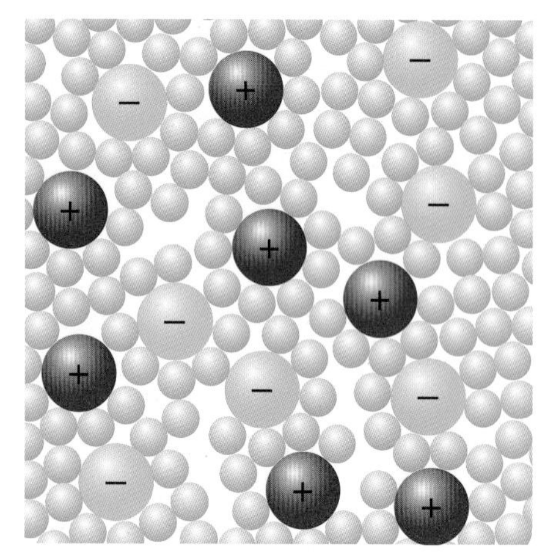

Figure 11.4 When an ionic substance is dissolved in water, the positive and negative ions move around separately, surrounded by water molecules.

The **solubility** of a solute in water, or any other solvent, is usually given in grams of solute per 100 grams of water (or solvent) at that temperature.

A **saturated solution** contains the maximum amount of solute that will dissolve at that temperature.

Some molecular compounds, including sucrose, oxygen and carbon dioxide, are also soluble in water. However, giant covalent structures such as silica (sand) are insoluble in water, as are many covalent compounds like the alkanes and other hydrocarbons.

What is solubility?

The **solubility** of a solute is the maximum amount of it that will dissolve fully in a particular mass of solvent (usually 100 g) under given conditions. For example, if 38 g of sodium chloride (but no more) will fully dissolve in 100 g of water at 20 °C, the solubility of sodium chloride in water at that temperature is 38 g per 100 g of water.

A **saturated solution** is one in which the maximum amount of solute (at that temperature) is dissolved. If there is more solute present, an equilibrium is set up between the solution and the excess solute. The solute particles move into and out of solution all the time, but the amount of solute in the solution stays constant. Different amounts of solute will dissolve in a solvent at different temperatures. If the temperature changes, then so does the equilibrium position.

Generally, the solubility of solids decreases as the temperature falls. Imagine a hot solution containing a dissolved solid that is not quite saturated. As the solution cools, it will reach a temperature when it becomes saturated. If the temperature falls further, some of the solute will come out of the solution. Solid crystals will start to form, with more and more crystals being produced as the solution cools further. If the cooling takes place slowly, large crystals will form as the solute particles take their place in the growing crystal lattice.

Figure 11.5 One way to form crystals is by cooling a saturated solution.

Under special conditions, supersaturated solutions can form. These have more solute dissolved in them than their solubility at that temperature should allow. Supersaturation can occur when a saturated solution is cooled in a very clean glass container, where there are no particles on which the crystals can form. Under these conditions, crystal formation does not take place and the solute stays in solution. However, if a speck of dust falls into the supersaturated solution, this provides a starting point for crystal formation and a mass of crystals will form very quickly.

Solubility curves

A **solubility curve** shows the solubility of a solute at different temperatures.

One useful way of showing how the solubility of a solute in a solvent changes with temperature is by drawing **solubility curves**. Look at the graphs in Figure 11.6. These graphs, or solubility curves, show the solubilities of two different solid solutes – sodium chloride and potassium nitrate – at temperatures between 0 °C and 100 °C. For example, at 27 °C, the solubility of both sodium chloride and potassium nitrate is 38 g per 100 g of water.

From these curves it is easy to see that changes in temperature have a very different effect on the solubility of the two solutes. While the solubility of potassium nitrate rises steeply with temperature, that of sodium chloride remains nearly the same throughout the temperature range.

Solubility curves can also be used to predict the temperature at which crystals of the solid solute will start to form as a solution is cooled. Imagine a solution containing 50 g of potassium nitrate per 100 g water. At 50 °C, the solution is unsaturated. The solubility decreases as the solution cools, and it becomes saturated at around 34 °C. Below this temperature, there will be excess solute present and crystals will start to form.

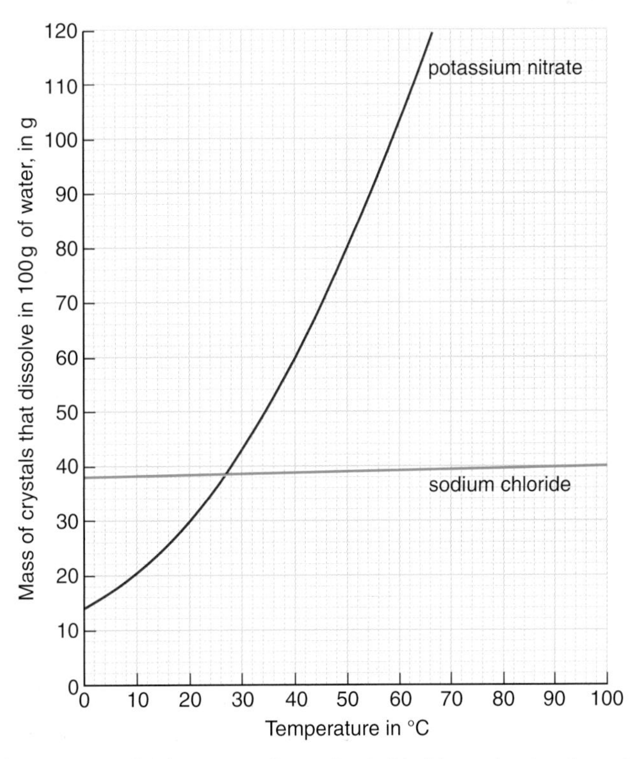

Figure 11.6 Solubility curves for sodium chloride and potassium nitrate

6. Explain the following terms:
   a) solute, solvent and solution
   b) solubility
   c) saturated solution.
7. Look at the solubility curves of sodium chloride and potassium nitrate in Figure 11.6.
   a) Explain what the curves show.
   b) Estimate how much of each of these two substances dissolves in 100 g of water at 20 °C and at 80 °C.
   c) Is the difference in solubility between 20 °C and 80 °C the same for potassium nitrate and sodium chloride?
   d) Explain how you would make a saturated solution of each of these two substances at room temperature.
8. Table 11.2 shows the results of an investigation to find out how temperature affects the solubility of potassium chlorate.
   a) Plot a graph of these results. Draw a smooth curve to show the pattern in the results.

Temperature in °C	Solubility (grams of potassium chlorate per 100 g of water)
0	5.5
10	5.6
20	8.0
30	10.5
40	15.0
50	20.0
60	26.2
70	32.0
80	41.0

Table 11.2 The solubility of potassium chlorate at different temperatures

   b) Describe the pattern in the results.
   c) Use your knowledge of particle theory to explain the shape of the curve.
   d) Estimate the amount of solute that would dissolve in 100 g of water at 90 °C.

Figure 11.7 When you heat water, gases come out of the solution and form bubbles that you can see.

Solubility of gases

Just as solids can dissolve in liquids, so can gases. Different gases dissolve in water to different extents. Some, such as hydrogen chloride, are extremely soluble. Others, like oxygen and carbon dioxide, are only slightly soluble. However, even the slight solubility of these gases is very important. Fish and other aquatic animals depend on dissolved oxygen, and aquatic plants use dissolved carbon dioxide for photosynthesis.

The effect of temperature on the solubility of gases in water and other solvents is very different from – in fact opposite to – that for solids. As the temperature rises, gases become less and less soluble and bubble out of solution. When you heat a beaker of tap water to about 50 °C, bubbles appear on the glass in contact with the water. These bubbles are dissolved gases coming out of solution.

The fact that gases become less soluble in water at higher temperatures has important consequences for animals that live in water. As the water loses its oxygen, aquatic animals cannot survive.

Pressure and gas solubility

Increasing the pressure of a gas increases its solubility. Carbonated or 'fizzy' water, also known as soda water or sparkling water, is plain water into which carbon dioxide gas has been dissolved under pressure. The higher the pressure and the lower the temperature, the more carbon dioxide will dissolve. This process of making carbonated water and

Figure 11.8 Both temperature and pressure have an effect on the solubility of gases in water.

other fizzy soft drinks is called carbonation. As the carbon dioxide dissolves in the water, some carbonic acid (H_2CO_3) is formed.

When a fizzy drink is opened, the pressure of the gas on the surface of the liquid suddenly becomes lower. This allows gases to come out of solution, and lots of bubbles are produced. If the fizzy drink is ice cold, most of the gas stays in solution. But if the fizzy drink is warm, even more effervescence (or fizz) occurs as the carbon dioxide quickly comes out of solution. This can cause problems for the unwary!

Alcoholic drinks that are naturally fizzy – such as champagne – also contain dissolved carbon dioxide. In this case, the fizz comes from carbon dioxide produced by a fermentation reaction. The gas in champagne is pressurised by keeping the bottle closed during the later stages of fermentation. As the amount of carbon dioxide in the bottle increases, so does the pressure.

Figure 11.9 Naturally sparkling mineral waters contain carbon dioxide that dissolved in them during higher-pressure conditions underground. As with other fizzy drinks, the carbon dioxide bubbles out of solution and escapes when the pressure is reduced.

9 Table 11.3 shows data for the solubility of ammonia gas (NH_3) in water between 0 °C and 100 °C.

Temperature in °C	Solubility (grams of ammonia per 100 g of water)
0	91.5
10	68.9
20	52.0
30	42.0
40	33.5
50	28.0
60	21.5
70	18.0
80	13.1
90	9.9
100	8.1

Table 11.3

a) Plot a solubility curve for these results.
b) Describe the shape of the graph.
c) State and explain the main difference between this graph and the solubility curves for solids.
d) Estimate the mass of ammonia gas that would dissolve in 1000 g of water at 25 °C.
e) Assuming 17 g (1 mole) of ammonia has a volume of approximately 24 000 cm^3 at 25 °C, what volume of ammonia dissolves in 1000 g of water at 25 °C?

10 a) What is soda water?
b) What is another name for soda water?
c) What is the name of the weak acid in soda water?
d) What happens when you open a bottle of warm soda water?
e) Does the same thing happen when you open a bottle of very cold soda water? Explain the difference.

11.3 What is hard water?

Figure 11.10 As the map shows, water in the central, eastern and southern areas of England is hardest. In the Lake District, Scotland, Northern Ireland and the west of England and Wales, water is usually soft.

Hard water contains dissolved compounds, usually of calcium or magnesium. Hard water forms a 'scum' with soap.

Soft water has no (or very few) calcium or magnesium ions dissolved in it.

Hard water is water that has dissolved minerals in it, particularly calcium and magnesium compounds. These come from minerals in the soil and rocks over which the water has flowed. The dissolved compounds are in the form of ions: calcium (Ca^{2+}), magnesium (Mg^{2+}), carbonate (CO_3^{2-}), hydrogencarbonate (HCO_3^-) and sulfate (SO_4^{2-}). **Soft water** does not contain much of any of these ions.

Most freshwater sources contain calcium and magnesium ions in varying quantities. Calcium ions usually enter the water from either limestone or chalk (calcium carbonate), or from gypsum (calcium sulfate). As you can see in Figure 11.10, water in the north-western regions of the UK is usually soft because there is not much chalk or limestone (calcium carbonate) in the soil. Another reason is that, in these regions, the water used is mainly surface water, and contact time between surface water and the ground is relatively short. In contrast, in the south-east of England the soil is very chalky and most of the water is pumped from underground, where it has had longer contact with the earth. Hence, the water is hard in this area.

Figure 11.11 Many mineral waters are hard water. Many people prefer the taste to soft water.

Figure 11.12 Hard or soft water? You can't tell by looking.

Hard water is beneficial to health, but it does cause some problems.

The advantages to hard water are:

- it helps to form stronger teeth and bones;
- it is ideal for brewing certain types of beer and whisky;
- it is considered tastier to drink.

One simple way to determine whether water is hard or soft is the lather test. Soft water produces a lather very easily when mixed with soap, while hard water produces little or no lather. Hard water reacts with soap to form scum, so more soap is needed to form a lather.

11 a) Explain what is meant by the term 'hard water'.
b) What ions are present in hard water?

12 a) Would you expect the water supply in south-east England to provide hard or soft water?
b) Would you expect the water supply in Scotland to provide hard or soft water?
c) If you had to wash some delicate clothes in soap flakes, what difference would you see if you washed them in south-east England compared with Scotland?

Temporary hardness is hardness in water that can be removed by boiling the water. In contrast, **permanent hardness** is not removed by boiling.

Temporary hardness can be removed by boiling. It is caused by calcium hydrogencarbonate in the water. Boiling turns the calcium hydrogencarbonate into calcium carbonate. This is insoluble, so it precipitates (forms a solid), leaving water that is less hard. However, this process can also produce limescale (see page 210).

$$Ca(HCO_3)_2(aq) \rightarrow CaCO_3(s) + CO_2(g) + H_2O(l)$$

calcium hydrogencarbonate → calcium carbonate + carbon dioxide + water

Permanent hardness cannot be removed by boiling. It is caused by calcium and magnesium sulfates and chlorides in the water. This can only be removed using a water softener or an ion exchange column (see page 211).

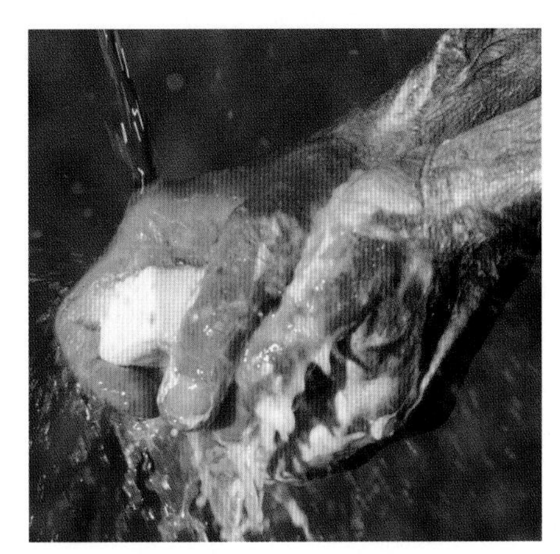

Figure 11.13 Soft water easily forms a lather with soap.

Problems caused by hard water

For many everyday uses, hard water and soft water are both suitable. But hard water is not as good as soft water for washing, because of the difficulty of producing a lather. Laundering clothes using hard water needs a lot more soap and detergent. This makes using hard water more expensive, and the use of extra soap and detergent can have a greater impact on the environment.

Hard water also makes it more difficult to keep bathrooms looking clean. When calcium and magnesium ions in hard water react with the long-chain fatty acid anions (negative ions) in soap, a scum is produced. This is hard to rinse away, and leaves ugly marks on surfaces such as baths and hand basins.

The reactions are complicated, but can be summarised as:

$Ca^{2+}(aq)$	+	$2X^{-}(aq)$	→	$CaX_2(s)$
calcium ions in hard water		long chain anions in soap		'scum' – insoluble calcium salts of anions in soap

Limescale

Limescale is the hard substance deposited on the inside of pipes or containers in which hard water has been heated.

Hard water also causes **limescale**. When hard water is heated, the minerals dissolved in it which cause temporary hardness come out of solution and form a hard deposit called limescale. This is mainly calcium carbonate. You will probably have seen limescale on the inside of electric kettles, where it often covers the heating element with a light-coloured flaky crust. This crust reduces the rate of heat transfer between the heating element and the water, so the kettle takes longer to boil.

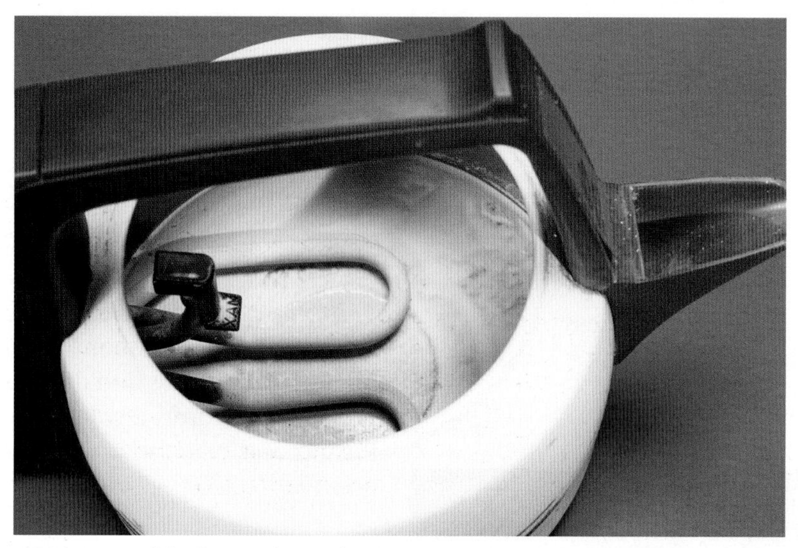

Figure 11.14 Most of the limescale in a kettle is caused by temporary hardness.

Limescale can also clog pipes, water taps, shower heads, boilers and washing machines – anywhere where hot water is used. As well as making them work less efficiently, it can cause blockages and other damage that is expensive to repair.

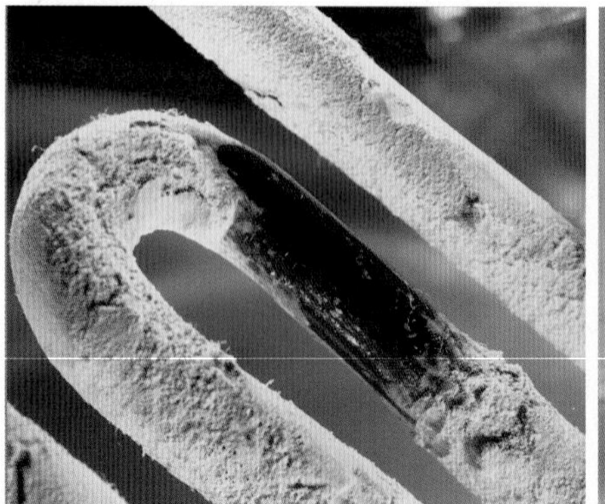

Figure 11.15 Limescale is mainly calcium carbonate. It is hard, white, and causes maintenance problems.

13 a) What is meant by temporary hardness in water?
b) Explain the chemical principles involved in removing temporary hardness from water by boiling.
c) What is limescale and why is it produced on the insides of kettles?

14 Look at the data for two different bottled waters in Table 11.4.
a) Which water do you think is softer? Explain how you arrived at your answer.
b) Which water is likely to be tastier?
c) Why might drinking this water also be good for your health?

Mineral composition in mg/l	Evian	Badoit
Calcium (Ca^{2+})	78	190
Magnesium (Mg^{2+})	24	85
Sodium (Na^{+})	5	150
Potassium (K^{+})	1	10
Hydrogencarbonate (HCO_3^{-})	357	1300
Chloride (Cl^{-})	4.5	40
Sulfate (SO_4^{2-})	10	40
Fluoride (F^{-})	3.8	1
Silica (SiO_2)	13.5	35

Table 11.4 Mineral waters contain different amounts of dissolved minerals.

Softening hard water

There are two main ways of softening hard water: adding sodium carbonate or using an ion exchange column.

Adding sodium carbonate (washing soda or bath salts)

Figure 11.16 Bath salts contain sodium carbonate plus colorants and perfumes.

The addition of sodium carbonate softens hard water. The calcium and magnesium ions in hard water react with sodium carbonate to form a precipitate of calcium carbonate and magnesium carbonate. Sodium sulfate is also formed, which remains in solution as a soluble salt and has no effect on the hardness of water.

$$Na_2CO_3(aq) + CaSO_4(aq) \rightarrow Na_2SO_4(aq) + CaCO_3(s)$$

sodium carbonate + calcium sulfate → sodium sulfate + calcium carbonate

Using an ion exchange column

An ion exchange column works on the principle of exchanging the calcium and magnesium ions in hard water for sodium or hydrogen ions when the water passes through the column. This process is called ion exchange.

An ion exchange column is an inexpensive way of softening household water. The column contains beads of resin to which are attached sodium ions or hydrogen ions. As the hard water flows through the column, calcium and magnesium ions are absorbed onto the surface of the beads,

Figure 11.17 This ion exchange water column removes the hardness from water.

displacing the sodium or hydrogen ions. Eventually the resin becomes exhausted when all the original ions have been displaced. It is then re-charged by flushing through with salt water or acid and discarding the effluent. This displaces the calcium and magnesium ions that have been absorbed onto the resin and replaces them with sodium or hydrogen ions again. This re-charging process can be repeated many times.

15 a) What is the relative formula mass of the compound that is washing soda ($Na_2CO_3.10H_2O$)?
b) Write a chemical equation for the reaction between washing soda and magnesium sulfate in hard water.
c) What mass of washing soda is needed to treat 50 dm^3 (50 litres) of hard water that contains 0.002 moles of magnesium sulfate per dm^3?
d) What is the insoluble product produced when washing soda reacts with magnesium sulfate in hard water?

16 a) Explain with a simple diagram how ion exchange resins can be used to soften hard water.
b) How can water softening ion exchange resins be regenerated when they become exhausted?

17 Explain briefly some of the problems that can arise from using hard water as a domestic water supply for washing and laundering.

11.4 Water for life

Figure 11.18 Washing soda is hydrated sodium carbonate, $Na_2CO_3.10H_2O$.

Good quality water is essential for all forms of life. Water plays an essential part in our metabolism. All the biochemical reactions in our bodies only work in watery solutions. Our blood is a suspension of cells in salty oxygenated water. Our lungs have a wet inside surface to dissolve oxygen gas.

Our bodies function well if we drink two litres of fresh water a day. Drinking water must:

- be low in dissolved salts, although we can tolerate some dissolved minerals like those in hard water;
- have a low level of microorganisms.

Microorganisms in water can cause various human diseases such as typhoid, cholera and dysentery.

In addition to drinking water, we also use an enormous amounts of fresh water every day for washing, sanitation and other uses (Table 11.5) with little thought about how much we use. Of course, there is a great deal of water around: 70% of the Earth is covered with ocean, but this salt

Activity	Amount of water used
Bath	120 litres per bath
Dishwasher	75 litres per load
Flushing the toilet	9 litres per flush
Lawn watering with hosepipe	15 litres per minute
Washing machine	136 litres per load
Shower	9 litres per minute

Table 11.5 The way we use water in the UK

water is useless for drinking and preparing food. So where does the clean, drinkable water in our homes come from?

The water supply to our homes

Producing clean water to use in our homes requires several stages of processing.

- First, the water is drawn from the source, which may be a well, river, lake or reservoir. Water from underground wells is usually very clean and needs the least processing.
- If the water is taken from the river or lake, it first passes through filtration screens to remove any large-size debris, such as sticks, stones and weeds. The filters also prevent fish from getting into the water treatment equipment. In the water treatment plant it is passed through finer filters to remove algae, insects and any similar-sized impurities.
- A chemical called a coagulant is then added. This makes the remaining impurities clump together so that most sink down and are removed.
- Filtration through beds of fine sand removes any remaining particles. The sand filters are cleaned frequently to get rid of the impurities collected.
- The cleaned water is then disinfected by adding chlorine, which kills bacteria and other microorganisms. The chlorine also keeps the water free from microorganisms on its journey to your tap. The acidity of the water is also adjusted at this stage.
- The clean and disinfected water is held in closed storage tanks until needed. The quality of the water is tested regularly to ensure that it is safe to drink.

In Britain, the water supply system is run by various private companies, for example Severn Trent Water. These companies charge users a fixed quarterly fee or provide them with a water meter to pay for what they use. The water supplied has to meet very strict standards of purity and healthiness (see Table 11.6).

18 Use the information in Table 11.5 to estimate how much water would be used in one week by a family of four in the UK.

19 Draw and annotate a diagram of the water treatment process from reservoir to clean tap water in your home.

20 a) Why is the permitted level of lead in drinking water lower than that of iron?
b) Why are *E. coli* bacteria not permitted in drinking water?

Impurity	Permitted concentration in micrograms per dm^3 of water
Escherichia coli (*E. coli*) bacteria	0
Acidity	Minimum pH of 6.5
Acrylamide	0.10
Aluminium	200
Arsenic	10
Benzene	1.0
Chromium	50
Copper(II)	2.0
Cyanide	50
Fluoride	1.5
Iron	200
Lead	25
Manganese	50
Mercury	1.0
Nickel	20
Nitrate	50
Sodium	200
Pesticides	0.50

Table 11.6 Permitted levels of drinking water impurities in the UK (from The Water Supply (Water Quality) Regulations 2000)

Activity – The Camelford incident

Read the account below of the Camelford water contamination incident.

A serious incident of accidental water contamination took place in the town of Camelford, Cornwall, in 1988. Some 20 000 residents in Camelford and the surrounding area were exposed to drinking water contaminated with high levels of dissolved aluminium, at up to 3000 times the legal safe limit.

The accident happened when a contractor's lorry arrived at the water treatment plant at Camelford loaded with aluminium sulfate. The chemical was intended for water treatment, but it was dumped into the wrong tank and went straight into the water supply.

Residents immediately complained about odd tastes in the water and other effects including skin irritation and plumbing corrosion. The water company initially reassured them that the water was safe to drink. Two days after the incident the water supply was cleaned by flushing through.

Over the next few months several hundred residents complained of medical symptoms including skin rashes, arthritic pains, sore throats, memory loss and exhaustion. These complaints were investigated, and in 1989 and 1991 official reports concluded that there was no convincing evidence of long-term harm to health caused by drinking the contaminated water. This conclusion angered many residents.

Then, in 2006, a resident of Camelford was found to have died from a rare form Alzheimer's disease. Normally, this form of the disease only occurs in people who have relatives that have also been affected – which was not so in this case. However, a high level of aluminium was found in the brain tissue of the Camelford resident. Exposure to aluminium has been associated with an increased risk of developing Alzheimer's disease.

Set up a 'public enquiry' about the Camelford incident. In the enquiry, try to decide if there is enough evidence of serious harm caused by the incident, and if so whether anyone should be held responsible. You may also want to make recommendations for future action (such as whether or not to offer medical testing to residents).

1. First, you will need to form a group of three people:
   - a spokesperson for the water company;
   - a spokesperson for the residents of Camelford;
   - a chairperson to weigh up the arguments from the two sides.
2. The chairperson should hear a 2-minute submission from each of the spokespeople and then ask each of them three more questions. After hearing the arguments, the chairperson should make a statement about his or her conclusions.
3. Finally, write your own brief report as a member of the enquiry, giving your own conclusions and the reasons for them.

Water filters for purity

Figure 11.19 There are many types of water filters on sale. They can be used to improve the clarity, taste and quality of tap water.

Although tap water in the UK is safe to drink, some people like to use water filters to purify their drinking water further. The important material in most water filters is activated charcoal. This material is effective in removing chlorine (which some people find has an unpleasant taste) and organic chemicals. Most ionic compounds, such as sodium compounds and nitrates, are not removed by the activated charcoal.

Activated charcoal is a form of carbon. It is made by passing oxygen through the charcoal at very high temperatures, producing minute pores

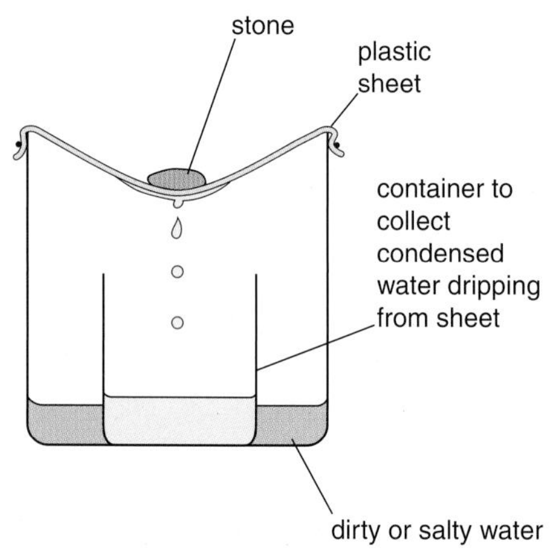

Figure 11.20 This simple solar still will make a small amount of drinking water from dirty water, without boiling.

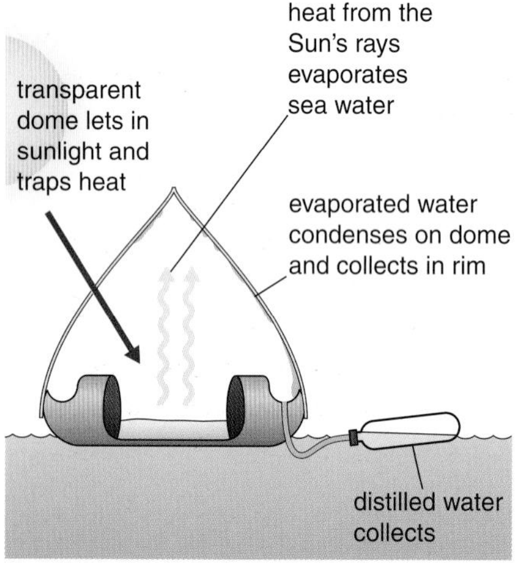

Figure 11.21 A survival still. A man survived a month at sea by using a still like this after his boat had capsized.

in the material. These increase the surface area of the charcoal, which can have a surface area of 1000 square metres per gram. When water is passed through the filter, some molecules attach themselves to the surface of the carbon or become trapped in the pores.

One problem with activated carbon filters is that any trapped organic material on the filters provides an excellent place for bacteria to grow. In order to overcome this, manufacturers put silver in the charcoal filter to prevent the growth of bacteria.

Distilled water

Pure water is made by distillation. Distilled water is used in applications where the mineral content of tap water would cause unacceptable levels of impurity or unwanted precipitates to form. Distillation involves boiling the water and then condensing the vapour (Section 2.1). This is very effective in producing pure, clear water free of dissolved salts. However, if the water is contaminated with water-soluble organic compounds such as ethanol or ethanoic acid, these may remain with the water during distillation. Fractional distillation must be used to remove these compounds.

The energy costs of boiling water are very high. This means that distilled water is expensive to produce. Distilled water was once used to top up car batteries and fill steam irons. Today, de-ionised water from an ion exchange system is used for these purposes because it is cheaper.

For the same reason, distillation is not a sensible way to make drinking water from salt water unless 'free' solar heating is available. Figures 11.20 and 11.21 show two devices that use solar heating to produce drinking water by evaporating and condensing impure water.

21 'Charcoal-filtered drinking water is better to drink.'
a) Explain why this advertising claim might be correct.
b) If you were worried about organic compounds such as pesticide residues in your water supply, would using an activated charcoal water filter help?
c) If your concerns were about nitrates from fertilisers, would the same water filter also help to remove these?

22 Describe the process of making distilled water from tap water. Describe what you would do and see at each stage.

Summary

- ✓ The **water cycle** transfers water continuously from ocean to land and back again by evaporation and precipitation.
- ✓ Water is a very good solvent for many ionic compounds and some molecular compounds.
- ✓ **Solubility** is usually measured in grams of solute per 100 g of solvent (g/100 g).
- ✓ A **solubility curve** shows the solubility of a solute in a particular solvent at different temperatures.
- ✓ A **saturated solution** will not dissolve any more solute. As a saturated solution cools, solid crystallises out.
- ✓ Carbon dioxide is dissolved in water under pressure to make carbonated or fizzy drinks.
- ✓ Oxygen dissolved in water is essential for aquatic life. Warm water has a lower concentration of dissolved oxygen than cold water.
- ✓ **Hard water** contains dissolved calcium and magnesium ions from rocks through which the water has passed. These ions react with the fatty acid anions in soap to form insoluble scum. **Soft water** has no (or very few) calcium or magnesium ions dissolved in it.
- ✓ Hard water causes increased costs because more soap is needed and **limescale** damages heating systems.
- ✓ **Temporary hardness** in water is caused by calcium hydrogencarbonate. It can be removed by boiling.
- ✓ **Permanent hardness** in water is caused by calcium and magnesium sulfates and chlorides. Boiling cannot remove permanent hardness; it can only be removed by using a water softener.
- ✓ Calcium and magnesium compounds cause limescale deposits in heating systems and kettles.
- ✓ Sodium carbonate (washing soda) acts as a water softener (removes hardness) by precipitating out the calcium and magnesium ions.
- ✓ Ion exchange columns act as water softeners by removing the calcium and magnesium ions and replacing them with sodium and hydrogen ions.
- ✓ The water supply has to be filtered through sand to remove fine particles and chlorinated to kill bacteria before it is fit for drinking.
- ✓ Charcoal can be used in home water filters to remove dissolved organic compounds and chlorine from tap water. Silver is added to water filters to prevent bacteria growing on them.
- ✓ Pure water can be made by distillation.

EXAMQUESTIONS

1 a) i) What is meant by hard water? *(2 marks)*
ii) Where does the 'hardness' come from? *(2 marks)*

b) Explain the reaction that happens to remove temporary hardness by boiling water. *(4 marks)*

c) Explain how hard water forms scum with soap. *(2 marks)*

d) Explain why limescale is deposited in kettles in hard water areas. *(3 marks)*

e) Explain, using equations, how sodium carbonate (washing soda) can reduce the cost of soap in hard water areas. *(4 marks)*

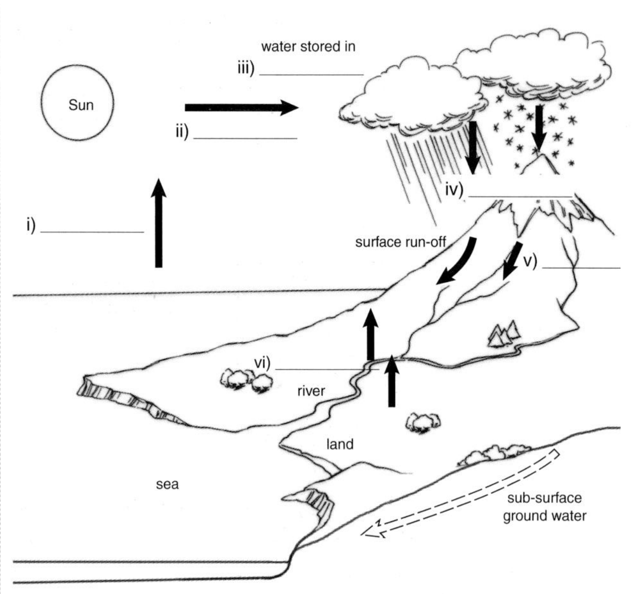

Figure 11.22

EXAMQUESTIONS

❷ a) Figure 11.22 shows the water cycle. Write out the missing descriptions for labels i) to vi). *(6 marks)*

b) Describe the processes that drive the water cycle. *(4 marks)*

c) Global warming is increasing air temperatures and the strength of winds. What effect will this have on the water cycle? *(4 marks)*

d) Use the data in Table 11.7 to comment on the amount of the world's water that is 'cycling' at any one time. *(4 marks)*

Reservoir	Volume of water in 1000s of km³	Per cent of total
Oceans	1370	97.25
Ice caps and glaciers	29	2.05
Ground water	9.5	0.68
Lakes	0.125	0.01
Soil moisture	0.065	0.005
Atmosphere	0.013	0.001
Streams and rivers	0.001 7	0.000 1
Biosphere	0.000 6	0.000 04

Table 11.7 The volume and per cent of water stored in different parts of the water cycle

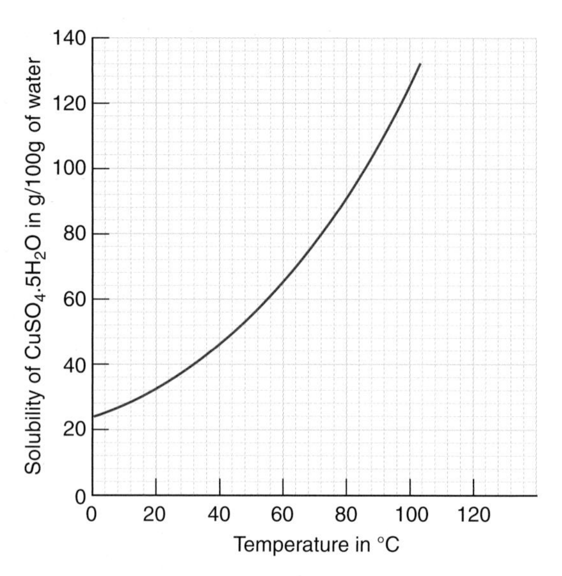

Figure 11.23 Solubility curve for copper sulfate

❸ Data for the solubility of copper sulfate crystals ($CuSO_4.5H_2O$) in water is given in Figure 11.23.

a) From the graph shown in Figure 11.23, what is the solubility of copper sulfate at:

i) 30 °C *(1 mark)*

ii) 70 °C? *(1 mark)*

b) Describe the trend in the graph as fully as possible. *(3 marks)*

c) A mixture containing 250 g of copper sulfate crystals dissolved in 500 g of water at 70 °C was allowed to cool. At what temperature would crystals start to appear? *(2 marks)*

❹ Read the passage below and then answer the questions which follow.

'The bends' were first recorded in 1841. Early deep-sea divers experienced strange symptoms when they came back to the surface, such as rashes, coughing, dizziness, unconsciousness and an inability to bend joints (hence the phrase 'the bends').

When people move from a high-pressure environment to one of low pressure, bubbles of gas sometimes form in their bloodstream. These cause the symptoms of the bends. Divers breathe pressurised air, which is mostly nitrogen. At the high pressures in deep water, nitrogen dissolves in the body's tissue fluids and blood rather than being exhaled. If a diver returns to the surface too quickly, the rapid decrease in pressure makes the dissolved nitrogen turn back into bubbles of gas. The way to avoid this is to ascend slowly from deep water, allowing the nitrogen to come out of solution steadily and be expelled from the body naturally.

a) What are the symptoms of the bends? *(2 marks)*
b) What gas causes the bends? *(1 mark)*
c) What is the effect of pressure on the solubility of a gas in water? *(2 marks)*
d) What is the effect of temperature on the solubility of a gas in water? *(2 marks)*
e) Describe how you would explain to junior school pupils why the bends occur, using a bottle of fizzy pop as a visual aid. *(4 marks)*

5 a) Describe how the ion exchange column in Figure 11.24 is used in the treatment of hard water. *(8 marks)*
b) Describe how you would regenerate the ion exchange resin when its sodium ions become depleted. *(4 marks)*

6 a) Draw the apparatus you would use to distil salt water to obtain pure water. *(5 marks)*

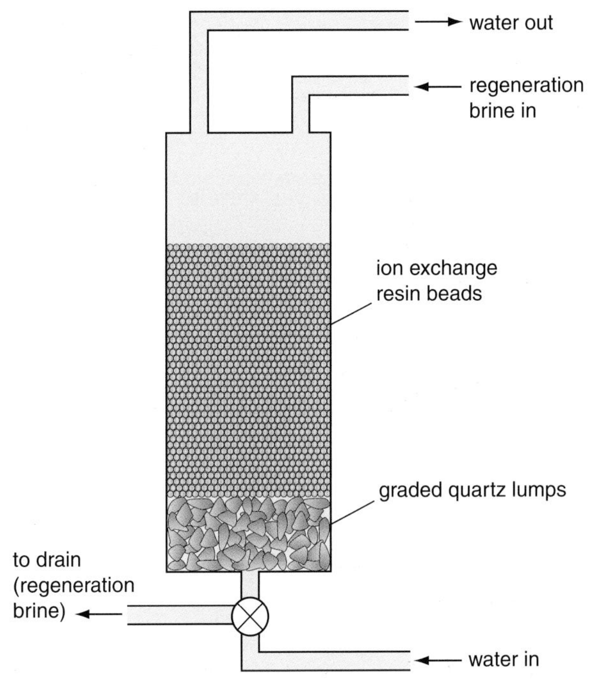

Figure 11.24 The cross-section of an ion exchange column for water purification

b) Explain how the apparatus separates pure water from muddy, salty water. *(3 marks)*
c) Describe and explain the additional process that must be used to purify water contaminated with ethanol. *(3 marks)*
d) Why is this process not used to produce drinking water in Britain? *(2 marks)*

Chapter 12
How much energy is involved in chemical reactions?

At the end of this chapter you should:

- ✓ know that energy is normally measured in joules, but that some food labels use calories and 1 calorie = 4.2 joules;
- ✓ know how to measure the energy transferred in chemical reactions by using calorimetry;
- ✓ understand that different foods and fuels store and provide different amounts of energy;
- ✓ know that carbohydrates, fats and oils provide relatively large amounts of energy;
- ✓ understand that the intake of too much food can lead to obesity;
- ✓ be able to consider the social, economic and environmental consequences of using fuels;
- ✓ be able to interpret simple energy level diagrams;
- ✓ understand what is meant by activation energy;
- ✓ understand how catalysts affect activation energy;
- ✓ understand that energy must be supplied to break bonds whereas energy is released when bonds form;
- ✓ be able to calculate the energy transferred in reactions.

Figure 12.1 Our bodies need to store energy from our food. However, these girls are eating too much food that is high in fat and are not burning it up through exercise. The result is that too much fat has been stored and they have become obese.

12.1 Energy from foods and fuels

Our bodies obtain the energy we need from our food. We also need to use energy in our homes and in industry, where it is supplied by fuels from crude oil (see Chapter 2). But how much energy does a food or fuel contain?

It is not possible to measure the amount of energy produced by a food or fuel directly. You have to measure the heat energy transferred when the food or fuel burns. Usually this is done by using the food or fuel to heat a known mass of water and measuring the rise in temperature.

We can compare the energy released per gram of different foods and fuels using the method illustrated in Figure 12.2.

1. Take similar amounts of each food or fuel.
   - Place liquid fuels in a spirit burner.
   - Place solid fuels on a small, clean upturned food tin.
2. Find the total mass of the fuel plus its container.
3. Place a known amount of water, say 250 g, in a tin can or metal calorimeter.
4. Burn the fuel or food so that the water is heated. Extinguish the fuel before the water reaches 70 °C (this minimises the heat loss). Take care to protect the burner from draughts of air.
5. Measure the temperature rise of the water.
6. Find the final mass of the spirit burner or the upturned tin plus unused fuel at the end of the experiment.

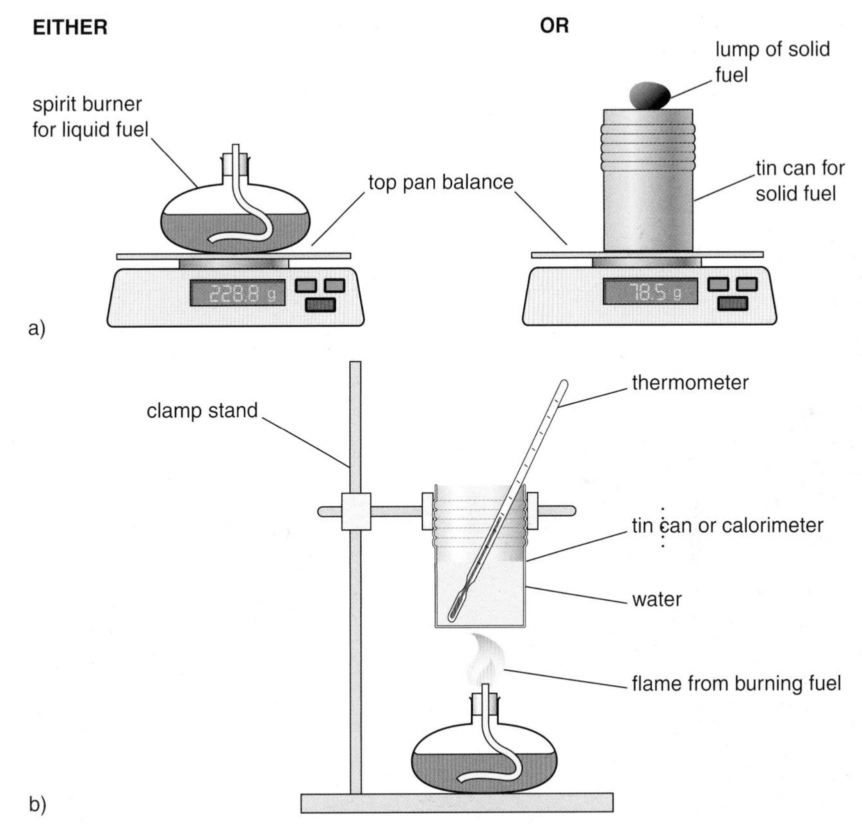

Figure 12.2 a) Use a top pan balance to find the initial mass of the apparatus plus fuel. b) Use the fuel to heat a known mass of water. Record the temperature rise of the water. The diagram shows the liquid fuel being burned.

A **calorimeter** is the apparatus used to measure the heat transferred when a substance burns. It can be a simple tin can or a metal container.

When energy changes are measured in the laboratory, a piece of apparatus called a **calorimeter** is used. This is placed so as to ensure that as much energy as possible from the burning reaction goes into heating the water, rather than being lost to the surroundings. This means that the measured temperature rise accurately represents the energy released.

1. Look at the diagram of the calorimeter in Figure 12.2 and compare it with the diagram in Figure 12.3. Explain why the results for the energy value of the food or fuel will be more accurate using the apparatus in Figure 12.3.

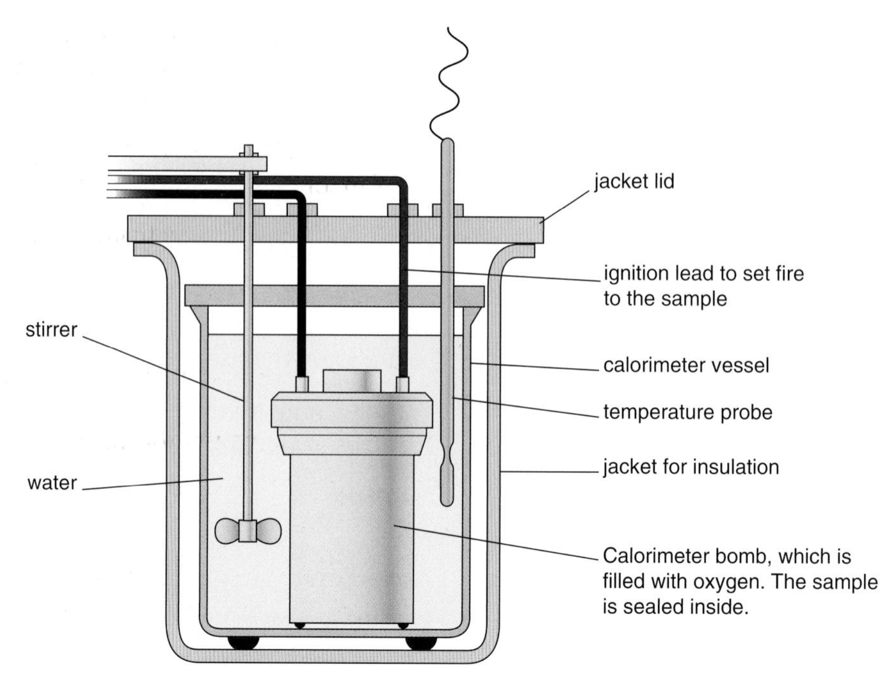

Figure 12.3 The energy transferred when a food or fuel burns can be measured more accurately using a calorimeter like this.

Calculating the energy change from the temperature rise

We can use the results of calorimeter experiments to measure the amount of energy transferred during burning or other reactions.

The amount of energy required to raise the temperature of 1 kg of water by 1 °C is 4.2 kJ. Scientists describe the amount of energy required to raise the temperature of 1 kg a substance by 1 °C as its specific heat capacity. So the specific heat capacity of water in 4.2 J/kg/°C. Different materials have different specific heat capacities.

In a calorimeter experiment, if we know the mass of water in the calorimeter and the rise in temperature, then we can use the value for the specific heat capacity of water to calculate the energy transferred to the water in the calorimer. So,

total energy transferred	=	mass of water	×	4.2	×	change in temperature
(in kJ)		(in kg)		(kJ/kg/°C)		(in °C)

Notice that the mass is in kilograms, because the value for specific heat capacity is per kg.

Here is an example of a calculation to find the amount of energy transferred when a particular fuel is burned.

Example

a) When 180 cm^3 of methane is burned, the heat produced is used to raise the temperature of 100 g (0.10 kg) of water from 20 °C to 35.9 °C. Calculate the energy released.

energy released = mass of water × 4.2 × change in temperature
= 0.10 × 4.2 × 15.9
= 6.68 kJ

b) How much energy is released per mole of methane? (Assume 1 mole of methane = 24 000 cm^3.)

180 cm^3 of methane releases 6.68 kJ

so 1 cm^3 of methane releases $\frac{6.68}{180}$ kJ

so 1 mole of methane (24 000 cm^3) releases $\frac{24\,000 \times 6.68}{180}$ = 891 kJ

c) Write an equation for this reaction and the energy change.
The equation is:

$CH_4 + 2O_2 \rightarrow CO_2 + 2H_2O$ Energy change = −891 kJ/mol

The energy change is described as −891 kJ/mol because the reaction is exothermic and heat is lost.

Fuel	Formula	Energy released in kJ /mol
Ethanol	C_2H_5OH	1370
Propanol	C_3H_7OH	2020
Butanol	C_4H_9OH	2670
Pentanol	$C_5H_{11}OH$	3320
Methane	CH_4	890
Ethane	C_2H_6	1560
Propane	C_3H_8	2220
Butane	C_4H_{10}	2880
Pentane	C_5H_{12}	3510

Table 12.1 The energy produced by burning different fuels

2 a) Using the data in Table 12.1, plot a graph to compare the energy output per mole of ethanol, propanol, butanol and pentanol. On the *x*-axis, plot the number of carbon atoms in the molecule.
b) Explain what the trend (pattern) in the graph shows.
c) Predict the energy output per mole for burning methanol (CH_3OH).

3 Sam says, 'It is better to compare the energy output per gram of a fuel than the energy output per mole. Output per mole is not a fair test for comparing fuels.'
a) Calculate the energy output per gram for methane, ethane, propane, butane and pentane. Record the results in a table with the energy output per mole.
b) Use your table to write a paragraph commenting on whether what Sam says is valid.

Measuring the energy from reactions in solution

The energy produced by chemical reactions in solution can be found by using a calorimeter. The reagents are mixed in an insulated container and the temperature change of the solution is measured.

For example, the energy released in a displacement reaction such as:

$$Fe(s) + CuSO_4(aq) \rightarrow FeSO_4(aq) + Cu(s)$$

can be measured using the apparatus in Figure 12.4.

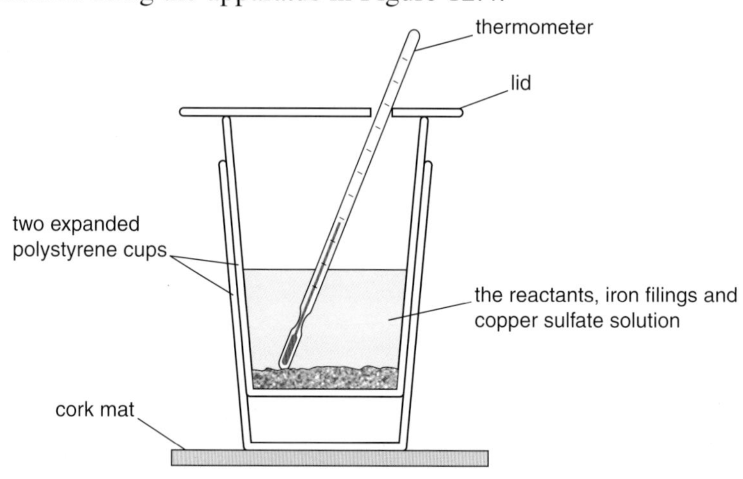

Figure 12.4 Measuring the energy released in a displacement reaction. Careful insulation and draught-proofing increase the accuracy of the method.

The method involves the following steps:

1 Collect the apparatus shown in Figure 12.4.
2 Allow all reactants to reach room temperature and take this as the starting temperature.
3 Measure out a known volume of 2.0 mol/dm^3 copper sulfate solution. Place this in the polystyrene cup and take its temperature.
4 Add an excess of iron filings (about 5 g) and place the lid on the cup.
5 Stir the mixture and take the temperature every minute until the maximum has been reached and the temperature starts to fall again.
6 Protect the container from draughts and keep the lid in place between measurements to prevent energy loss from the reacting mixture.
7 Plot a graph of the temperature measurements against time (Figure 12.5). To find the maximum temperature, draw a smooth curve through the plotted points and read off the maximum value from the curve.

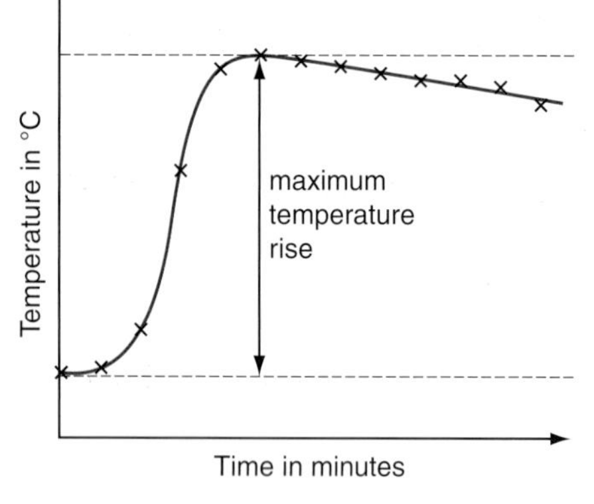

Figure 12.5 A graph showing how to find the maximum temperature reached by a reacting mixture

4 When 8.0 g of ammonium nitrate is dissolved in water, there is a drop in temperature. The energy change for the reaction is 1787 joules.
a) Calculate the energy change when 1 mole (80 g) of ammonium nitrate dissolves.
b) Write an equation for the reaction followed by an energy change value.

As with burning reactions, we can use the measurements taken during reactions between solutions to calculate the energy changes in the reactions.

Example

When 120 cm^3 of 2.0 mol/dm^3 hydrochloric acid is neutralised with 120 cm^3 of sodium hydroxide solution there is a temperature rise of 13.5 °C. Calculate the energy change per mole for the neutralisation of hydrochloric acid by sodium hydroxide.

The equation for the reaction is:

$$HCl(aq) + NaOH(aq) \rightarrow NaCl(aq) + H_2O(l)$$

In 120 cm^3 of 2.0 mol/dm^3 HCl there are $\frac{120}{1000} \times 2.0$ moles = 0.24 moles HCl.

As the solutions are mostly water, we can say:

$$\text{volume of water heated} = 120 + 120 = 240\ cm^3$$

1000 cm^3 of water has a mass of 1 kg, so 240 cm^3 of water has a mass of $\frac{240}{1000} = 0.24$ kg.

$$\underset{\text{(in kJ)}}{\text{energy gained by water}} = \underset{\text{(in kg)}}{\text{mass of water}} \times \underset{\text{(kJ/g/°C)}}{4.2} \times \underset{\text{(in °C)}}{\text{temperature change}}$$

$$= 0.24 \times 4.2 \times 13.5$$

$$= 13.6\ kJ$$

So 0.24 moles of HCl releases 13.6 kJ when neutralised.

Therefore 1 mole of HCl releases $\frac{13.6}{0.24}$ kJ of energy = 56.7 kJ.

As the reaction is exothermic, we can write:

$$HCl(aq) + NaOH(aq) \rightarrow NaCl(aq) + H_2O(l) \qquad \text{Energy change} = -56.7\ kJ$$

5 The results below were obtained in an experiment to find the energy released during the combustion of ethanol (C_2H_5OH).

Mass of ethanol and burner before experiment	= 208.8 g
Mass of ethanol and burner after experiment	= 208.5 g
Mass of ethanol burned	=
Initial temperature	= 20 °C
Final temperature	= 36.5 °C
Temperature rise	=
Mass of water	= 100 g

a) Copy and complete the results.
b) Write a balanced equation for the reaction.
c) Calculate the energy released when 1 mole of ethanol is burned.
d) Is the reaction endothermic or exothermic?
e) How well does your calculated value fit with the accepted value in Table 12.1 on page 223?

12.2 Food for fuel and energy storage

Energy is normally measured in joules (J) or kilojoules (kJ). Sometimes energy values on food packets are given in **calories**. 1 calorie = 4.2 joules

Our bodies need energy to maintain body temperature, for movement and for growth of new tissues. Different foods produce different amounts of energy (Table 12.2). Often the energy content is given on the food packet. This is usually in units of kilojoules per 100 g (kJ/100 g). Sometimes an older unit, the **calorie** (or kilocalorie), is used.

Food	Energy value in kJ/100 g
Bread	980
Butter, margarine	3000
Chicken (no skin)	621
Lamb	1110
Milk (full fat)	272
Green beans	19
Cooking oil	3700
Boiled egg	612
Fried fish	982
Carrots	98
Boiled potatoes	322
Chips	980
Apples	196
Pasta	1450
Breakfast cereal	1500

Table 12.2 The energy values of various foods

Food with a high proportion of fat or oil has the highest energy content. Foods high in fat and oil include items that we don't usually think of as fatty, such as crisps, cooked meat and salad dressings. Foods with a high proportion of carbohydrate, such as bread and pasta, also have a high energy value.

If you take in more energy in food than you use up, you will put on weight. The excess energy is stored as fat. This fat accumulates under the skin and in the walls of blood vessels. Fat is a long-term energy store. Some fat is also needed to keep us warm, because the layer of fat under the skin acts as an insulting layer.

Obesity is an abnormally high body weight (relative to a person's height), caused by an accumulation of fat. This accumulation results from taking in more energy from food than is used up.

Some people have more body fat than others, and there is no 'ideal' amount. However, too much body fat is unhealthy. The deposits formed in blood vessels can cause heart disease. If you have a large amount of body fat, you may be obese. **Obesity** is becoming more common, especially in modern Western countries, and is responsible for many health problems.

Activity – Controlling energy intake

The amount of energy needed each day by people with different occupations is shown in Figure 12.6.

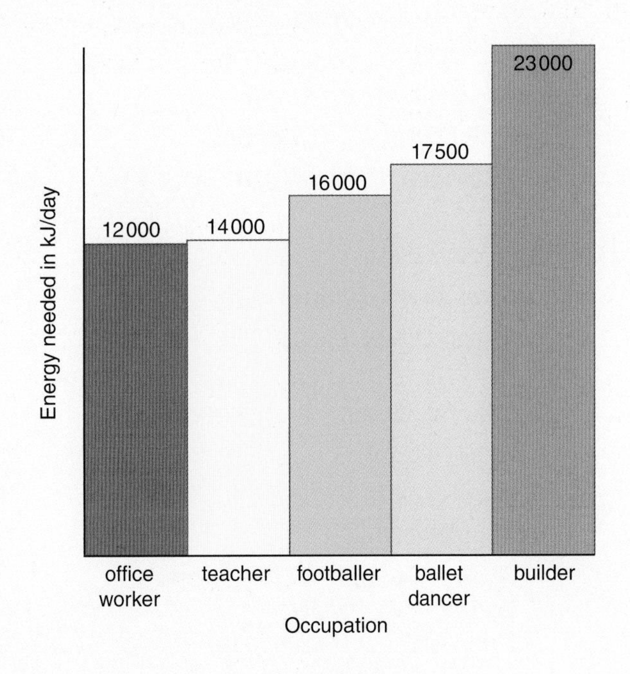

Figure 12.6 The energy needed each day by different people

Tables 12.3 and 12.4 show the food intake at breakfast and lunch of an office worker, a ballet dancer and a builder. In addition to these meals, they each eat snacks totalling 3000 kJ.

Person	Food eaten	Energy in the food in kJ
Office worker	2 thick slices of bread (100 g)	980
	Butter (25 g)	750
	2 large boiled eggs (200 g)	1224
	Breakfast cereal (200 g)	3000
	Milk (100 g)	272
		Total = 6226 kJ
Ballet dancer	2 apples (200 g)	392
	Glass of milk (200 g)	544
		Total = 936 kJ
Builder	2 thick slices of bread (100 g)	980
	Butter (25 g)	750
	2 large boiled eggs (200 g)	1224
	Breakfast cereal (200 g)	3000
	Milk (100 g)	272
		Total = 6226 kJ

Table 12.3 The breakfast foods of an office worker, a ballet dancer and a builder

Person	Food eaten	Energy in the food in kJ
Office worker	Fried lamb steak (200 g)	2220
	Oil for cooking lamb (20 g)	740
	Pasta (100 g)	1450
	Bread (100 g)	980
	Butter (25 g)	750
	Large glass of milk (300 g)	816
		Total = 6956 kJ
Ballet dancer	Chicken (200 g)	1242
	Green beans (200 g)	38
	Carrots (100 g)	98
		Total = 1378 kJ
Builder	Fried fish (200 g)	1964
	Chips (100 g)	980
	Carrots (200 g)	196
	2 apples (200 g)	392
	Chicken for sandwich for tea break in afternoon (100 g)	621
	2 thick slices of bread (100 g)	980
	Butter (25 g)	750
	Milk for tea (50 g)	136
		Total = 6019 kJ

Table 12.4 The lunch foods of an office worker, a ballet dancer and a builder

1. What types of food give the least energy per gram? Explain how you arrived at your answer.
2. Which person is likely to become obese? Explain why.
3. Write a short note giving each person advice about their energy intake. Include advice on what foods they should eat for their evening meal.

6. a) Give one example from Table 12.2 of a food that contains mainly i) protein, ii) carbohydrate, iii) fat.
   b) Work out the approximate amounts of energy we get per 100 g from these different types of food.
7. Would grilling or deep fat frying be better for reducing the energy content of food? Explain your answer.
8. The nutrition label on a jam roly poly packet provides this information:

   100 g contains: energy 335 kcal, protein 2.8 g, carbohydrate 51.9 g, fats 12.5 g, fibre 4.3 g, sodium 0.2 g

Figure 12.7 Jam roly poly is a high-energy food, eaten as a dessert.

   a) Calculate the energy content of jam roly poly in kJ per 100 g.
   b) Which of the food groups present in jam roly poly make it a high-energy food? Explain your answer.

12.3 Energy level diagrams

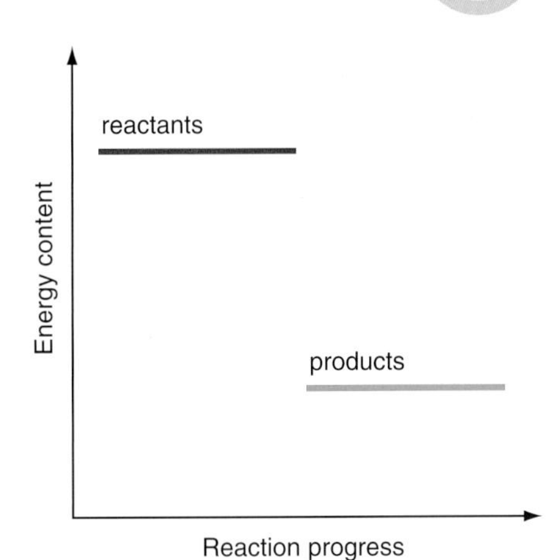

Figure 12.8 A simple energy level diagram where the products have less energy than the reactants

When a chemical reaction occurs, energy changes take place at the same time. In any reaction mixture, the reactants and products contain chemical energy. The products of the reaction have a different energy content from the reactants. For example, when we burn a fuel, energy is released. This energy comes from the reactants, so the products formed in the reaction have less energy than the reactants.

An energy level diagram (Figure 12.8) can be used to show the difference in energy content between the reactants and the products.

Exothermic reactions

In exothermic reactions, some energy is transferred to the surroundings, usually causing heating. As a result, the energy content of the products is less than that of the reactants. This means the energy change for an exothermic reaction is always negative (Figure 12.9).

An example of an exothermic reaction is ethanol burning in air.

$$C_2H_5OH(l) + 3O_2(g) \rightarrow 2CO_2(g) + 3H_2O(g)$$

ethanol + oxygen → carbon dioxide + water

Activation energy is the energy required by the reactant particles for a reaction to happen.

In Figure 12.9, E_A is the energy required to start the reaction going. This is called the **activation energy**. In the ethanol example, the energy from a match or a spark is needed to get the burning reaction started.

Once started, an exothermic reaction gives out enough energy to keep itself going. In some exothermic reactions, the activation energy is very large (for example burning charcoal), but they are still exothermic. In other reactions that happen spontaneously, the activation energy may be very small.

Figure 12.10 It takes a lot of effort to get a car moving, but once it is moving it tends to stay moving. This is rather like activation energy, which is the energy needed to get a reaction going.

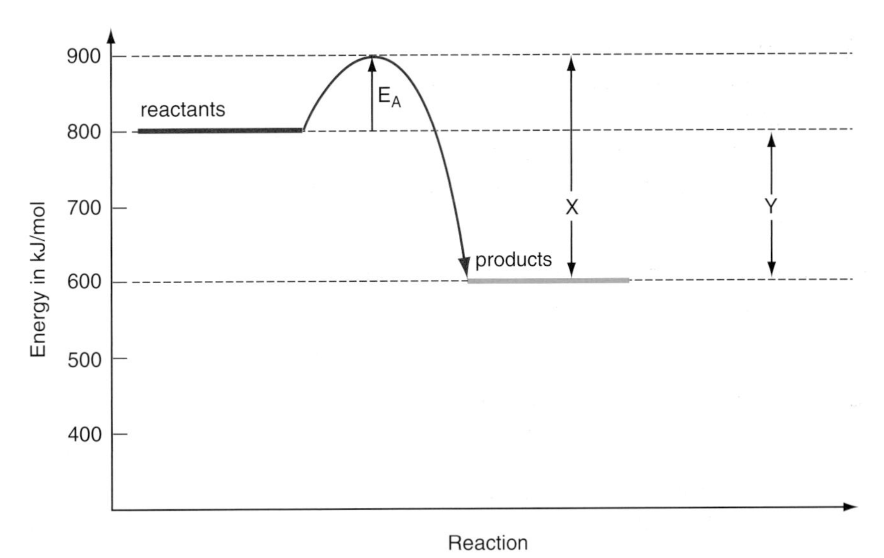

Figure 12.9 The energy level diagram for an exothermic reaction

Endothermic reactions

In endothermic reactions, energy is transferred to the reacting mixture from the surroundings. This energy is taken in during the reaction and stored as chemical energy in the products. In this case, the energy content of the products is greater than the energy content of the reactants. This means the energy change for an endothermic reaction is always positive (Figure 12.11). Overall, the products contain more energy than the reactants.

An example of an endothermic reaction is solid hydrated barium hydroxide reacting with solid ammonium nitrate.

$$Ba(OH)_2.8H_2O(s) + 2NH_4NO_3(s) \rightarrow Ba(NO_3)_2(aq) + 2NH_3(g) + 10H_2O(l)$$

hydrated barium hydroxide + ammonium nitrate → barium nitrate + ammonia + water

> **9** a) Explain what is meant by an endothermic reaction.
> b) Give one example of an endothermic reaction and explain why this reaction is endothermic.
> c) Explain what is meant by an exothermic reaction.
> d) Give one example of an exothermic reaction and explain why this reaction is exothermic.

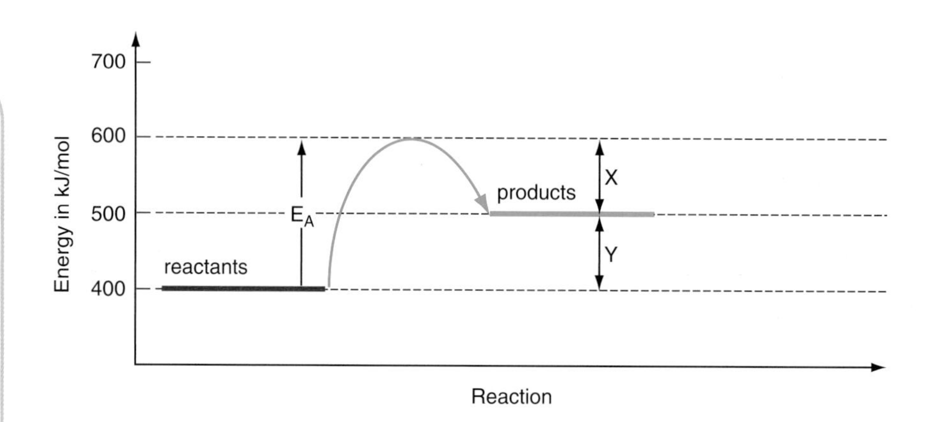

Figure 12.11 The energy level diagram for an endothermic reaction

Calculating energy changes from energy level diagrams

In the energy level diagram for an endothermic reaction (Figure 12.11), Y is the total amount of energy transferred in from the surroundings by the end of the reaction.

Y is the change in the energy content during the reaction – that is, the difference between the energy of the reactants and that of the products. So this is the amount of energy put in to the reactants (E_A), minus the energy transferred out during the reaction (X).

Therefore, Y = change in energy content
and Y = E_A (activation energy taken in) − X (energy transferred out during the reaction)

The change in energy content (Y) is measured in kilojoules (kJ), or kJ/mol if it relates to one mole of a substance.

From the graph in Figure 12.11, E_A = 200 kJ and X = 100 kJ,
so

$$Y = E_A - X$$
$$= 200 - 100 = +100\,kJ$$

Using the equation $Y = E_A - X$ for an exothermic reaction (Figure 12.9):

$Y = E_A$ (activation energy taken in) $-$ X (energy transferred out during the reaction.

From the graph in Figure 12.9, $E_A = 100\,kJ$ and $X = 300\,kJ$,

so $Y = 100\,kJ - 300\,kJ = -200\,kJ$

10 a) What is E_A the symbol for?
b) Explain what kJ/mol means.

11 Look at Figure 12.12.

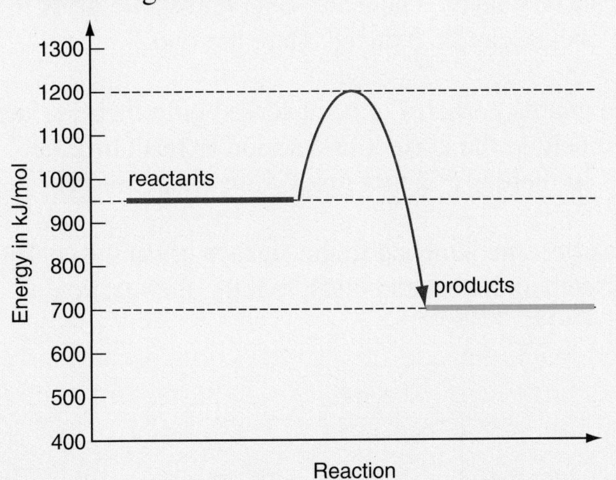

Figure 12.12 The energy level diagram for an exothermic reaction

a) How much energy is stored in the reactants?
b) How much energy is stored in the products?
c) How much energy must first be put in to make the reaction happen?
d) What is the change in energy content during the reaction (in kJ/mol)?

12 Look at Figure 12.13. Explain as fully as you can what this energy level diagram tells you about the reaction.

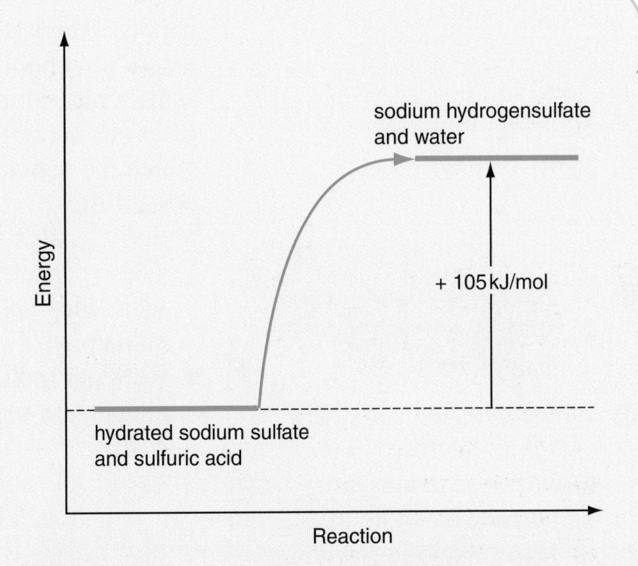

Figure 12.13 The energy level diagram for making sodium hydrogensulfate

13 The reaction between hydrogen gas and chlorine gas is exothermic with an energy change of $-184\,kJ$ per mole of hydrogen and a low activation energy.
a) Sketch an energy level diagram for this reaction.
b) Write a balanced equation including energy data.

14 When 0.6 g of octane C_8H_{18} (petrol) was burned it produced 29 kJ of energy. What is the energy change for the combustion of octane in kJ/mol?

12.4 The effect of catalysts on activation energy

As we have already seen (page 229), most reactions require a 'kick' to get them started. For example, when methane burns in air, the reaction needs a match or a spark to get it started, and the reaction between hydrogen and chlorine can be started using ultraviolet light. This 'kick' to get the reaction started is the activation energy. This energy is needed to give the reactant particles sufficient energy to react.

Collisions between particles determine how quickly most reactions happen (Chapter 7). A collision between reactant particles does not result in a reaction unless the collision is energetic enough for the particles to reach the activation energy level. In any reaction mixture, some particles will have more energy than others. The more energetic particles may be able to react if they happen to collide.

Catalysts

A catalyst (Section 7.3) can provide a different way for the reaction to happen. The new pathway has a lower activation energy than the normal reaction without a catalyst. This will make the reaction faster at the same temperature. Many catalysts are solids, which tend to take part in reactions using their surface layer. There are several possible ways in which the activation energy can be reduced. Here are two possibilities.

- Solid catalysts can enable particles to be absorbed onto their surface. These particles are held in the correct orientation to react together when they collide, so there is a higher proportion of successful collisions.
- Reactant molecules become attached to the surface of catalysts. The attachment may stretch and weaken bonds, making the reaction easier.

15 a) What is activation energy?
b) Why is activation energy needed for a reaction?

16 Where does the activation energy come from in the following reactions:
a) burning a candle in air;
b) petrol exploding with air in a car engine;
c) gas igniting automatically on a central heating pilot light;
d) lighting a match?

17 Catalytic converters in cars oxidise carbon monoxide in exhaust fumes to carbon dioxide using oxygen from the air.

$$2CO(g) + O_2(g) \xrightarrow{\text{catalyst}} 2CO_2(g)$$

Explain simply using a diagram how the catalyst surface might do this.

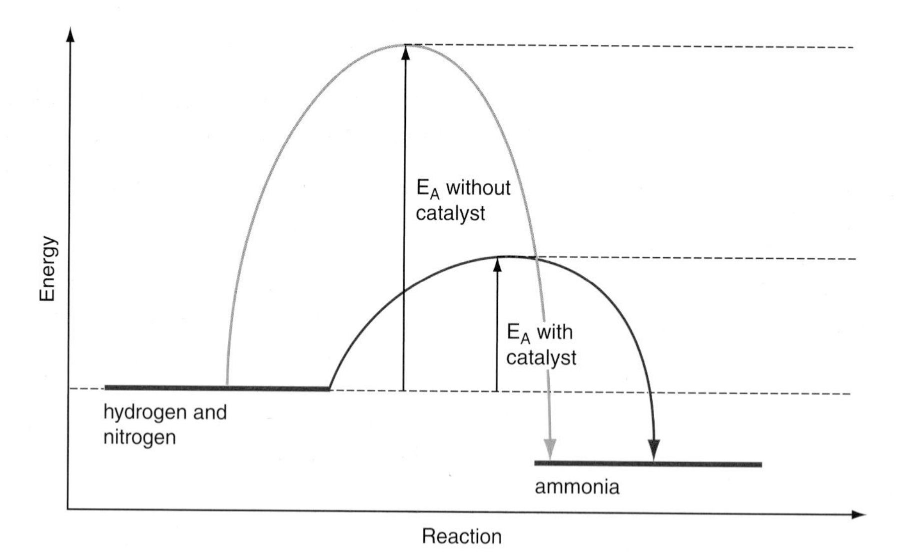

Figure 12.14 The catalyst lowers the activation energy so more of the collisions result in a reaction. This makes a reaction more likely to happen and therefore the rate of reaction is faster.

12.5 Bond breaking and bond making

Bond breaking is endothermic (requires energy) whereas bond making is exothermic (releases energy).

For a reaction to happen, existing bonds between atoms in the reactants need to be broken. This takes in energy, as energy is needed to break the bonds between atoms. When new chemical bonds form to make the products, energy is released.

Why are reactions exothermic or endothermic?

In any reaction, energy must be supplied to break the chemical bonds between atoms, and energy is then released when chemical bonds form in the products. If more energy is given out than is put in, the reaction is exothermic.

For example, when methane is burned, the chemical bonds in methane and oxygen are broken and new bonds are formed in the products, carbon dioxide and water:

$$CH_4 + O_2 \rightarrow CO_2 + H_2O$$

methane + oxygen → carbon dioxide + water

The chemical bonds in methane and oxygen are weaker than the chemical bonds between atoms in carbon dioxide and water. This means more energy is released when the products form than was put in when bonds in the reactants were broken. So the reaction is exothermic.

Figure 12.15 Hydrocarbon gases make good portable fuel supplies. Here a blow torch is being used in plumbing.

In endothermic reactions, more energy is needed to break bonds in the reactants than is released when bonds form between atoms in the products. Hence, energy is taken in overall. This energy is stored in the bonds of the products.

For example, in photosynthesis, plants take in carbon dioxide and water but need to break these molecules apart to make the products:

$$6CO_2 + 6H_2O \xrightarrow{\text{chlorophyll}} C_6H_{12}O_6 + 6O_2$$

carbon dioxide + water → glucose + oxygen

This is done using large amounts of energy from sunlight and with help from chlorophyll. The broken-up molecules of carbon dioxide and water recombine to make glucose and oxygen. Much less energy is given out when these molecules form than is taken in to rip the reactants apart. So lots of energy is stored in the products of this reaction – which is the whole point of photosynthesis.

Figure 12.16 Green plants take in energy to complete the changes needed for photosynthesis.

18 Draw energy level diagrams to represent:
a) the reaction between methane and oxygen;
b) the photosynthesis reaction that converts carbon dioxide and water to oxygen and glucose.
On each diagram, mark:
- the energy content of reactants;
- the energy content of products;
- the activation energy;
- the energy transferred by the reaction.

19 Suggest three ways in which chemical bonds can be broken.

20 Explain what the following information means in as much detail as you can.

$$2H_2(g) + O_2(g) \rightarrow 2H_2O(g)$$
Energy change = −242 kJ

21 a) Write a word equation and a balanced chemical equation for the thermal decomposition of copper carbonate. (Hint: Use the information from the data sheet on page 257.)
b) Draw an energy level diagram for this reaction.

For any chemical bond between two atoms, the **bond energy** is the amount of energy needed to break the bond and separate the atoms. Bond energies are measured in kilojoules per mole of bonds broken.

How do chemical bonds get broken?

The energy to break chemical bonds and start a reaction can sometimes come from the particles themselves. In other cases, energy input is needed in the form of heat, electricity or even light to start a reaction.

Some chemical changes, such as thermal decomposition, do not involve collisions. For example, when solid copper carbonate is heated, the more energetic vibrations of the atoms cause it to decompose. Copper carbonate becomes copper oxide plus carbon dioxide gas. The bonds in copper carbonate store less energy than those in the products (copper oxide and carbon dioxide) so this is an endothermic change. Enzymes and other catalysts make it easier for chemical bonds to break in some substances and reduce the energy needed for reactions to occur (Section 12.4).

Calculating energy changes from bond energies

Bond breaking requires energy, whereas bond making releases energy. By careful experiments it is possible to measure (in kJ/mol) the energy required to break a bond. This will be the same as the energy released when the bond is made. By simple 'accountancy' – that is, adding and subtracting the **bond energies** – you can work out how much energy is required in total to break up the reactants into atoms, and the energy released when these atoms combine to form the products. By doing this, it is possible to calculate the overall energy change for a reaction.

Bond	Energy required to break one mole of bonds in kJ/mol	Bond	Energy required to break one mole of bonds in kJ/mol
C–C	346	H–Cl	431
C=C	611	Br–Br	193
C–H	413	H–Br	366
C=O	803	N–H	390
O=O	497	N≡N	945
O–H	464	Cl–Cl	242
H–H	436		

Table 12.5 The bond energies of various bonds

Using the bond energies in Table 12.5, we can work out the energy changes in some reactions.

Examples

1 Hydrogen burns to form water. What is the energy change in the reaction?

$$2H_2(g) + O_2(g) \rightarrow 2H_2O(g)$$

a) Energy needed to break bonds in the reactants:
$= 2(H-H) + (O=O)$
$= (2 \times 436) + 497$
Energy input $= 1369$ kJ

b) Energy released by making bonds in the products:
$= 4\ (O–H)$
$= 4 \times 464$
Energy output $= 1856$ kJ

$\therefore$ Energy change in the reaction = energy input − energy output
$= +1369 - 1856 = -487$ kJ

2 What is the energy change for the following reaction?

$$H_2(g) + Cl_2(g) \rightarrow 2HCl(g)$$

a) Energy needed to break bonds in the reactants:
$= (H-H) + (Cl-Cl)$
$= 436 + 242$ kJ/mol
Energy input $= +678$ kJ

b) Energy released by making bonds in the products:
$= 2(H-Cl)$
$= 2 \times 431$
Energy output $= 862$ kJ

$\therefore$ Energy change $= +678 - 862$ kJ $= -184$ kJ

22 Use the bond energies in Table 12.5 to calculate the energy change for the number of moles in the equation below for when methane burns in air.

$$CH_4(g) + 2O_2(g) \rightarrow CO_2(g) + 2H_2O(g)$$

23 a) Why do you think that nitrogen (N_2) is such a stable molecule?
b) Calculate the energy change for the formation of ammonia (NH_3) in the reaction below.

$$N_2(g) + 3H_2(g) \rightarrow 2NH_3(g)$$

c) Is this reaction endothermic or exothermic?

Summary

- ✓ The energy transferred in a chemical reaction can be calculated by measuring the temperature rise in a known mass of solution.
- ✓ It is important to insulate the reacting mixture when measuring the energy transferred in chemical reactions.
- ✓ A **calorimeter** is the apparatus used to measure the heat transferred during a reaction.
- ✓ The unit of energy is the joule (J). Sometimes food energy data is given in **calories**. 1 calorie = 4.2 joules.
- ✓ Different foods and fuels store different amounts of energy.
- ✓ Fats, oils and carbohydrates store relatively large amounts of energy.

- ✓ Taking in too much food, which provides more energy than your body needs, can lead to **obesity**.
- ✓ The energy changes in a reaction can be represented by an energy level diagram.
- ✓ **Activation energy** is the energy required by reactant particles to enable them to react.
- ✓ Catalysts work by providing a new reaction pathway with a lower activation energy.
- ✓ Energy must be supplied to break bonds whereas energy is released when bonds are formed.
- ✓ In an exothermic reaction, bond making releases more energy than the bond breaking took in.
- ✓ In an endothermic reaction, bond making releases less energy than the bond breaking requires.
- ✓ It is possible to calculate the energy transferred in a reaction if you know the **bond energies** involved.

EXAMQUESTIONS

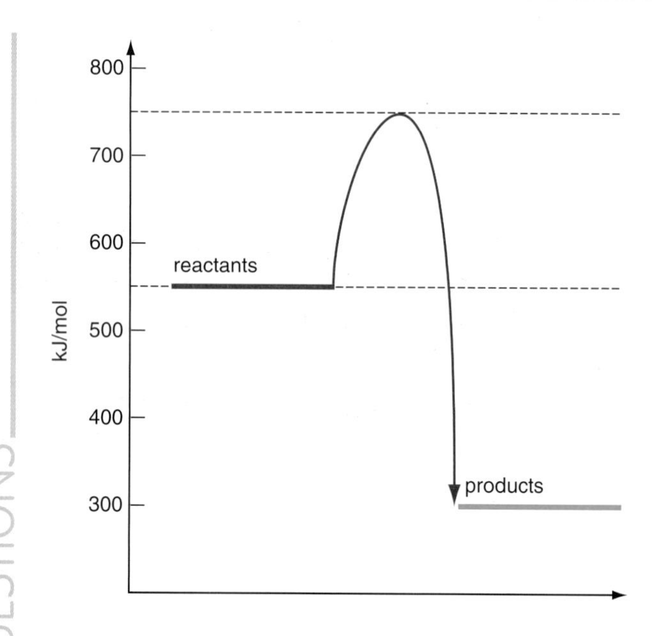

Figure 12.17 An energy level diagram for an exothermic reaction

1 a) Copy the energy level diagram in Figure 12.17 and mark on it:
   i) the energy change in the reaction; *(1 mark)*
   ii) the activation energy. *(1 mark)*

b) Explain why the energy change has a minus value in this case. *(2 marks)*

c) From the diagram calculate the value of the energy change in kJ. *(2 marks)*

d) State what the *x*-axis represents in this diagram. *(2 marks)*

e) State what the *y*-axis represents in this case. *(2 marks)*

2 a) Use words from the box below to complete the following paragraph.

make	break	takes in
broken	bonds	reactant
products	gives out	

For a chemical reaction to take place, the bonds holding the __________ substances together must be __________. This process __________ energy. When the __________ of the reaction are formed, new __________ form between the atoms. This process __________ energy. *(6 marks)*

b) Explain the difference between endothermic and exothermic reactions in terms of the energy changes in bond breaking and bond formation. *(4 marks)*

3 'Producer gas' contains carbon monoxide, hydrogen and nitrogen. It is made by passing air and steam over hot coke (carbon). It is used as an industrial fuel.

First a blast of air is blown through the coke. Then a blast of steam is used to produce the gas.

The two reactions involved are:

$$C(s) + O_2(g) \rightarrow CO_2(g)$$
Energy change = −394 kJ/mol

$$C(s) + H_2O(g) \rightarrow CO(g) + H_2(g)$$
Energy change = +137 kJ/mol

Use this data to show that the energy

transferred in the first reaction will keep the overall producer gas process going to make the carbon monoxide and hydrogen. *(6 marks)*

4 The apparatus shown in Figure 12.18 was used to study the reaction between iron filings and copper sulfate solution.

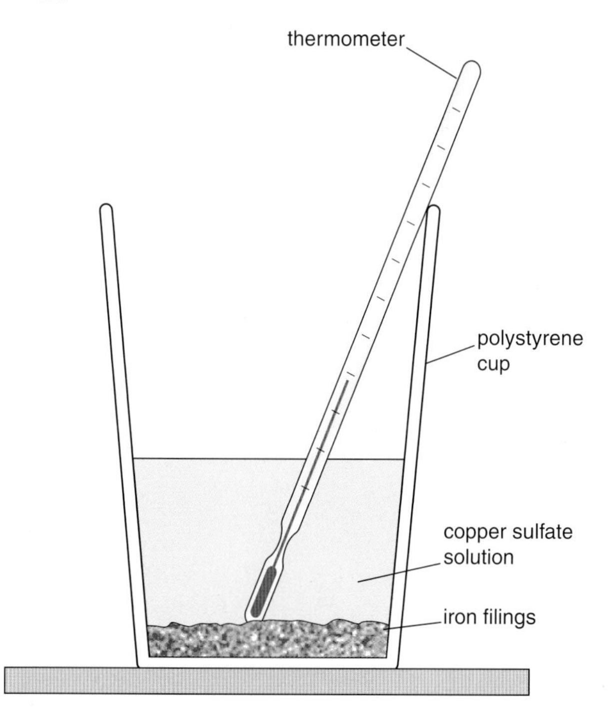

Figure 12.18 Apparatus for the displacement reaction between copper sulfate solution and iron filings

The reactants were excess iron filings and 100 cm^3 of 0.25 mol/dm^3 copper sulfate solution.

The reaction mixture was stirred until the blue colour of the copper sulfate had disappeared.

A maximum temperature rise of 8.3 °C was recorded.

a) Write an equation for the reaction. *(2 marks)*

b) What would you see during the reaction? *(2 marks)*

c) When the reaction mixture heats up it takes in the same energy as heating up 100 g of water. Work out the energy change per mole of copper sulfate. *(6 marks)*

5 Use the bond energies in Table 12.5 on page 234 to answer parts a) and b) below.

a) What is the energy change for the reaction between hydrogen and bromine?

$$H_2(g) + Br_2(g) \rightarrow 2HBr(g)$$

(4 marks)

b) Show that the value for the combustion of propane in Table 12.1 agrees with the value for the reaction calculated from bond energies.

$$C_3H_8(g) + 5O_2(g) \rightarrow 3CO_2(g) + 4H_2O(g)$$

(6 marks)

Chapter 13
How do we identify and analyse substances?

At the end of this chapter you should:

- ✓ understand that a range of chemical tests can be used to detect and identify elements and compounds;
- ✓ know how flame tests and reactions with sodium hydroxide solution can be used to test for various cations;
- ✓ know the tests for common anions – carbonate, chloride, bromide, iodide, sulfate and nitrate;
- ✓ be able to interpret the results of these chemical tests and identify various cations and anions;
- ✓ be able to evaluate the advantages and disadvantages of instrumental methods of analysis;
- ✓ appreciate the importance of modern computers and other electronic devices in increasing the speed, accuracy and sensitivity of modern instrumental analysis;
- ✓ be able to interpret the results of instrumental analysis and draw conclusions;
- ✓ understand the benefits of modern instrumental analysis in industry, environmental monitoring and forensic science;
- ✓ understand that different methods such as mass spectrometry, emission spectroscopy, absorption spectroscopy and nuclear magnetic resonance are suitable for identifying elements and compounds.

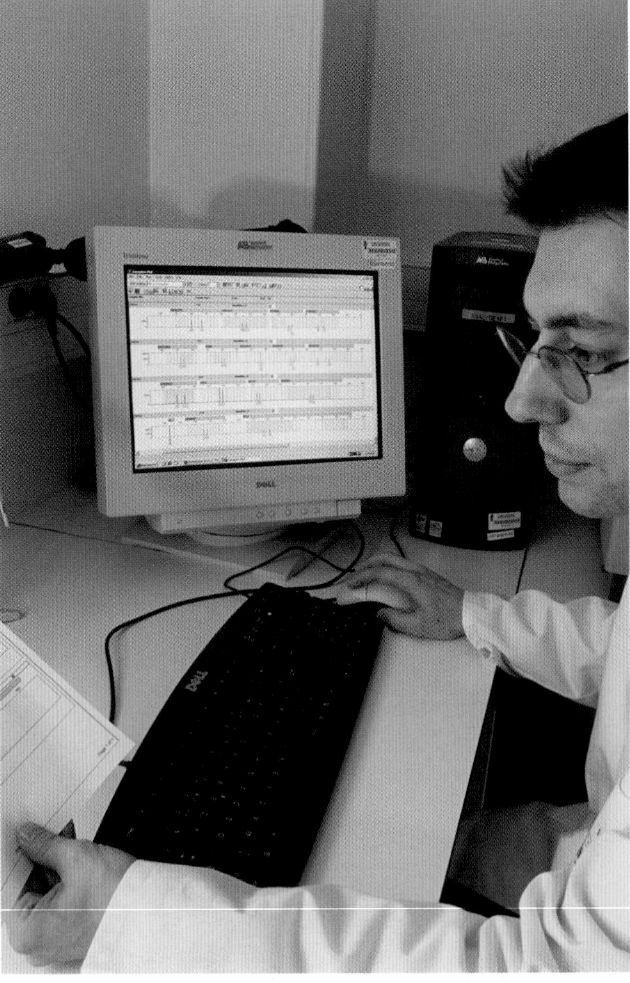

Figure 13.1 Chemists use a wide range of techniques and instruments to analyse materials in our food, in the environment and at the scene of a crime. Their instruments can work with tiny samples of material. A computer is used to control the instruments and record data.

13.1 Identifying substances by heating them

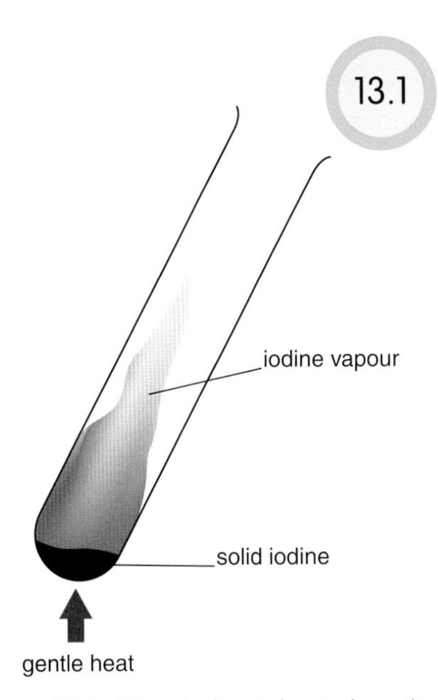

Figure 13.2 When iodine is heated gently, the dark grey solid sublimes (changes directly from solid to vapour), producing a purple vapour. This is a very distinctive change.

When a substance is heated, its temperature will increase and the sample may change in various ways. For example, a solid may:

- expand;
- melt;
- react chemically.

Sometimes, the change that happens when a substance is heated is so distinctive and so informative that it identifies the substance immediately (Figures 13.2 and 13.3). If you try any of the tests in this chapter, you must get the approval of your teacher first and always wear eye protection.

Heating ionic (metal) compounds

Some ionic compounds decompose on heating. These reactions can sometimes be used to identify the substance being heated.

The equations below summarise what happens when copper(II) carbonate and zinc carbonate are heated. The distinctive colour changes (shown below the formulae) help to identify both of these substances.

copper(II) carbonate	→	copper(II) oxide	+	carbon dioxide
$CuCO_3(s)$	→	$CuO(s)$	+	$CO_2(g)$
green		black		colourless

zinc carbonate	→	zinc oxide	+	carbon dioxide
$ZnCO_3(s)$	→	$ZnO(s)$	+	$CO_2(g)$
white		yellow when hot, white when cold		colourless

Green copper(II) carbonate forms black copper(II) oxide on heating. When white zinc carbonate is heated, hot yellow zinc oxide is produced, which turns white on cooling.

The reactions above are examples of thermal decomposition (Section 1.5). Notice how similar the equations are. In each case, the metal carbonate decomposes to a solid metal oxide and carbon dioxide.

$$\text{metal carbonate} \xrightarrow{\text{heat}} \text{metal oxide} + \text{carbon dioxide}$$

Figure 13.3 Zinc oxide is white when cold, but yellow when hot. This is a very distinctive change.

The action of heat on metal carbonates can be related to the reactivity series. The reactions are summarised in Table 13.1.

Other common metal compounds, including sulfates, sulfides and chlorides, are more stable than carbonates and do not decompose so readily on heating.

Notice from Table 13.1 that the metal carbonates fall into three groups.

1 The most reactive metals at the top of the reactivity series, such as sodium and potassium, have stable carbonates which do not decompose on heating with a Bunsen burner.

Metal	Action of heat on the carbonate
K Na	Stable
Ca Mg Al Zn Fe Pb Cu	Decompose to oxide + carbon dioxide, e.g. $MgCO_3 \rightarrow MgO + CO_2$
Ag Au	Carbonates are too unstable to exist

Table 13.1 The action of heat on metal carbonates

2 Metals in the middle of the reactivity series, from calcium to copper, form less stable carbonates which decompose on heating to give the metal oxide.

3 The least reactive metals at the bottom of the reactivity series, like silver and gold, are so unreactive that they cannot form carbonates even at room temperature.

❶ When green copper carbonate is heated, it produces a black solid, A, and a colourless gas, B. When A is heated in hydrogen gas, it forms an orange-brown solid, C, which conducts electricity, and a vapour, D, which condenses to a colourless liquid.

a) What are the names of substances A, B, C and D?

b) Write word equations for:

i) copper carbonate forming A and B on heating;

ii) A reacting with hydrogen to form C and D.

c) Write balanced chemical equations for the reactions in b) parts i) and ii).

❷ a) How is the decomposition of metal carbonates related to the reactivity series?

b) Suppose you were given unlabelled samples of three white carbonates – sodium carbonate, magnesium carbonate and zinc carbonate. How would you identify which was which?

Figure 13.4 This welder is using an oxy-acetylene welding torch. Acetylene (C_2H_2) mixes with oxygen at the end of the torch and burns completely with a smokeless bright blue flame. The products of burning are carbon dioxide and water.

Organic compounds are compounds containing carbon, hydrogen and sometimes other non-metals.

Heating organic (carbon) compounds

Most of the carbon compounds that occur naturally, and that we use in fuels, foods, pharmaceuticals and plastics, come from crude oil or plants. Other important sources of carbon compounds are coal and natural gas.

The original sources of all these carbon compounds were living organisms. For this reason, carbon compounds are often called **organic compounds**. Most organic compounds contain carbon and hydrogen, often combined with oxygen and sometimes with other non-metals.

When organic compounds such as natural gas (methane), Calor gas (propane) and acetylene (Figure 13.4) are heated in a plentiful supply of air or oxygen, they burn completely. In these cases, carbon in the organic compounds is converted to carbon dioxide and hydrogen is converted to water. For example:

propane in Calor gas + oxygen in air → carbon dioxide + water

$$C_3H_8 + 5O_2 \rightarrow 3CO_2 + 4H_2O$$

If there is too little oxygen for an organic compound to burn fully, it may burn incompletely, producing a mixture of carbon (soot and smoke), carbon monoxide, carbon dioxide and water.

However, a lot of solid organic compounds, including many plastics, do not catch fire and burn easily. On heating, they just melt and char, leaving a black deposit of carbon. It is often possible to conclude that a compound is organic if it burns or chars on heating.

Figure 13.5 This photo shows the charred remains of plastics after a fire. Plastics are organic materials.

Detecting carbon and hydrogen in organic compounds

More definite evidence for organic compounds can be obtained by heating them and testing for the carbon dioxide and water produced when they burn (Figure 13.6).

The formation of carbon dioxide and water shows that the compound must contain carbon and hydrogen.

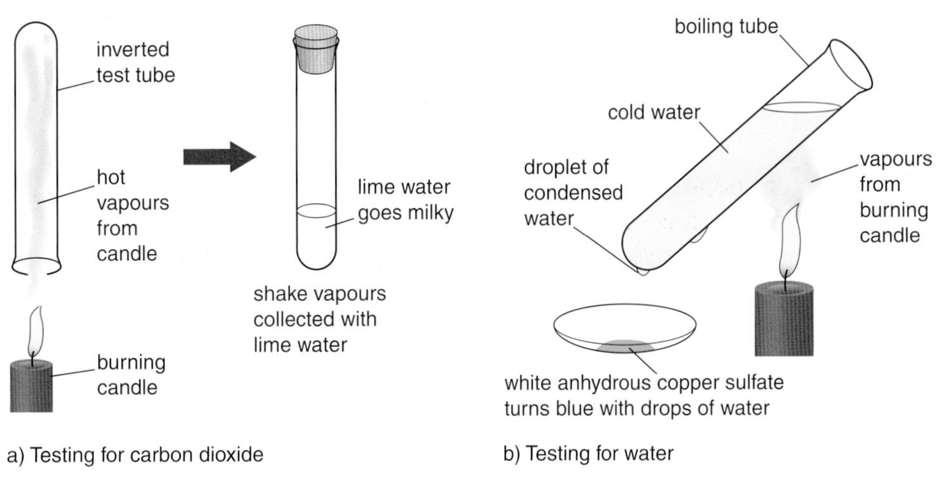

Figure 13.6 Testing for carbon dioxide and water when an organic compound burns

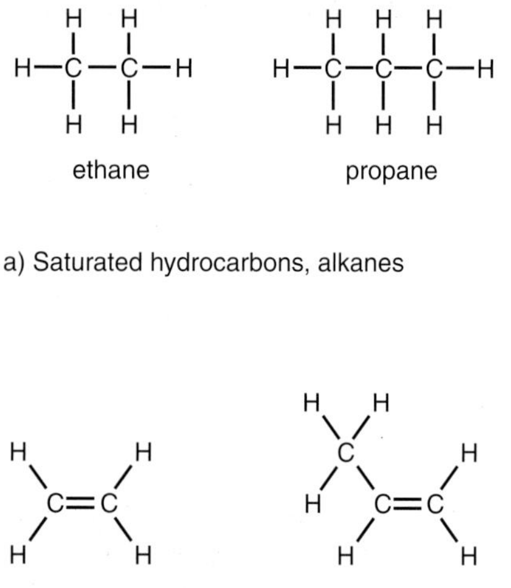

ethane propane

a) Saturated hydrocarbons, alkanes

ethene propene

b) Unsaturated hydrocarbons, alkenes

Figure 13.7 The two main groups of hydrocarbons

There are many carbon compounds that contain only carbon and hydrogen. These compounds are called hydrocarbons (Section 2.1). There are two main groups of hydrocarbons:

- saturated hydrocarbons or alkanes like ethane and propane (Figure 13.7). In these hydrocarbons, all the carbon atoms are joined by single bonds (Section 2.2).
- unsaturated hydrocarbons or alkenes like ethene and propene (Figure 13.7). These hydrocarbons have molecules with at least one double carbon–carbon bond (Section 2.2).

Alkenes are much more reactive than alkanes owing to their double carbon–carbon bonds. Because of their double bonds, alkenes decolorise yellow-orange bromine water but alkanes do not. So, this reaction can be used to identify and test for unsaturated organic compounds containing double carbon–carbon bonds (Section 3.5).

3 a) What is meant by the term 'organic compound'?
b) Describe the tests you would carry out to show that ethanol is an organic compound.

4 Look at the structures of hexene, cyclohexene and cyclohexane in Figure 13.8.

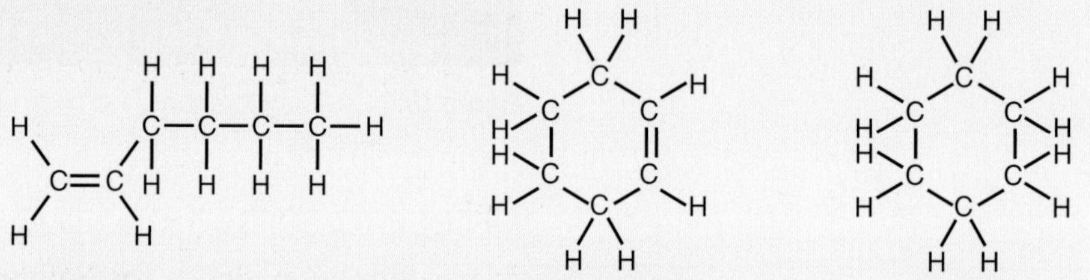

hexene cyclohexene cyclohexane

Figure 13.8 The structural formulae of hexene, cyclohexene and cyclohexane

a) Which of these compounds:
i) are saturated organic compounds;
ii) are unsaturated organic compounds;
iii) have the same molecular formula?
b) Describe the test you would carry out to show which of these compounds is/are unsaturated.

13.2 Finding the formulae of organic compounds

You should recall from Section 5.3 that it is possible to determine the empirical formula of a compound by doing experiments to find the masses of the different elements in a given mass of the compound.

For example, 8.0 g of an oxide of titanium (Ti) was found to contain 4.8 g of titanium and 3.2 g of oxygen (Ti = 48, O = 16). So, we can calculate the empirical formula of this oxide along the lines shown in Table 13.2.

	Ti	O
Masses reacting	4.8 g	3.2 g
Mass of 1 mole	48 g	16 g
∴ Moles reacting	0.1	0.2
Ratio of moles	1	2
	⇒ Formula is TiO_2	

Table 13.2 Finding the formula of titanium oxide

With organic compounds, it is difficult to find the masses of carbon and hydrogen in a sample directly. However, if an organic compound is burned completely in oxygen, the masses of carbon and hydrogen in the compound can be calculated from the masses of carbon dioxide and water produced.

Using relative atomic masses (H = 1, C = 12, O = 16), we can calculate that:

12 g of carbon combine with 32 g of oxygen to produce 44 g of carbon dioxide

and

2 g of hydrogen combine with 16 g of oxygen to produce 18 g of water.

$$\therefore \quad \frac{12}{44} \text{ or } \frac{3}{11} \text{ parts of carbon dioxide are carbon}$$

and

$$\frac{2}{18} \text{ or } \frac{1}{9} \text{ parts of water are hydrogen.}$$

So, if 2.4 g of a substance is burned completely and it produces 6.6 g of carbon dioxide and 5.4 g of water,

$$\text{the mass of carbon in 2.4 g of the substance} = 6.6 \times \frac{3}{11} = 1.8\text{ g}$$

and

$$\text{the mass of hydrogen in 2.4 g of the substance} = 5.4 \times \frac{1}{9} = 0.6\text{ g}$$

The formula of the compound can now be worked out as shown in Table 13.3.

	C	H
Masses reacting	1.8 g	0.6 g
Mass of 1 mole	12.0 g	1.0 g
∴ Moles reacting	0.15	0.6
Ratio of moles	1	4
	⇒ Formula = CH_4	

Table 13.3 Finding the formula of methane

Activity – Finding the formula of an organic liquid

Figure 13.9 shows the apparatus which Laura and Ali set up to find the formula of an organic liquid. The liquid was burned in a small spirit burner. The water produced was absorbed in anhydrous copper sulfate ($CuSO_4$). The carbon dioxide was absorbed in soda lime (a mixture of sodium hydroxide and calcium oxide).

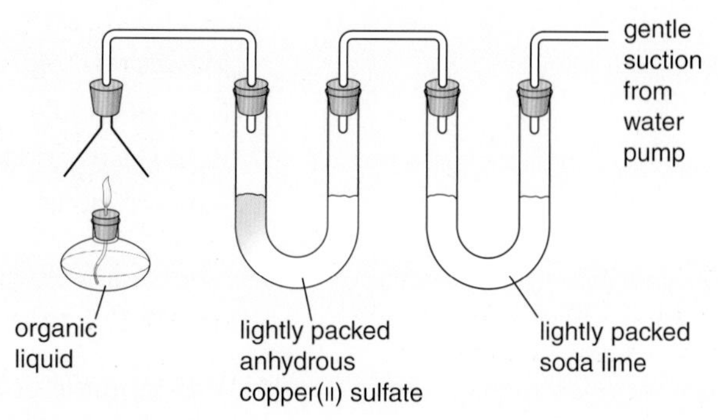

Figure 13.9 Finding the masses of carbon dioxide and water produced when an organic liquid burns

Here are the results of Laura and Ali's experiment.

Mass of spirit burner + organic liquid before the experiment	= 281.9 g
Mass of spirit burner + organic liquid after the experiment	= 281.2 g
Mass of U tube + anhydrous copper sulfate before the experiment	= 304.5 g
Mass of U tube + anhydrous copper sulfate after the experiment	= 305.4 g
Mass of U tube + soda lime before the experiment	= 315.2 g
Mass of U tube + soda lime after the experiment	= 317.4 g

1. During the experiment, the anhydrous copper sulfate reacts with any water produced to form hydrated copper sulfate ($CuSO_4.5H_2O$).
   a) What colour change occurs in the copper sulfate during the experiment?
   b) Write a chemical equation for the reaction of copper sulfate during the experiment.
2. Gentle suction was used in the experiment. Why was this?
3. Use Laura and Ali's results to calculate:
   a) the mass of organic liquid burned;
   b) the mass of water produced;
   c) the mass of carbon dioxide produced.
4. Calculate:
   a) the mass of carbon in the carbon dioxide produced;
   b) the mass of hydrogen in the water produced. (Remember that the mass of carbon is 3/11ths the mass of carbon dioxide and the mass of hydrogen is 1/9th the mass of water.)
5. The mass of carbon plus the mass of hydrogen which you have calculated in question **4** should now tell you that the organic liquid is a hydrocarbon. Why can you come to this conclusion?
6. Using the masses of carbon and hydrogen from question **4**, calculate the empirical formula of the hydrocarbon along the same lines as the calculation in Table 13.3.
7. In question **6**, you should have found that the empirical formula of the organic liquid is CH_2. This gives the simplest whole number ratio for the atoms of carbon and hydrogen in the organic liquid.
   a) The molecular formula of the organic liquid could be C_2H_4, C_3H_6, C_4H_8, etc. Why could the molecular formula of the organic liquid not be CH_2? (Hint: How many bonds must C and H atoms form?)
   b) The relative molecular mass of the organic liquid is 70. What is the molecular formula of the organic liquid?
8. Draw one possible structural formula for the organic liquid.

13.3 Identifying cations

We can identify different substances by testing for the atoms in them as we have done for carbon and hydrogen in organic compounds. We can also identify substances using tests for the ions which they contain.

The best way to identify a substance is to find a property or a reaction which is characteristic of that substance and is easy to observe.

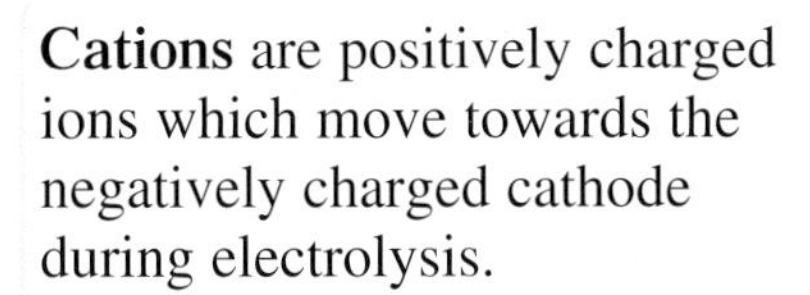

Cations are positively charged ions which move towards the negatively charged cathode during electrolysis.

Anions are negatively charged ions which move towards the positively charged anode during electrolysis.

Two of the best tests for **cations** (metal ions) use flame tests and sodium hydroxide solution. Remember that you must have your teacher's permission before trying any of the tests in this chapter and you must always wear eye protection.

Flame tests

When substances are heated strongly, the electrons in them absorb extra energy. Scientists say that the electrons are 'excited'. The excited electrons soon release this excess energy as light and become stable again. Different cations emit different colours of light when substances are heated. The distinctive flame colours produced by six cations are shown in Table 13.4.

Cation in the substance	Flame colour
K^+ (potassium)	Lilac
Na^+ (sodium)	Yellow
Li^+ (lithium)	Red
Ca^{2+} (calcium)	Orange-red
Ba^{2+} (barium)	Pale green
Cu^{2+} (copper(II))	Blue-green

Table 13.4 The flame colours of some cations

Figure 13.10 Carrying out a flame test on a compound containing potassium

Flame tests for the ions in Table 13.4 can be carried out using the following method.

Dip a nichrome wire in concentrated hydrochloric acid and then heat it in a pale blue Bunsen flame until it gives no colour to the flame. The wire is now clean. Dip it in concentrated hydrochloric acid again, and then in the substance to be tested. Heat the wire in the Bunsen and note the flame colour (Figure 13.10).

Remember that the colours in flame tests are produced by metal ions in compounds. Don't confuse the flame colours in these tests with the colour of the flame when an element burns in air. In some cases, such as sodium, the colour is the same, but in others it is not. For example, magnesium has a very bright white flame when it burns in air, but magnesium compounds do not produce any colour during flame tests.

Tests with sodium hydroxide solution

The hydroxides of all metals (except those in Group 1) are insoluble. These insoluble hydroxides form as solid precipitates when sodium hydroxide solution, containing hydroxide ions, is added to a solution containing metal ions (Figure 13.11). For example, when sodium hydroxide solution is added to copper sulfate solution, a blue precipitate of copper hydroxide is produced.

copper sulfate solution	+	sodium hydroxide solution	→	copper hydroxide precipitate	+	sodium sulfate solution
$Cu^{2+}(aq) + SO_4^{2-}(aq)$	+	$2Na^+(aq) + 2OH^-(aq)$	→	$Cu(OH)_2(s)$	+	$2Na^+(aq) + SO_4^{2-}(aq)$

The sulfate ions ($SO_4^{2-}(aq)$) and the sodium ions ($2Na^+(aq)$) take no part in the reaction and are left behind in the solution as sodium sulfate. Because of this, they are described as spectator ions (Section 8.1). The only reaction that takes place involves copper(II) ions ($Cu^{2+}(aq)$) reacting with twice as many hydroxide ions ($2OH^-(aq)$) to form a solid precipitate of copper hydroxide. The sulfate ions and sodium ions are the same after the reaction as they were before the reaction. So, they can be cancelled in the above equation and the reaction can be summarised as:

$$Cu^{2+}(aq) + 2OH^-(aq) \rightarrow Cu(OH)_2(s)$$

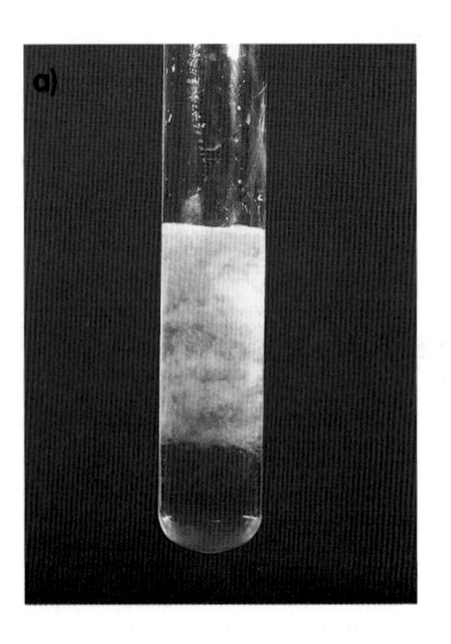

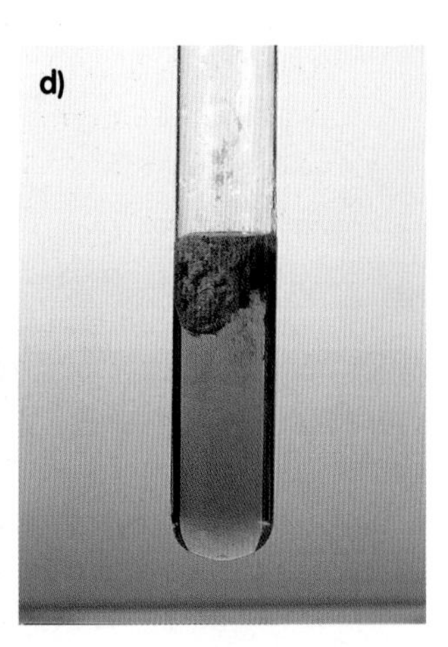

Figure 13.11 Precipitates of a) magnesium hydroxide, b) aluminium hydroxide, c) copper (II) hydroxide and d) iron (III) hydroxide. These precipitates form when sodium hydroxide solution is added to a solution of the relevant cation.

Activity – Testing for metal ions with sodium hydroxide solution

Table 13.5 shows what happens when:

- a little sodium hydroxide solution and
- then excess sodium hydroxide solution

is added to solutions of some common cations.

Look carefully at Table 13.5 and answer the following questions.

1. Write a chemical equation for the reaction of sodium hydroxide solution with magnesium sulfate solution to form a precipitate of magnesium hydroxide.
2. Use your equation from question **1** to write a summary equation for the precipitation of magnesium hydroxide which could go in Table 13.5.

3 a) Which cations in Table 13.5 give no precipitate with sodium hydroxide solution?
b) How can you tell which of these cations is which, using flame tests?

4 a) Which cations give a white precipitate which does not dissolve with excess sodium hydroxide?
b) How can you tell which of these cations is which using flame tests?

5 a) Which cations give a coloured (non-white) precipitate with sodium hydroxide solution?
b) How can you tell the difference between these cations from the colour of their hydroxide?

6 The ammonium cation, NH_4^+, can also be identified using sodium hydroxide solution. In this case the ammonium compound, containing NH_4^+ ions, should be warmed with sodium hydroxide solution. The NH_4^+ ions react with hydroxide ions (OH^-) to produce water and ammonia (NH_3), which has a very pungent smell and is the only common alkaline gas.
a) How can damp litmus paper identify ammonia and ammonium compounds?
b) Copy and complete the following equation for the reaction of ammonium ions with hydroxide ions.

$NH_4^+ + OH^- \rightarrow$ ________ + ________

7 Look at the tablets and their container in Figure 13.12. Describe how you would test the tablets to check whether they contain Fe^{2+} or Fe^{3+} ions.

Figure 13.12

Cation in solution	Three drops of NaOH(aq) added to 3 cm³ of solution of cation	10 cm³ of NaOH(aq) added to 3 cm³ of solution of cation
Potassium, K^+	No precipitate	No precipitate
Sodium, Na^+	No precipitate	No precipitate
Calcium, Ca^{2+}	A white precipitate of $Ca(OH)_2$ forms $Ca^{2+} + 2OH^- \rightarrow Ca(OH)_2$	White precipitate remains
Magnesium, Mg^{2+}	A white precipitate of $Mg(OH)_2$ forms	White precipitate remains
Aluminium, Al^{3+}	A white precipitate of $Al(OH)_3$ forms $Al^{3+} + 3OH^- \rightarrow Al(OH)_3$	White precipitate dissolves to give a colourless solution
Iron(II), Fe^{2+}	A green precipitate of $Fe(OH)_2$ forms $Fe^{2+} + 2OH^- \rightarrow Fe(OH)_2$	Green precipitate remains
Iron(III), Fe^{3+}	A brown precipitate of $Fe(OH)_3$ forms $Fe^{3+} + 3OH^- \rightarrow Fe(OH)_3$	Brown precipitate remains
Copper(II), Cu^{2+}	A blue precipitate of $Cu(OH)_2$ forms $Cu^{2+} + 2OH^- \rightarrow Cu(OH)_2$	Blue precipitate remains

Table 13.5 Testing for metal ions with sodium hydroxide solution

5. a) How would you carry out a flame test on a sample of chalk?
   b) What colour would you expect from the flame test?
   c) Why must you wear eye protection when you are carrying out a flame test?
   d) What causes the flame colours from certain cations?
6. a) A garden fertiliser is thought to contain copper(II) sulfate. Describe two tests you could do to show that it contains Cu^{2+} ions.
   b) How would you show that the fertiliser also contains an ammonium compound?
7. A metal of fairly low density has become coated with a layer of oxide. Describe the tests that you would do to decide whether the metal is aluminium, magnesium or calcium. (Hint: The oxides are not soluble in water, but they react with dilute hydrochloric acid to form solutions of the metal chlorides.)

13.4 Identifying anions

Test for carbonates (CO_3^{2-})

Most carbonates decompose on heating to give a metal oxide and carbon dioxide. The carbon dioxide will turn lime water milky. This provides some evidence that the substance heated may be a carbonate, but the test is not foolproof.

Some carbonates, such as sodium carbonate and potassium carbonate, don't decompose on heating. In addition, some organic compounds may burn or char when heated, producing carbon dioxide. So, a more specific test for carbonates is needed.

In order to make sure that a substance is a carbonate, add dilute acid to the solid. If the substance is a carbonate, carbon dioxide is produced which turns lime water milky (Figure 13.13).

carbonate + acid → carbon dioxide + water

$$CO_3^{2-}(s) + 2H^+(aq) \rightarrow CO_2(g) + H_2O(l)$$

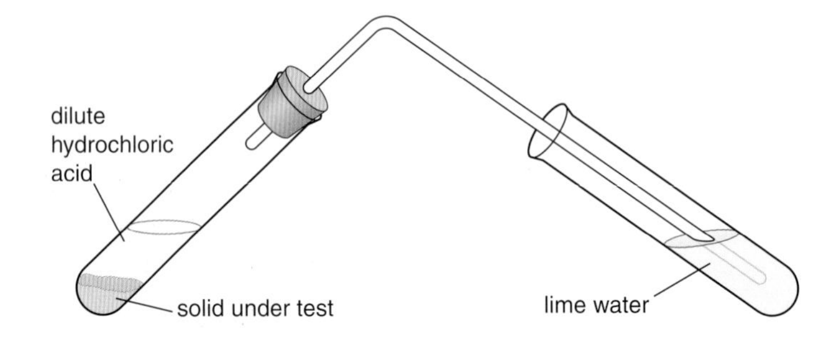

Figure 13.13 Testing for carbonate using dilute acid and lime water

Test for halides (Cl^-, Br^-, I^-)

Precipitates form when silver nitrate solution is added to solutions of chlorides, bromides and iodides mixed with dilute nitric acid (Figure 13.14).

Chlorides (Cl^-) give a white precipitate of silver chloride which turns purple-grey in sunlight.

silver ion	+	chloride	→	silver chloride
$Ag^+(aq)$	+	$Cl^-(aq)$	→	$AgCl(s)$

Bromides (Br^-) give a cream precipitate of silver bromide which turns green-yellow in sunlight.

silver ion	+	bromide	→	silver bromide
$Ag^+(aq)$	+	$Br^-(aq)$	→	$AgBr(s)$

Iodides (I^-) give a yellow precipitate of silver iodide which does not change in sunlight.

silver ion	+	iodide	→	silver iodide
$Ag^+(aq)$	+	$I^-(aq)$	→	$AgI(s)$

Figure 13.14 Testing for halides with silver nitrate solution. Chlorides give a white precipitate (left), bromides give a cream precipitate (centre) and iodides give a yellow precipitate (right).

Test for sulfate (SO_4^{2-})

A thick white precipitate of barium sulfate forms when dilute hydrochloric acid and then barium chloride solution are added to a solution of a sulfate.

barium ion	+	sulfate	→	barium sulfate
$Ba^{2+}(aq)$	+	$SO_4^{2-}(aq)$	→	$BaSO_4(s)$

Test for nitrate (NO_3^-)

In the Activity at the end of Section 13.3, we found that ammonia (NH_3) could easily be identified because it is the only common gas which turns damp red litmus paper blue.

This reaction of ammonia also opens up a test for nitrate ions. When a nitrate is heated with sodium hydroxide solution plus a little aluminium powder, ammonia is produced. In the presence of sodium hydroxide solution, aluminium can reduce nitrate ions (NO_3^-) to ammonia (NH_3), which turns damp red litmus paper blue (Figure 13.15).

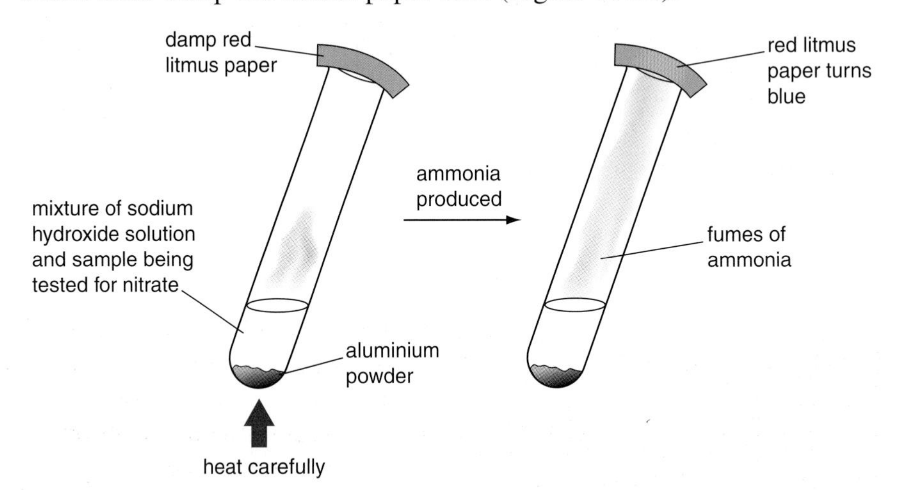

Figure 13.15 Testing for a nitrate

Don't confuse the test for ammonium ions (NH_4^+) with the test for nitrate ions (NO_3^-).

Ammonium compounds produce ammonia on heating with sodium hydroxide solution. Nitrates only produce ammonia on heating with sodium hydroxide solution plus aluminium powder.

The key tests for common anions are summarised in Table 13.6.

Anion	Test	Positive result
Carbonate, CO_3^{2-}	Add dilute acid to the solid being tested and pass any gases produced through lime water	Carbon dioxide is produced which turns lime water milky
Chloride, Cl^- Bromide, Br^- Iodide, I^-	Add dilute nitric acid, then silver nitrate solution to the solution being tested	White precipitate with Cl^- Cream precipitate with Br^- Yellow precipitate with I^-
Sulfate, SO_4^{2-}	Add dilute hydrochloric acid, then barium chloride solution to the solution being tested	Thick white precipitate
Nitrate, NO_3^-	Add sodium hydroxide solution plus a little aluminium powder to the solid being tested. Then, heat carefully	Ammonia is produced which turns damp red litmus paper blue

Table 13.6 Key tests for common anions

8. Two fertilisers are labelled 'sulfate of potash' and 'nitrate of potash'.
   a) How would you check which was which?
   b) Which metal ion (cation) do you think is present in these fertilisers?
   c) One of the fertilisers is thought to contain an iron salt. How would you investigate this?

9. Hamish heated a solid, X, and passed the gases produced through lime water. The lime water turned milky and Hamish concluded that X was a carbonate.
   a) Could X be a carbonate?
   b) Is X definitely a carbonate?
   c) Explain your answer to part b).

10. Ammonium chloride, potassium nitrate and ammonium nitrate are all used in fertilisers.
   a) How would you show that ammonium chloride contains ammonium ions (NH_4^+)?
   b) How would you show that potassium nitrate contains nitrate ions (NO_3^-), but no ammonium ions?
   c) How would you show that ammonium nitrate contains both ammonium ions and nitrate ions? (Hint: When ammonium compounds are heated with excess sodium hydroxide solution, all the ammonium ions are converted to ammonia gas.)

13.5 Instrumental methods of analysis

Figure 13.16 This prototype 'walk through' spectrometer can detect traces of explosives left in the air that passes over a person who has handled explosives, and is being tested in some airports.

In the last hundred years, the work of analytical chemists has been revolutionised by the development of very sophisticated measuring instruments. These include:

- mass spectrometers (Section 5.2), which measure the relative masses of different particles;
- spectroscopes, which measure the wavelengths of light and other electromagnetic radiations emitted and absorbed by different substances.

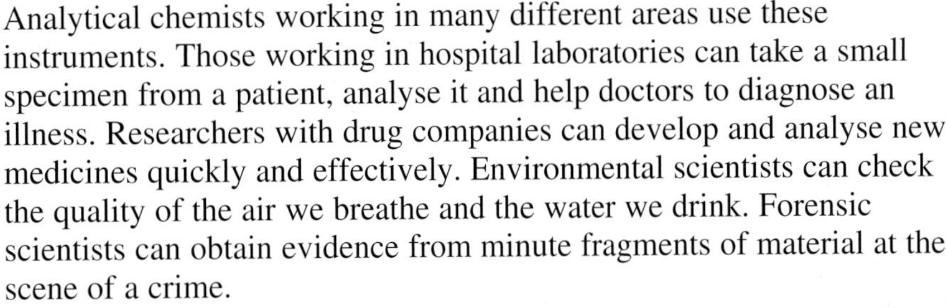

Analytical chemists working in many different areas use these instruments. Those working in hospital laboratories can take a small specimen from a patient, analyse it and help doctors to diagnose an illness. Researchers with drug companies can develop and analyse new medicines quickly and effectively. Environmental scientists can check the quality of the air we breathe and the water we drink. Forensic scientists can obtain evidence from minute fragments of material at the scene of a crime.

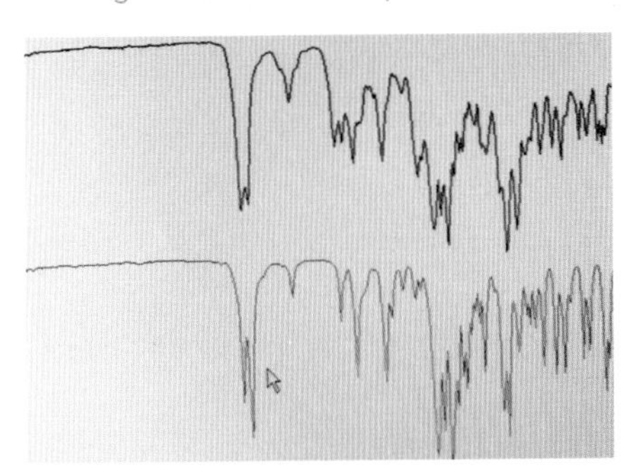

Figure 13.17 Using infra-red spectroscopy, forensic scientists can check whether a sample of material contains drugs such as heroin by comparing the sample's spectrum (black) with that of heroin (blue).

Modern instrumental methods of chemical analysis have taken over from traditional laboratory tests because they are:

- faster – many samples can be analysed in a short time. This is important in making rapid security checks at airports and in industry when materials require regular, frequent monitoring;
- more sensitive – this allows analysis of tiny fragments of material and the detection of small amounts in a huge sample. This is important for drug testing in sport and in testing materials at the scene of a crime;
- more accurate – human error is reduced as a result of automation involving computer analysis and recording.

The remarkable development in modern instrumental methods could not have happened without the rapid progress in technologies such as electronics and computing.

Computers, printers and other electronic devices have speeded up the whole process of chemical analysis and the storage of data. For example, the analysis of an oil slick used to take weeks of painstaking laboratory experiments. Today, the same analysis can be carried out in a few hours. Computers have also speeded up the interpretation of complex data requiring detailed calculations and the search of information previously stored in databases.

11 a) State three advantages of using modern instruments to detect and analyse substances.
b) Name two kinds of instrument now used by analytical chemists.
c) Give two examples of the use of instrumental analysis.

12 In December 1978, four American and two Russian space probes landed on Venus. They had on board a range of instruments, including mass spectrometers, spectroscopes and gas chromatograms. Because of weight limitations on the probe, the mass spectrometers had only low resolution. A particle of relative mass 64.0 was identified, but the resolution could not show whether the particle was SO_2 or S_2.
a) Why are instrumental methods of analysis important in space exploration?
b) Why could the mass spectrometers not show whether a particle on Venus was SO_2 or S_2? (S = 32.0, O = 16.0)
c) High-resolution mass spectrometers can measure relative masses to three decimal places. (O = 15.995, S = 31.972). Explain how a high-resolution mass spectrometer could determine whether a particle was SO_2 or S_2.

13.6 More about instrumental methods

The instruments used by analytical chemists can detect the presence of certain elements and compounds. Some instruments, such as emission and absorption spectroscopes, are suited to identifying elements. Other instruments, such as infra-red and NMR spectroscopes, are normally used to identify compounds. Mass spectrometers can be adapted to identify elements or compounds.

Figure 13.18 A scientist using an absorption spectroscope to analyse a water sample

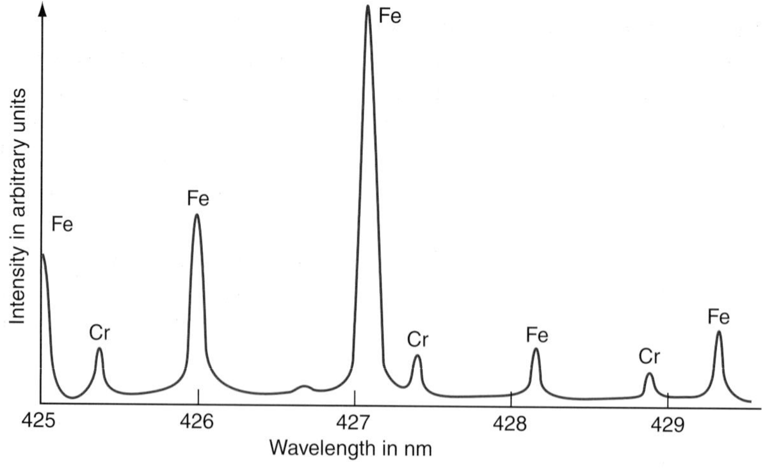

Figure 13.19 Absorption spectroscopy is used in the steel industry to monitor the amounts of different elements in steels to control the alloy's quality as it is being made. This print-out shows the presence of chromium along with iron in a steel alloy.

Table 13.7 gives more information about the instruments used, what they identify and a short explanation of how they work. You are not required to remember the details of how these instruments work.

Instrument used	What does the instrument identify?	How does it work?	Example of the application
Mass spectrometer (see Section 5.2)	Elements and compounds (from particles as small as atoms to complex molecules like proteins)	Positive ions of elements and compounds pass through a magnetic field where they are deflected. Lighter particles are deflected more than heavier particles. So, the mass of an ion can be determined by measuring how much it is deflected.	Checking that the additives present in food are permitted
Emission spectroscope (visible, infra-red or ultraviolet)	Elements	When substances are heated strongly, their electrons are 'excited' to higher energy levels. The electrons soon release this energy as visible, infra-red (IR) or ultraviolet (UV) radiations. (This is what happens in flame tests, Section 13.3.) By identifying the wavelength or frequency of the radiation, it is possible to identify the elements in the material being analysed.	Analysing the composition of elements in the Sun and other stars
Absorption spectroscope (visible and ultraviolet)	Elements	Strong visible and UV light pass through a sample of the substance. Electrons in the sample absorb radiation of a particular wavelength and are excited to higher energy levels. By comparing the incoming and outgoing radiation it is possible to identify the elements in the sample.	Detecting elements in steel; pollution monitoring to identify heavy metals
Infra-red spectroscope	Organic compounds (by identifying the bonds in compounds)	These involve the same technique as absorption spectroscopes but with infra-red. IR waves are absorbed by electrons in the bonds of molecules. Different bonds absorb energy at different wavelengths, so bonds such as C–H, C–O and C=C can be identified.	Identifying drugs in forensic testing
Nuclear magnetic resonance spectroscope (often called 'NMR')	Organic compounds (by identifying the different groups of atoms containing hydrogen)	Nuclei of hydrogen atoms in organic compounds can act as small magnets. Radio waves of specific frequencies are absorbed by the nuclei of hydrogen atoms as they move into line with a strong magnetic field. The amount of energy absorbed corresponds to the number of hydrogen atoms in the same type of bond.	Analysing the molecular structure of new pharmaceutical products
Gas–liquid chromatograph	Volatile liquids	The liquid is allowed to vaporise into a stream of inert gas passing through a long, thin tube packed with porous material. The time it takes a vaporised substance to reach the end of the tube is used to identify it.	Analysing the different components in petrol

Table 13.7 The instruments used in modern chemical analysis

Activity – Interpreting the NMR and mass spectrometer print-outs for ethanol

Figure 13.20 shows a simplified NMR print-out for ethanol, CH_3CH_2OH. The different absorption peaks correspond to hydrogen atoms in different groups in the molecules of ethanol. The height of each peak is roughly proportional to the number of hydrogen atoms in these groups. The highest peak in the print-out corresponds to the three hydrogen atoms in the CH_3 group in ethanol.

1. Draw the structural formula for ethanol showing all the bonds between atoms.
2. Why are there three peaks in the NMR print-out for ethanol?
3. The relative heights of the peaks are roughly in the ratio 1 : 2 : 3. Why is this?
4. Sketch the NMR print-out and show on your sketch which peak belongs to which hydrogen atoms in ethanol.

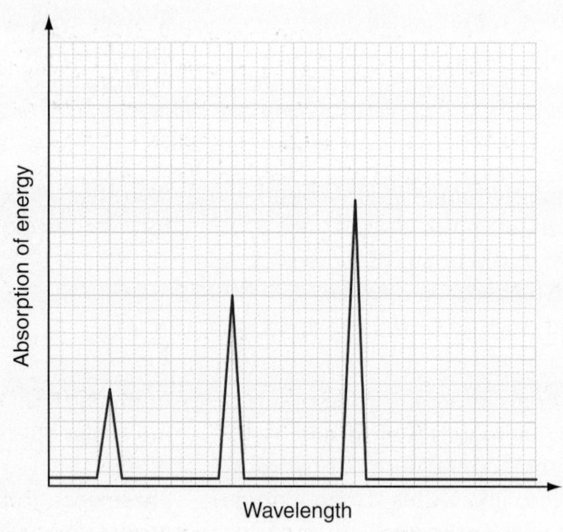

Figure 13.20 A simplified NMR print-out for ethanol, CH_3CH_2OH

Mass spectrometry can often be used to identify the complete structure of an organic compound. This is because some molecules break up into smaller fragments in the mass spectrometer. This means that several other peaks appear on the print-out in addition to the peak corresponding to the molecule itself. Look at the simplified mass spectrometer print-out for ethanol in Figure 13.21.

5. What is the relative molecular mass of ethanol? (C = 12, H = 1, O = 16)

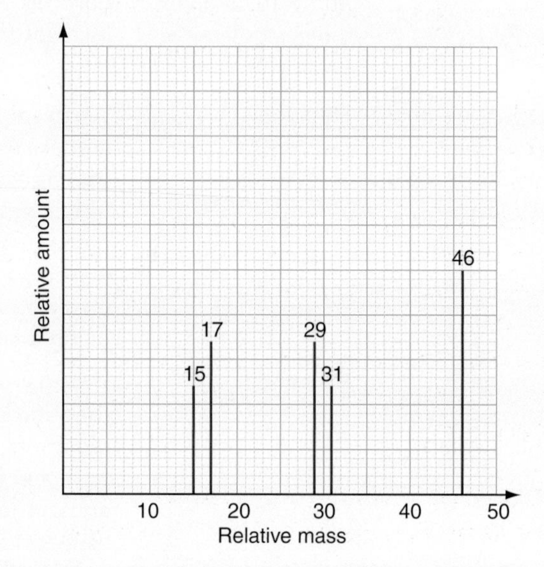

Figure 13.21 A simplified mass spectrometer print-out for ethanol

6. Which peak in Figure 13.21 corresponds to a molecule of ethanol?
7. The peak at relative mass 15 units corresponds to the CH_3 group. What fragments of the ethanol molecule do the other three peaks correspond to?

Summary

- ✓ Sometimes the change that occurs to a substance when it is heated is so distinctive that it identifies the substance immediately.
- ✓ The carbonates of metals from calcium to copper in the reactivity series decompose on heating to form a metal oxide and carbon dioxide.
- ✓ **Organic compounds** are compounds containing carbon, hydrogen and sometimes other non-metals.
- ✓ Organic compounds burn or char when heated in air. They often form carbon dioxide and water at the same time.
- ✓ Unsaturated organic compounds containing a double carbon–carbon bond can be identified by shaking with yellow bromine water, which they decolorise.
- ✓ It is possible to determine the empirical formula of an organic compound by burning a known mass of the compound.
- ✓ **Cations** are positively charged ions which move towards the negatively charged cathode during electrolysis.
- ✓ Metal ions (cations) can be identified by flame tests or by reaction with sodium hydroxide solution.
- ✓ **Anions** are negatively charged ions which move towards the positively charged anode during electrolysis.
- ✓ There are a number of tests for anions:
  - Carbonates react with acids to form carbon dioxide.
  - Halides give precipitates with silver nitrate solution.
  - Sulfates give a precipitate with barium chloride solution.
  - Nitrates react with aluminium powder and sodium hydroxide solution on heating to form ammonia, which turns damp red litmus paper blue.
- ✓ Instrumental methods of analysis are fast, accurate and so sensitive that very little material is required. Laboratory methods are slow in comparison and require larger amounts of material.
- ✓ The development of modern instrumental methods has been aided by the rapid developments in computers and electronics.
- ✓ Instrumental methods have brought very significant benefits to society, including more efficient monitoring of industrial products, more accurate medical diagnoses, improved access to forensic information and more systematic checks of our food and the environment.

EXAMQUESTIONS

1 A white powder, P, dissolves in water to produce a colourless solution. This solution gives a white precipitate with sodium hydroxide solution and a precipitate with a mixture of silver nitrate solution and dilute nitric acid. Which three of the following substances could be P?

A aluminium chloride
B calcium nitrate
C calcium iodide
D copper(II) chloride
E magnesium carbonate
F magnesium bromide *(3 marks)*

2 Copy and complete the sentences below by using words from the box.

atoms	electrons	energy	heat
infra-red	ions	protons	visible

In flame tests, the colours arise because the __________ of the flame causes __________ in metal __________ to move from lower to higher __________ levels. These 'excited' ions lose energy by emitting __________ radiation. *(5 marks)*

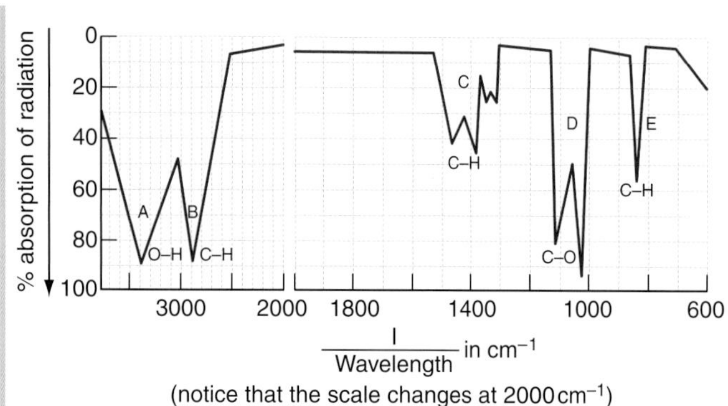

Figure 13.22 A simplified infra-red spectrum for ethanol

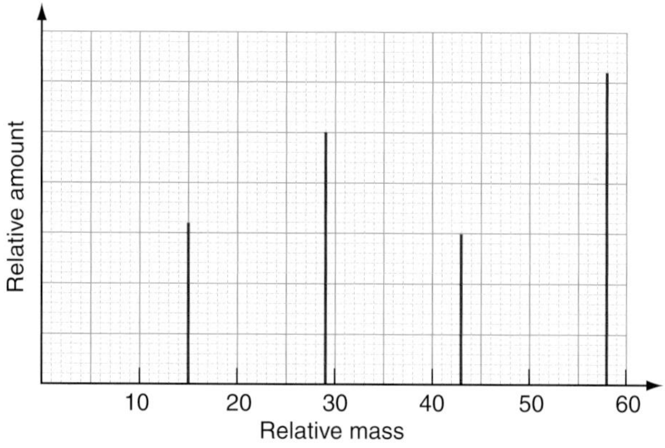

Figure 13.23 The mass spectrometer print-out for hydrocarbon Y

EXAMQUESTIONS

3 Figure 13.22 shows a simplified infra-red spectrum for ethanol (C_2H_5OH). Infra-red spectra show the percentage absorption of radiation on the vertical axis against the reciprocal of wavelength horizontally. The bonds in ethanol associated with the different absorption peaks are shown in Figure 13.22 and labelled A, B, C, D and E.

a) Why are the units along the horizontal axis cm^{-1}? *(2 marks)*

b) i) Draw the structural formula of ethanol. *(2 marks)*

ii) Which bonds in ethanol are not identified in its infra-red spectrum? *(1 mark)*

c) i) Draw the structural formula of propane (C_3H_8). *(2 marks)*

ii) Use the peak labels A–E to explain what the infra-red spectrum of propane will look like. *(2 marks)*

d) i) Draw the structural formula of propanol (C_3H_7OH). *(1 mark)*

ii) How do you think the infra-red spectrum of propanol will compare with that of ethanol? *(1 mark)*

4 Figure 13.23 shows the simplified mass spectrometer print-out for the hydrocarbon Y. The print-out shows the relative masses of the main fragments and the whole molecule. (C = 12, H = 1)

a) What is the relative molecular mass of Y? *(1 mark)*

b) The peak at relative mass 15 corresponds with the CH_3 group. What do the peaks at 29 and 43 represent? *(2 marks)*

c) i) Draw the structure of Y. *(2 marks)*

ii) What is the name of Y? *(1 mark)*

5 Which instrumental method would you use for each of the following investigations?

a) Checking the colourings in food dyes. *(1 mark)*

b) Identifying the alkanes in a new brand of petrol. *(1 mark)*

c) Identifying the isotopes in liquid bromine. *(1 mark)*

d) Detecting the O–H group in a medicine. *(1 mark)*

e) Monitoring the composition of alloys of brass. *(1 mark)*

Data sheet

1. Reactivity series of metals

Potassium
Sodium
Calcium
Magnesium
Aluminium
Carbon
Zinc
Iron
Tin
Lead
Hydrogen
Copper
Silver
Gold
Platinum

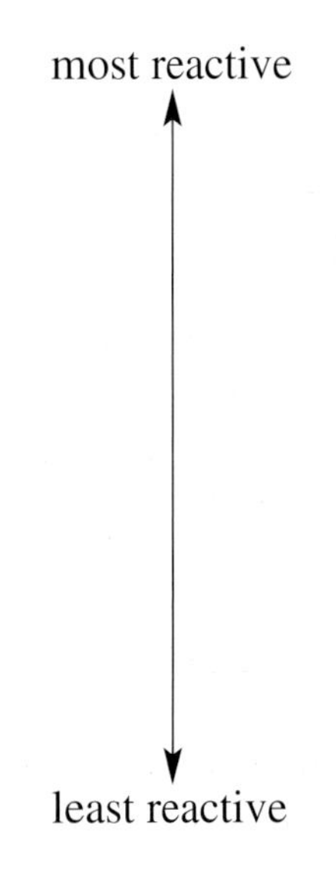

(elements in italics, though non-metals, have been included for comparison)

2. Formulae of some common ions

Positive ions

Name	Formula
Hydrogen	H^{+}
Sodium	Na^{+}
Silver	Ag^{+}
Potassium	K^{+}
Lithium	Li^{+}
Ammonium	NH_4^{+}
Barium	Ba^{2+}
Calcium	Ca^{2+}
Copper(II)	Cu^{2+}
Magnesium	Mg^{2+}
Zinc	Zn^{2+}
Lead	Pb^{2+}
Iron(II)	Fe^{2+}
Iron(III)	Fe^{3+}
Aluminium	Al^{3+}

Negative ions

Name	Formula
Chloride	Cl^{-}
Bromide	Br^{-}
Fluoride	F^{-}
Iodide	I^{-}
Hydroxide	OH^{-}
Nitrate	NO_3^{-}
Oxide	O^{2-}
Sulfide	S^{2-}
Sulfate	SO_4^{2-}
Carbonate	CO_3^{2-}

Key

relative atomic mass
atomic symbol
atomic name
proton number

1	2											3	4	5	6	7	0
							1 **H** hydrogen 1										4 **He** helium 2
7 **Li** lithium 3	9 **Be** beryllium 4											11 **B** boron 5	12 **C** carbon 6	14 **N** nitrogen 7	16 **O** oxygen 8	19 **F** fluorine 9	20 **Ne** neon 10
23 **Na** sodium 11	24 **Mg** magnesium 12											27 **Al** aluminium 13	28 **Si** silicon 14	31 **P** phosphorus 15	32 **S** sulfur 16	35.5 **Cl** chlorine 17	40 **Ar** argon 18
39 **K** potassium 19	40 **Ca** calcium 20	45 **Sc** scandium 21	48 **Ti** titanium 22	51 **V** vanadium 23	52 **Cr** chromium 24	55 **Mn** manganese 25	56 **Fe** iron 26	59 **Co** cobalt 27	59 **Ni** nickel 28	63.5 **Cu** copper 29	64 **Zn** zinc 30	70 **Ga** gallium 31	73 **Ge** germanium 32	75 **As** arsenic 33	79 **Se** selenium 34	80 **Br** bromine 35	84 **Kr** krypton 36
85 **Rb** rubidium 37	88 **Sr** strontium 38	89 **Y** yttrium 39	91 **Zr** zirconium 40	93 **Nb** niobium 41	96 **Mo** molybdenum 42	**Tc** technetium 43	101 **Ru** ruthenium 44	103 **Rh** rhodium 45	106 **Pd** palladium 46	108 **Ag** silver 47	112 **Cd** cadmium 48	115 **In** indium 49	119 **Sn** tin 50	122 **Sb** antimony 51	128 **Te** tellurium 52	127 **I** iodine 53	131 **Xe** xenon 54
133 **Cs** caesium 55	137 **Ba** barium 56	139 **La** lanthanum 57	178 **Hf** hafnium 72	181 **Ta** tantalum 73	184 **W** tungsten 74	186 **Re** rhenium 75	190 **Os** osmium 76	192 **Ir** iridium 77	195 **Pt** platinum 78	197 **Au** gold 79	201 **Hg** mercury 80	204 **Tl** thallium 81	207 **Pb** lead 82	209 **Bi** bismuth 83	[209] **Po** polonium 84	[210] **At** astatine 85	[222] **Rn** radon 86
[223] **Fr** francium 87	[226] **Ra** radium 88	[227] **Ac*** actinium 89	[261] **Rf** rutherfordium 104	[262] **Db** dubnium 105	[266] **Sg** seaborgium 106	[264] **Bh** bohrium 107	[277] **Hs** hassium 108	[268] **Mt** meitnerium 109	[271] **Ds** darmstadtium 110	[272] **Rg** roentgenium 111	Elements with atomic numbers 112–116 have been reported but not fully authenticated						

*The Lanthanides (atomic numbers 58–71) and the Actinides (atomic numbers 90–103) have been omitted.

Cu and **Cl** have not been rounded to the nearest whole number.

Index